Barron's Regents Exams and Answers

Three-Year Sequence for High School Mathematics (Course I)

Sequential Integrated Mathematics

LESTER W. SCHLUMPF

Retired Principal
John Adams High School, Queens

Former Lecturer in Mathematics, Queens College

LAWRENCE S. LEFF

Assistant Principal
Mathematics Supervisor

Franklin D. Roosevelt High School, Brooklyn

Barron's Educational Series, Inc.

Three-Year Sequence for High School Mathematics (Course I)

All inquiries should be addressed to:
Barron's Educational Series, Inc.
250 Wireless Boulevard
Hauppauge, New York 11788

ISBN 0-8120-4144-5

PRINTED IN THE UNITED STATES OF AMERICA

345 100 987654321

Contents

Regents Examinations and Answers

Preface

A helpful word to the student:

This book is designed to strengthen your understanding and mastery of the ninth year mathematics curriculum as exemplified in the New York State syllabus for the Three-Year Sequence in High School Mathematics—Course I. The book has been specifically written to assist you in preparing for the Regents examination covering this syllabus. It has been revised to adapt to changes made in this syllabus in the 1988–1989 school year that were first tested on the June, 1989 Regents examination.

Special features include:

Complete sets of questions from 10 previous Regents examinations in this subject. Attempting to solve these will make you familiar with the topics tested on the examination and with the degree of difficulty you are expected to master in each topic. Solving the questions on many tests will provide drill, improve your understanding of the topics, and increase your confidence as the nature and the language of the questions become more familiar.

Solutions to all Regents questions with step-by-step explanations of the solutions. Careful study of the solutions and explanations will improve your mastery of the subject. Each explanation is designed to show you how to apply the facts and principles you have learned in class. Since the explanation for each solution has been written with emphasis on the reasoning behind each step, its value goes far beyond the application to that particular question. You should read the explanation of the solution even if you have answered the question correctly. It gives insight into the topic that may be valuable when answering a more difficult question on the same topic on the next test you face.

A Practice Section at the front of the book consisting of questions taken from former Regents, each with a completely explained step-by-step solution. The questions are classified into 31 topic groups.

A specially prepared Self-Analysis Chart following each Regents. The chart is keyed to the Practice Section questions to direct you to concentrated study on those topics where you discover you have weaknesses. You will notice that Self-Analysis Charts for Regents given prior to June, 1989 do not contain any entries for topic groups numbered 26 through 31. This is because such topics were not included in Course I before that time. However, the Practice Section of this book does provide questions on these topics in addition to those on Regents given on and after June, 1989. These Practice Questions were taken from the August, 1989 Course I Regents or were adapted from similar questions that appeared on Course II and Course III Regents, where topic groups 26 through 31 were formerly included.

Follow the procedures in the section entitled "How to Use This Book"—*and watch that mark grow.*

How to Use This Book

This book has a built-in program to identify your areas of strength and weakness, and to guide you in concentrating your study on those topics where you most need assistance. The first part of the book consists of Practice Exercises. The section contains 189 questions, all selected from former Regents. The questions are classified into 31 topics, and a completely explained solution is provided for each question.

The questions in the first 25 topic groups are all from former Course I Regents examinations. They have been selected as typical, both in type and degree of difficulty, of those asked on the examinations. Topic groups numbered 26 through 31 cover material that was added to Course I when the Regents Board rearranged the sequential mathematics courses and included questions on these topics in Course I beginning with the June, 1989 Regents. The Practice Exercises selected for topic groups numbered 26 through 31 were obtained from the August, 1989 Course I examination and from adaptations of questions that appeared in earlier Course II and III Regents when these topics were part of Courses II and III.

The second part of the book contains 10 of the most recent Regents examinations with completely explained solutions for each question. A Self-Analysis Chart has been constructed for each examination. By following the procedure below you can use it to locate your strong and weak points, and to direct you to the proper topics in the Practice Exercises where you need additional study:

1. Do a complete Regents examination, answering *all* questions (even though you will have a choice on the actual Regents).
2. Compare your answers with those in the explained solutions.
3. On the Self-Analysis Chart, find the topic under which each question is classified, and enter the number of points you earned if you answered it correctly.
4. Obtain your percentage for each topic by dividing your earned points by the number of test points on that topic, carrying the division to two decimal places.
5. If you are not satisfied with your percentage on any topic, turn to that topic in the Practice Exercises in the front of the book and answer all the questions there, comparing your results with the explained solutions.
6. If you still need additional practice on a particular topic, locate appropriate questions in other Regents examinations by using

their Self-Analysis Charts to see which questions are listed for this topic. Regents examinations given prior to June, 1989 do not have topic groups 26 through 31 included in their Self-Analysis Charts since these topics were not part of the course prior to that time. For topic groups 26 through 31, you will find additional practice questions in the Regents examinations of June, 1989 and thereafter, and also in the Practice Exercises section in the front of this book.

7. Repeat the first six steps, using other complete Regents examinations, until you are satisfied with your percentages in all the topic classifications.

Test Techniques — Practice Makes Perfect

TAKING MATH EXAMS

Success on any examination, including those in the field of mathematics, is dependent on three factors:

1. Competence in the subject matter tested.
2. "Test wiseness," that is, understanding what to expect as to the nature of the test and the conditions under which it is to be given.
3. Emotional attitude or degree of self-confidence in taking the examination.

This book is designed to improve your effectiveness in all three of the above factors. To improve your mathematics competence, practice by answering a number of the past Regents examinations and afterward studying the explained solutions to the questions. It is advisable to concentrate first on the most recent Regents examinations since questions on topic groups 26 through 31 appear only on tests given in June, 1989 and thereafter, when these topics were added to Course I. If you practice answering a Regents examination given prior to June, 1989, it is suggested that you supplement the Regents questions by also answering one question chosen from each of the topic groups numbered 26 through 31 that appear in the Practice Exercises section in the front of this book. This procedure will give you experience with a total set of questions as close as possible to the ones you will have to answer when you take your own Regents examination. Follow the directions in the section on "How to Use This Book" to focus your study on those topics where you need the most attention.

Any nervousness you may have will disappear and your self-confidence will grow as your record on the Self-Analysis Charts improves and as you become familiar with the nature of the Regents examination through practicing on several of them from prior years.

You can also improve your "test wiseness" by developing a practice strategy that will build habits that can serve you well when you take the actual Regents examination. Some suggestions for such a strategy follow.

TIPS FOR PRACTICE

How to Handle Individual Problems

1. **READ** the problem through the **FIRST TIME** to get a *general idea* of what you are asked to find; determine what question is being asked.

2. **READ** the problem through a **SECOND TIME** to *identify* and *relate* information or data.
3. **READ** the problem through a **THIRD TIME** to pick out specific facts and decide on a method for solving the problem.
4. **DRAW A SKETCH OR DIAGRAM,** if appropriate. Some problems give very little information, whereas others give a great deal. A sketch or diagram will help you identify and relate the *pertinent* information.
5. **ESTIMATE THE ANSWER.**
6. **WRITE AN EQUATION** to translate into symbols any statement in the question indicating that two quantities are equal.
7. **SOLVE THE EQUATION.**
8. **CHECK THE ANSWER BY COMPARING IT WITH YOUR ESTIMATE.**
9. **REREAD THE QUESTION** to make sure you have answered what was asked.

How to Practice Effectively and Efficiently.

1. Do not spend too much time on one question if you cannot come up with a method to be used or if you cannot complete the solution. Instead, put a slash through the number of any question you cannot complete. When you have completed as many questions as you can you will be able to return quickly to the unanswered questions and try them again.
2. After trying the unanswered questions again, check the answer key for the entire test.
3. Circle the number of each question you answered incorrectly.
4. Study the explained solutions for those questions you answered incorrectly. (If the solution employs a formula or rule you do not know, write it on a piece of paper and attach the paper to the inside of your review book cover.)
5. Enter the points for your correct answers on the Self-Analysis Chart following the Regents you tried, and follow the procedure in the section on "How to Use This Book" to direct your remedial study.

LET'S PRACTICE

Part I. Let's do your first practice Regents together. Allow yourself between one hour and one and one-half hours to do Part I. Follow the steps outlined under "Tips for Practice." Answer *all* 35 of the Part I questions even though you will be required to choose only 30 of them on the actual Regents—those on which you believe you can get correct answers. By completing all 35 questions on the practice exams and using the Self-Analysis Charts you will be able to establish a pattern of your strong and weak areas. You may find, for instance, that you usually solve the probability questions correctly but have difficulty with questions on logical implication. Remember your Self-Analysis Chart results, and you will know which 30 questions to choose on the actual Regents.

On the actual Regents, if you have difficulty in finding at least 30 questions on which you are reasonably sure of your ability, it is advisable to choose some multiple-choice questions on which you can eliminate one or more of the choices as obviously incorrect. There is no penalty for guessing, but do not guess until you have first tried to solve the question. If you can validly eliminate some choices, your guess from among the remaining choices will stand a better chance of being correct. Practice this technique as you answer all the Part I questions on practice exams so that you can use it to advantage if you have to resort to it in choosing the questions you will answer on the actual exam.

Part II. Spend one and one-half hours doing *all* seven Part II questions. Follow the same steps as you did in Part I.

On the actual Regents you will have to complete only four of the seven Part II questions—those on which you believe you can score the highest (partial credit is allowed on Part II for solutions that are not completely correct if major parts are accurate).

THE ACTUAL REGENTS

In the weeks before the Regents, you should plan to spend one-half hour preparing each night. It is better to spread out your preparation time this way than to prepare for, say, three hours in one evening.

On the night before the Regents, read over the formulas you have attached to the review book cover. If you have prepared each night for a month or so, this is a good time to study calmly for about an hour and go to bed reasonably early.

You will have three hours to do the Course I Math Regents. After you have practiced on a number of the past examinations, you will be able to complete the actual Regents in about one and one-half hours. Even so, spend the full three hours. It is better to be correct than to be fast.

Answering the Questions. Follow steps similar to those you used for the practice exams. Before going on to Part II, go back over Part I, making sure you have answered 30 questions. Go over any of the 30 you were unsure of (those whose numbers you have marked with a slash) and try them again. If you answered more than 30 questions, cross out the ones you do not wish to have counted.

At the end of the first hour of the exam, go on to Part II. Read all the questions completely and select the four question types you have had the most success with on practice exams according to your Self-Analysis Charts. Work out each problem on scrap paper, and when you feel confident about the solution, copy it into the answer booklet *in ink*. Show all work. If you cannot completely solve four questions, pick the four that will give you the potential for the most credit.

Time Schedule. The Course I Math Regents exam is a three-hour examination. Below is a suggested schedule to follow.

First Hour:
- Complete 30 Part I questions.
- Return to troublesome questions (those you have marked with slashes) and try again.
- In multiple-choice questions, rule out any obviously impossible choices and then guess if you absolutely cannot answer.

Second Hour:
- Complete four Part II questions.

Next 25 Minutes:
- Redo the 30 questions selected from Part I. Do not refer to the scrap paper because it is unlikely you will spot an arithmetic error by scanning your computation. Re-solve the problems instead; it is unlikely you will make exactly the same errors again.
- If any answers differ the second time around, redo these questions a third time. You still may not be able to decide which is the correct answer—in this case, choose the answer that seems more reasonable.

Next 25 Minutes:
- Redo the four Part II questions you chose.
- If you arrive at a different answer, redo the question a third time as you did with Part I.

Last 10 Minutes:
- Check that 30 Part I questions and four Part II questions are answered.
- Check that Part II answers are labeled and that you have included your scrap paper in the booklet.
- If you answered more than 30 Part I questions or more than four Part II questions, write *OMIT* on any you do not wish to be marked.
- Make sure all answers are *in ink* in the answer booklet.

Practice Exercises

This section of the book consists of 189 questions selected from former Regents examinations and classified into 31 topic groups. Solutions are provided for each of the questions with a step-by-step explanation given for each solution. The 31 topic groups are keyed to the Self-Analysis Charts that follow each of the 10 complete Regents examinations. This enables you to make use of these Practice Exercises to overcome any weaknesses that are revealed through the use of the Self-Analysis Charts. Topic groups 26 through 31 do not appear in the Self-Analysis Charts of Regents given prior to June, 1989, since these topics were not added to Course I until that time.

1. QUESTIONS ON NUMBERS (RATIONAL, IRRATIONAL); PERCENT

1. Which is an irrational number?

(1) 0 (2) $\frac{1}{2}$ (3) $\sqrt{3}$ (4) $\sqrt{4}$

2. Which represents an irrational number?

(1) $-\frac{4}{5}$ (2) π (3) $\sqrt{9}$ (4) 0

3. If p represents "x is a prime number," and q represents "x is an even number greater than 2," which of the following must be true?
(1) $p \rightarrow q$ (2) $p \vee q$ (3) $p \wedge q$ (4) $q \rightarrow \sim p$

4. What is 2.5% of 1,000?
(1) 2,500 (2) 250 (3) 25 (4) 2.50

5. If 30 students took an examination and 24 passed, what percent of the students passed the examination?

6. The top 12% of the class is placed on the honor roll. If 42 students are on the honor roll, how many students are in the class?

Solutions to Questions on Numbers (rational, irrational); Percent

1. An irrational number *cannot* be expressed in the form $\frac{p}{q}$ where p and q are integers.

Consider each choice in turn:

(1) 0: 0 can be expressed in the form $\frac{p}{q}$ as $\frac{0}{1}$ or $\frac{0}{3}$ or $\frac{0}{68}$, etc. Therefore, 0 is a rational number.

(2) $\frac{1}{2}$: $\frac{1}{2}$ is in the form $\frac{p}{q}$ and 1 and 2 are integers. Therefore, $\frac{1}{2}$ is a rational number.

(3) $\sqrt{3}$: $\sqrt{3}$ *cannot* be expressed as the ratio of two integers. As a decimal it would be a nonrepeating, nonterminating number, approximately equal to 0.1732 . . . Such a number *cannot* be expressed in the form $\frac{p}{q}$ where p and q are integers. Therefore, $\sqrt{3}$ is an irrational number.

(4) $\sqrt{4}$: $\sqrt{4} = 2$. 2 can be expressed in the form $\frac{p}{q}$ as $\frac{2}{1}$ or $\frac{6}{3}$ or $\frac{10}{20}$, etc. Therefore, $\sqrt{4}$ is a rational number.

The correct choice is **(3)**.

2. An irrational number is a number which *cannot* be written in the form $\frac{p}{q}$, where p and q are integers.

Consider each choice in turn:

(1) $-\frac{4}{5}$ is already in the form $\frac{p}{q}$ with $p = -4$ and $q = 5$. Therefore, $-\frac{4}{5}$ is a rational number.

(2) π is a nonterminating, nonrepeating decimal. Such a decimal *cannot* be written as the quotient of two integers. Therefore, π is an irrational number.

(3) $\sqrt{9} = 3$. 3 can be written in the form $\frac{p}{q}$ as $\frac{3}{1}$ with $p = 3$ and $q = 1$. Therefore, 3 is a rational number.

(4) 0 can be written in the form $\frac{p}{q}$ as $\frac{0}{1}$ with $p = 0$ and $q = 1$. Therefore, 0 is a rational number.

The correct choice is **(2)**.

3. A prime number is a number which is divisible only by itself and 1. Any even number greater than 2 is divisible by 2.

Consider each choice in turn:
(1) $p \rightarrow q$ is the *implication* that if p is true then q is also true. But if x is a prime number then it cannot also be divisible by 2. Hence, the implication $p \rightarrow q$ is false.
(2) $p \vee q$ is the *disjunction* of p and q. It states that either p is true or q is true or both are true. But if x is some number chosen at random, say 9, it may be neither a prime number nor an even number greater than 2. Therefore, the disjunction $p \vee q$ is not always true.
(3) $p \wedge q$ is the *conjunction* of p and q. It states that both p and q are true. But if x is a prime number it cannot also be a number which is even (that is divisible by 2) and greater than 2. Hence, the disjunction $p \wedge q$ can never be true.
(4) $q \rightarrow \sim p$ is the *implication* that if q is true the *negation* of p is true, that is, if q is true p is not true. If x is an even number greater than 2 it is certainly true that it cannot be a prime number. Hence, the implication $q \rightarrow \sim p$ is true.
The correct choice is **(4)**.

4. 2.5% of 1,000
2.5% can be represented as the decimal .025 (the percent sign stands for two decimal places; removing the percent sign thus requires that the decimal point be moved 2 places to the left).
To multiply a number by 1,000, the decimal point is moved 3 places to the right. Thus, $.025 \times 1{,}000 = 25$.
The correct choice is **(3)**.

5. Of the 30 students who took the examination, 24 passed.
The percent of students who passed equals the number who passed divided by the total number of students who took the test: $\frac{24}{30}$

Reduce the fraction to lowest terms by dividing both numerator and denominator by 6: $\frac{4}{5}$

Convert the fraction to a percent, either by remembering that $\frac{4}{5} = 80\%$ or by dividing 5 into 4.00 to get two decimal places, that is .80: 80%
The percent of students who passed the examination is **80**.

6. Let $x =$ the number of students in the class.
Since 12%, written as a decimal is 0.12, we have the equation: $0.12x = 42$

Clear decimals by multiplying both sides of the equation by 100:

$$100(0.12x) = 100(42)$$
$$12x = 4200$$

Divide both sides of the equation by 12:

$$\frac{12x}{12} = \frac{4200}{12}$$
$$x = 350$$

There are **350** students in the class.

2. QUESTIONS ON PROPERTIES OF NUMBER SYSTEMS

1. If a and b are integers, which statement is *always* true?

(1) $a - b = b - a$ (2) $a + b = b + a$ (3) $\frac{a}{b} = \frac{b}{a}$
(4) $a + 2b = b + 2a$

2. Which of the following is illustrated by the equation $x + y = y + x$?
(1) the commutative law for addition
(2) the associative law for addition
(3) the distributive law
(4) the identity property for addition

3. For what value of x is the expression $\frac{8}{x - 7}$ undefined?

4. For which value of x is the expression $\frac{x - 6}{x + 5}$ undefined?

(1) -6 (2) -5 (3) 5 (4) 6

5. Which expression is undefined or meaningless when $x = 3$?

(1) x^0 (2) x^{-3} (3) $\frac{1}{x - 3}$ (4) $\frac{1}{x + 3}$

Solutions to Questions on Properties of Number Systems

1. Consider each of the choices in turn:
(1) $a - b = b - a$ is a statement that subtraction is commutative. But this is not true of integers in general. For example, $8 - 3 \neq 3 - 8$.
(2) $a + b = b + a$ is a statement that addition is commutative. This is always true for any values of the integers a and b.

(3) $\frac{a}{b} = \frac{b}{a}$ is a statement that division is commutative. But this is not true of integers in general. For example, $\frac{4}{2} \neq \frac{2}{4}$.

(4) $a + 2b = b + 2a$ is a statement that the sum of one integer and double another is the same no matter which is doubled. This is not true in general. For example, $4 + 2(3) \neq 3 + 2(4)$ since $4 + 6 \neq 3 + 8$.

The correct choice is **(2)**.

2. The equation $x + y = y + x$ illustrates the fact that the sum of two quantities is the same no matter in which order they are added. This is the commutative law for addition.

Consider why the other choices must be ruled out. The associative law for addition states that $a + (b + c) = (a + b) + c$, that is, the sum of three quantities is the same no matter how they are grouped in pairs (associated) in the process. The distributive law states that $a(b + c) = ab + ac$, that is, in multiplying the sum of two numbers by a multiplier, the multiplier is "distributed" to each of the terms in the sum. The identity property for addition states that if the identity element, 0, is added to any number, the sum is identical to the original number, that is, $0 + a = a$ or $a + 0 = a$.

The correct choice is **(1)**.

3. Division by 0 is undefined. Hence, $\frac{8}{x - 7}$ is undefined if the denominator, $x - 7$, equals 0:

$$x - 7 = 0$$

Add 7 (the additive inverse of -7) to both sides of the equation:

$$\frac{\quad 7 = 7}{x \quad = 7}$$

The expression is undefined if $x = 7$.

4. Division by 0 is undefined. Hence, $\frac{x - 6}{x + 5}$ is undefined if the denominator, $x + 5$, equals 0:

$$x + 5 = 0$$

Add -5 (the additive inverse of 5) to both sides of the equation:

$$\frac{\quad -5 = -5}{x \quad = -5}$$

Note that the value of the numerator, $x - 6$, has no effect on the question of whether the expression is undefined. Choice (4) concerns $x = 6$. If 6 is substituted for x the expression becomes $\frac{6-6}{6+5}$ or $\frac{0}{11}$, which is equal to 0. The expression is thus defined to equal 0 when $x = 6$.

The correct choice is (2).

5. When $x = 3$:

(1) x^0 becomes 3^0 which is equal to 1. x^0 is equal to 1 for all values of x except $x = 0$.

(2) x^{-3} becomes 3^{-3} which is defined as $\frac{1}{3^3}$ or $\frac{1}{27}$.

(3) $\frac{1}{x-3}$ becomes $\frac{1}{3-3}$ or $\frac{1}{0}$. But $\frac{1}{0}$ is undefined or meaningless since division by 0 is undefined.

(4) $\frac{1}{x+3}$ becomes $\frac{1}{3+3}$ which is equal to $\frac{1}{6}$.

The correct choice is (3).

3. QUESTIONS ON OPERATIONS ON RATIONAL NUMBERS AND MONOMIALS

1. The product of $6x^3$ and $5x^4$ is

(1) $11x^{12}$ (2) $11x^7$ (3) $30x^{12}$ (4) $30x^7$

2. Express $\frac{15x^2}{-3x}$ in *simplest form.*

3. The expression $(3x^2y^3)^2$ is equivalent to

(1) $9x^4y^6$ (2) $9x^4y^5$ (3) $3x^4y^6$ (4) $6x^4y^6$

4. The value of 3^{-2} is

(1) -9 (2) -6 (3) $-\frac{1}{9}$ (4) $\frac{1}{9}$

5. The product of $(-2xy^2)(3x^2y^3)$ is

(1) $-5x^3y^5$ (2) $-6x^2y^6$ (3) $-6x^3y^5$ (4) $-6x^3y^6$

6. When $4x^3y^3$ is multiplied by $8xy^2$, the product is

(1) $12x^3y^6$ (2) $12x^4y^6$ (3) $32x^3y^6$ (4) $32x^4y^5$

Solutions to Questions on Operations on Rational Numbers and Monomials

1. $(6x^3)(5x^4)$

To multiply two monomials, first multiply their coefficients to get the coefficient of the product: $6 \cdot 5 = 30$

Multiply literal factors with the same base by adding the exponents of that base to get the exponent of the product: $x^3 \cdot x^4 = x^7$

Putting the two together: $(6x^3)(5x^4) = 30x^7$

The correct choice is (4).

2. $\frac{15x^2}{-3x}$

To divide two monomials, first divide their numerical coefficients to find the numerical coefficient of the quotient: $(15) \div (-3) = -5$

Divide the literal factors to find the literal factor of the quotient. Remember that powers of the same base are divided by subtracting their exponents: $(x^2) \div (x^1) = x^1$

Putting the two results together: $\frac{15x^2}{-3x} = -5x$

The fraction in simplest form is $\mathbf{-5x}$.

3. $(3x^2y^3)^2$ means $(3x^2y^3)(3x^2y^3)$

The numerical coefficient of the product is the product of the two numerical coefficients: $(3)(3) = 9$

In multiplying powers of the same base, the exponents are added to obtain the exponent for that base in the product: $(x^2y^3)(x^2y^3) = x^4y^6$

Hence: $(3x^2y^3)^2 = 9x^4y^6$

The correct choice is (1).

4. By the definition of a negative exponent, $x^{-n} = \frac{1}{x^n}$. Therefore, $3^{-2} = \frac{1}{3^2}$ or $\frac{1}{9}$.

The correct choice is (4).

5. $(-2xy^2)(3x^2y^3)$

To find the product of two monomials, first find the numerical coefficient of the product by multiplying the two numerical coefficients together: $(-2)(3) = -6$

Find the literal factor of the product by multiplying the literal factors together. The product of two powers of the same base is found by adding the exponents of that base: $(x^1y^2)(x^2y^3) = x^3y^5$

Putting the two results together: $(-2xy^2)(3x^2y^3) = -6x^3y^5$

The correct choice is (3).

6. Indicate the product as: $(4x^3y^3)(8xy^2)$

First find the numerical coefficient of the product by multiplying the two numerical coefficients together: $(4)(8) = 32$

Find the literal factor of the product by multiplying the literal factors together. The product of two powers of the same base is found by adding the exponents of that base: $(x^3y^3)(x^1y^2) = x^4y^5$

Putting the two results together: $(4x^3y^3)(8xy^2) = 32x^4y^5$

The correct choice is (4).

4. QUESTIONS ON OPERATIONS ON POLYNOMIALS

1. Subtract $4m - h$ from $4m + h$.

2. Find the sum of $2x^2 + 3x - 1$ and $3x^2 - 2x + 4$.

3. Perform the indicated operations and express the result as a trinomial:

$$3x(x + 1) + 4(x - 1)$$

4. Express the product $(2x - 7)(x + 3)$ as a trinomial.

5. If the product $(2x + 3)(x + k)$ is $2x^2 + 13x + 15$, find the value of k.

6. When $3x^3 + 3x$ is divided by $3x$, the quotient is
(1) x^2 (2) $x^2 + 1$ (3) $x^2 + 3x$ (4) $3x^3$

Solutions to Questions on Operations on Polynomials

1. Subtraction is the inverse operation of addition. Thus, to subtract we add the additive inverse of the subtrahend (expression to be subtracted) to the minuend (expression subtracted from). Therefore, to subtract $4m - h$ from $4m + h$, we add $-4m + h$ to $4m + h$:

Subtract: $\begin{array}{r} 4m + h \\ \underline{4m - h} \end{array}$ means add: $\begin{array}{r} 4m + h \\ \underline{-4m + h} \\ 2h \end{array}$

The difference is **$2h$**.

2. Write the second trinomial under the first, placing like terms in the same column:

$$\begin{array}{r} 2x^2 + 3x - 1 \\ \underline{3x^2 - 2x + 4} \end{array}$$

Add each column by adding the numerical coefficients algebraically and bringing down the literal factor:

$$5x^2 + \quad x + 3$$

The sum is **$5x^2 + x + 3$**.

3. $3x(x + 1) + 4(x - 1)$

Remove parentheses by applying the distributive law of multiplication over addition (multiply each term in the parentheses by the factor outside): $3x^2 + 3x + 4x - 4$

Combine like terms: $3x^2 + 7x - 4$

The trinomial is **$3x^2 + 7x - 4$**.

4. $(2x - 7)(x + 3)$

Apply the distributive law by multiplying each term of $(2x - 7)$ by $(x + 3)$: $2x(x + 3) - 7(x + 3)$

Again apply the distributive law by multiplying each term in the parentheses by the factor outside: $2x^2 + 6x - 7x - 21$

Combine like terms: $2x^2 - x - 21$

The trinomial is **$2x^2 - x - 21$**.

ALTERNATIVE SOLUTION: The product of two binomials may be found by multiplying each term of one by each term of the other, using a procedure analogous to that used in arithmetic to multiply multidigit numbers:

$$\begin{array}{r} 2x - 7 \\ \underline{x + 3} \\ 2x^2 - 7x \quad\quad \\ \underline{6x - 21} \\ 2x^2 - \quad x - 21 \end{array}$$

Combine like terms: $2x^2 - x - 21$

5. $(2x + 3)(x + k)$

Apply the distributive law by multiplying $(x + k)$ by each term of $(2x + 3)$: $2x(x + k) + 3(x + k)$

Again apply the distributive law by multiplying each term in parentheses by the factor outside: $2x^2 + 2kx + 3x + 3k$

Factor x from the two middle terms: $2x^2 + x(2k + 3) + 3k$

Since this product is equal to $2x^2 + 13x + 15$, the coefficients of x must be equal and the constant terms must be equal: $2k + 3 = 13 \qquad 3k = 15$

Solve either of these equations for k:

$$\begin{array}{rl} 2k + 3 &= 13 \\ -3 &= -3 \\ \hline 2k &= 10 \\ \frac{2k}{2} &= \frac{10}{2} \\ k &= 5 \end{array} \qquad \begin{array}{rl} 3k &= 15 \\ \frac{3k}{3} &= \frac{15}{3} \\ k &= 5 \end{array}$$

The value of k is **5**.

6. Indicate the division in fractional form: $\frac{3x^3 + 3x}{3x}$

Apply the distributive law by dividing each term of $3x^3 + 3x$ in turn by $3x$. Remember that in dividing powers of the same base, the exponents are subtracted. Also note that $3x \div 3x = 1$: $x^2 + 1$

The correct choice is **(2)**.

5. QUESTIONS ON SQUARE ROOT; OPERATIONS ON RADICALS

1. The expression $\sqrt{300}$ is equivalent to
(1) $50\sqrt{6}$ (2) $12\sqrt{5}$ (3) $3\sqrt{10}$ (4) $10\sqrt{3}$

2. Which is equivalent to $\sqrt{40}$?
(1) 20 (2) $2\sqrt{10}$ (3) $10\sqrt{2}$ (4) $4\sqrt{10}$

3. The expression $\sqrt{27} + \sqrt{12}$ is equivalent to
(1) $\sqrt{39}$ (2) $13\sqrt{3}$ (3) $5\sqrt{6}$ (4) $5\sqrt{3}$

4. The sum of $2\sqrt{3}$ and $\sqrt{12}$ is
(1) $4\sqrt{6}$ (2) $8\sqrt{3}$ (3) $3\sqrt{15}$ (4) $4\sqrt{3}$

5. The expression $\sqrt{50} + 3\sqrt{2}$ can be written in the form $x\sqrt{2}$. Find x.

Solutions to Questions on Square Root; Operations on Radicals

1. $\sqrt{300}$

Simplify $\sqrt{300}$ by finding two factors of the radicand, 300, one of which is the highest perfect square that divides into 300: $\sqrt{100(3)}$

Take the square root of the perfect square factor, 100, and place it outside the radical sign as the coefficient, 10: $10\sqrt{3}$

The correct choice is **(4)**.

2. $\sqrt{40}$

Factor out any perfect square factors in the radicand (the number under the radical sign): $\sqrt{4(10)}$

Remove the perfect square factor from under the radical sign by taking its square root and writing it as a coefficient of the radical: $2\sqrt{10}$

The correct choice is **(2)**.

3. $\sqrt{27} + \sqrt{12}$

We cannot add $\sqrt{27}$ and $\sqrt{12}$ in their present form because only *like radicals* may be combined. Like radicals have the same radicand (number under the radical sign) and the same index (here, understood to be 2, representing the square root). A radical can be simplified by finding two factors of the radicand, one of which is the highest possible perfect square.

$\sqrt{27} + \sqrt{12}$

STEP 1: Factor the radicands, using the highest possible perfect square in each: $\sqrt{9 \cdot 3} + \sqrt{4 \cdot 3}$

STEP 2: Simplify by taking the square root of the perfect square factor and placing it outside the radical sign as a coefficient of the radical: $3\sqrt{3} + 2\sqrt{3}$

STEP 3: Combine the like radicals by adding their coefficients to get the new coefficient: $5\sqrt{3}$

The given expression is equivalent to $5\sqrt{3}$.

The correct choice is **(4)**.

4. $2\sqrt{3} + \sqrt{12}$

Only *like radicals* can be added. Like radicals must have the same root (here both are square roots) and must have the same radicand (the number under the radical sign). Factor out any perfect square factor in the radicands: $2\sqrt{3} + \sqrt{4(3)}$

Remove the perfect square factors from under the radical sign by taking their square roots and writing them as coefficients of the radical: $2\sqrt{3} + 2\sqrt{3}$

Combine the like radicals by adding their numerical coefficients and writing the sum as the numerical coefficient of the common radical: $4\sqrt{3}$

The correct choice is (4).

5. $\sqrt{50} + 3\sqrt{2}$

Factor out any perfect square factors in the radicand (the number under the radical sign): $\sqrt{25(2)} + 3\sqrt{2}$

Remove the perfect square factors from under the radical sign by taking their square roots and writing them as coefficients of the radical: $5\sqrt{2} + 3\sqrt{2}$

Combine the like radicals by combining their coefficients and using the sum as the numerical coefficient of the common radical: $8\sqrt{2}$

$8\sqrt{2}$ is in the form $x\sqrt{2}$ with $x = 8$.

$x = 8$.

6. QUESTIONS ON EVALUATING FORMULAS AND EXPRESSIONS

1. If $b = -2$ and $c = 3$, find the value of $b^2 + c$.

2. Find the value of $6b^2 - 4a^2$ when $b = 2$ and $a = 1$.

3. If $x = 3$ and $y = -1$, find the value of $2x + y^2$.

4. Using the formula $V = x^2h$, find V when $x = 6$ and $h = 2$.

5. Given the formula $F = \frac{9}{5}C + 32$, find F when $C = 15$.

6. If $xy^2 = 18$, find x when $y = -3$.

Solutions to Questions on Evaluating Formulas and Expressions

1. $b^2 + c$ $\quad b = -2, c = 3$

Substitute -2 for b and 3 for c in the given expression: $(-2)^2 + 3$

Simplify: $4 + 3$

7

The value is 7.

2. $6b^2 - 4a^2$ $\quad b = 2 \quad a = 1$

Substitute 2 for b and 1 for a in the given expression:

$$6(2)^2 - 4(1)^2$$
$$6(4) - 4(1)$$
$$24 - 4$$
$$20$$

The value is **20.**

3. $2x + y^2$

Since $x = 3$ and $y = -1$, substitute these values for x and y respectively: $2(3) + (-1)^2$

Perform the indicated multiplications: $6 + 1$

Combine like terms: 7

The value is **7.**

4. $V = x^2h$

Since $x = 6$ and $h = 2$, substitute 6 for x and 2 for h in the formula: $V = (6)^2(2)$

Square 6: $V = 36(2)$

Perform the indicated multiplication: $V = 72$

$\mathbf{V = 72}$.

5. $F = \frac{9}{5}C + 32$

Substitute 15 for C in the formula: $F = \frac{9}{5}(15) + 32$

Divide the factor 5 out of numerator and denominator of the first term on the right side of the equation: $F = \frac{9}{\cancel{5}_1}(\cancel{15}^3) + 32$

Multiply 9 by 3: $F = 27 + 32$

Combine like terms: $F = 59$

$\mathbf{F = 59}$.

6. $xy^2 = 18$

Substitute -3 for y: $x(-3)^2 = 18$

Square -3: $9x = 18$

Divide both sides of the equation by 9: $\frac{9x}{9} = \frac{18}{9}$

$x = 2$

$\mathbf{x = 2}$.

7. QUESTIONS ON SIMPLE LINEAR EQUATIONS (INCLUDING PARENTHESES)

1. Solve for x: $4x = x + 21$

2. Solve for x: $8x = 2(x + 15)$

3. Solve for x: $3(x + 4) - x = 18$

4. Solve for x: $2(x + 3) = 12$

5. Solve for x: $3(2x - 5) = 9$

6. Solve for y: $6(y + 3) = 2y - 2$

Solutions to Questions on Simple Linear Equations (including parentheses)

1. $4x = x + 21$

Add $-x$ (the additive inverse of x) to both sides of the equation:

$$\begin{array}{r} -x = -x \\ \hline 3x = \quad 21 \end{array}$$

Divide both sides of the equation by 3:

$$\frac{3x}{3} = \frac{21}{3}$$

$$x = 7$$

The solution is **7**.

2. Remove parentheses by using the distributive property of multiplication on the right side:

$$8x = 2(x + 15)$$

$$8x = 2x + 30$$

Add $-2x$ (the additive inverse of $2x$) to both sides:

$$\begin{array}{r} -2x = -2x \\ \hline 6x = 30 \end{array}$$

Multiply both sides by $\frac{1}{6}$ (the multiplicative inverse of 6):

$$\frac{1}{6}(6x) = \frac{1}{6}(30)$$

The solution is $x = $ **5**.

$$x = 5$$

3. $3(x + 4) - x = 18$

Remove parentheses by using the distributive property of multiplication:

$$3x + 12 - x = 18$$

Combine like terms: $2x + 12 = 18$

Add -12 (the additive inverse of 12) to both sides of the equation:

$$\begin{array}{r} 2x + 12 = 18 \\ -12 = -12 \\ \hline 2x \quad = 6 \end{array}$$

Multiply both sides by $\frac{1}{2}$ (the multiplicative inverse of 2):

$$\frac{1}{2}(2x) = \frac{1}{2}(6)$$

$$x = 3$$

The value of x is 3.

4.

$$2(x + 3) = 12$$

Remove parentheses by applying the distributive law of multiplication over addition:

$$2x + 6 = 12$$

Add -6 (the additive inverse of 6) to both sides of the equation:

$$\begin{array}{r} -6 = -6 \\ \hline 2x \quad = 6 \end{array}$$

Divide both sides of the equation by 2:

$$\frac{2x}{2} = \frac{6}{2}$$

$$x = 3$$

The solution is 3.

5.

$$3(2x - 5) = 9$$

Apply the distributive law of multiplication over addition in order to remove parentheses:

$$6x - 15 = 9$$

Add 15 (the additive inverse of -15) to both sides of the equation:

$$\begin{array}{r} 15 = 15 \\ \hline 6x \quad = 24 \end{array}$$

Divide both sides of the equation by 6:

$$\frac{6x}{6} = \frac{24}{6}$$

$$x = 4$$

The solution is 4.

6.

$$6(y + 3) = 2y - 2$$

Remove parentheses by applying the distributive law of multiplication over addition:

$$6y + 18 = 2y - 2$$

Add -18 (the additive inverse of 18) and also add $-2y$ (the additive inverse of $2y$) to both sides of the equation:

$$\begin{array}{r} -2y - 18 = -2y - 18 \\ \hline 4y \quad = \quad -20 \end{array}$$

Divide both sides of the equation by 4:

$$\frac{4y}{4} = \frac{-20}{4}$$

$$y = -5$$

The solution is $y = -5$.

8. QUESTIONS ON LINEAR EQUATIONS CONTAINING DECIMALS OR FRACTIONS

1. Solve for y: $\frac{y}{3} + 2 = 5$

2. Solve for a: $\frac{a}{2} + \frac{a}{6} = 2$

3. Solve for h: $\frac{24}{h} = \frac{16}{4}$

4. Solve for x: $0.4x + 2 = 12$

5. Solve for x: $1.04x + 8 = 60$

6. Solve for n: $3n + 1.4n = 8.8$

Solutions to Questions on Linear Equations Containing Decimals or Fractions

1.

$$\frac{y}{3} + 2 = 5$$

To clear fractions in the equation, multiply each term by the least common denominator, in this case, by 3:

$$3\left(\frac{y}{3}\right) + 3(2) = 3(5)$$

Simplify:

$$y + 6 = 15$$

Add -6 (the additive inverse of 6) to both sides:

$$\begin{array}{r} -6 = -6 \\ \hline y \quad = 9 \end{array}$$

The solution for y is **9**.

2. To clear fractions in an equation, multiply each term on both sides by the

$$\frac{a}{2} + \frac{a}{6} = 2$$

lowest common denominator, in this case, 6:

$$6\left(\frac{a}{2}\right) + 6\left(\frac{a}{6}\right) = 6(2)$$

$$3a + a = 12$$

Combine like terms:

$$4a = 12$$

Multiply both sides by $\frac{1}{4}$ (the multiplicative inverse of 4):

$$\frac{1}{4}(4a) = \frac{1}{4}(12)$$

$$a = 3$$

The solution is $a = 3$.

3.

$$\frac{24}{h} = \frac{16}{4}$$

Reduce the fraction on the right side to lowest terms by dividing numerator and denominator by 4:

$$\frac{24}{h} = \frac{4}{1}$$

The equation is a *proportion.* In a proportion, the product of the means equals the product of the extremes. Therefore, cross-multiply:

$$4h = 24$$

Multiply each side by $\frac{1}{4}$ (the multiplicative inverse of 4):

$$\frac{1}{4}(4h) = \frac{1}{4}(24)$$

$$h = 6$$

The solution is **6.**

4.

$$0.4x + 2 = 12$$

To clear decimals, multiply each term on both sides of the equation by 10:

$$10(0.4x) + 10(2) = 10(12)$$

$$4x + 20 = 120$$

Add -20 (the additive inverse of 20) to both sides of the equation:

$$-20 = -20$$

$$4x \quad = 100$$

Multiply both sides of the equation by $\frac{1}{4}$ (the multiplicative inverse of 4):

$$\frac{1}{4}(4x) = \frac{1}{4}(100)$$

$$x = 25$$

The solution for x is **25.**

5.

$$1.04x + 8 = 60$$

Clear decimals by multiplying each term on both sides of the equation by 100:

$$100(1.04x) + 100(8) = 100(60)$$

$$104x + 800 = 6{,}000$$

Add -800 (the additive inverse of 800) to both sides of the equation:

$$\begin{aligned} -800 &= -800 \\ \hline 104x &= 5{,}200 \end{aligned}$$

Divide both sides of the equation by 104:

$$\frac{104x}{104} = \frac{5{,}200}{104}$$

$$\begin{array}{r} 50 \\ 104\overline{)5{,}200} \\ \underline{5\ 20} \end{array}$$

$$x = 50$$

The solution is **50.**

6. $$3n + 1.4n = 8.8$$

Clear decimals by multiplying each term on both sides of the equation by 10:

$$10(3n) + 10(1.4n) = 10(8.8)$$
$$30n + 14n = 88$$

Combine like terms:

$$44n = 88$$

Divide both sides of the equation by 44:

$$\frac{44n}{44} = \frac{88}{44}$$

$$n = 2$$

The solution is **2.**

9. QUESTIONS ON GRAPHS OF LINEAR FUNCTIONS (INCLUDING SLOPES)

1. Which point lies on the graph of $2x + y = 10$?
(1) (0,8) (2) (10,0) (3) (3,4) (4) (4,3)

2. A point on the graph of $x + 3y = 13$ is
(1) (4,4) (2) $(-2,3)$ (3) $(-5,6)$ (4) $(4,-3)$

3. What is the slope of the line whose equation is $4y = 3x + 16$?

4. What is the slope of the graph of $y = 2x + 3$?

5. The graph of $y = 3x - 4$ is parallel to the graph of
(1) $y = 4x - 3$ (2) $y = 3x + 4$ (3) $y = -3x + 4$
(4) $y = 3$

6. The equation of a line whose slope is 2 and whose y-intercept is -2 is
(1) $2y = x - 2$ (2) $y = -2$ (3) $y = -2x + 2$
(4) $y = 2x - 2$

Solutions to Questions on Graphs of Linear Functions (including slopes)

1. $2x + y = 10$
If a point lies on the graph of the equation above, its coordinates must satisfy the equation. Therefore, test each point in turn by substituting its coordinates for x and y to see if the equation is satisfied:

(1) (0,8): $2(0) + 8 \stackrel{?}{=} 10$
$0 + 8 \stackrel{?}{=} 10$
$8 \neq 10$ (0,8) does *not* check.
(2) (10,0): $2(10) + 0 \stackrel{?}{=} 10$
$20 + 0 \stackrel{?}{=} 10$
$20 \neq 10$ (10,0) does *not* check.
(3) (3,4): $2(3) + 4 \stackrel{?}{=} 10$
$6 + 4 \stackrel{?}{=} 10$
$10 = 10$✓ (3,4) is on the graph.
(4) (4,3): $2(4) + 3 \stackrel{?}{=} 10$
$8 + 3 \stackrel{?}{=} 10$
$11 \neq 10$ (4,3) does *not* check.
The correct choice is **(3)**.

2. If a point is on the graph of $x + 3y = 13$, then its coordinates must satisfy this equation.
Try each choice in turn by substituting the coordinates in the equation $x + 3y = 13$:

(1) (4,4): $4 + 3(4) \stackrel{?}{=} 13$
$4 + 12 \stackrel{?}{=} 13$
$16 \neq 13$ (4,4) is *not* on the graph.
(2) (−2,3): $-2 + 3(3) \stackrel{?}{=} 13$
$-2 + 9 \stackrel{?}{=} 13$
$7 \neq 13$ (−2,3) is *not* on the graph.
(3) (−5,6): $-5 + 3(6) \stackrel{?}{=} 13$
$-5 + 18 \stackrel{?}{=} 13$
$13 = 13$✓ (−5,6) *is* on the graph.

(4) $(4,-3)$: $4+3(-3) \stackrel{?}{=} 13$
$4-9 \stackrel{?}{=} 13$
$-5 \neq 13$ $(4,-3)$ is *not* on the graph.

The correct choice is (3).

3. $4y=3x+16$

If an equation of a straight line is in the form $y=mx+b$, then m represents its slope.

To get $4y=3x+16$ into the $y=mx+b$ form, divide each term by 4:

$$y=\frac{3}{4}x+4$$

Here, $m=\frac{3}{4}$.

The slope is $\frac{3}{4}$.

4. If the equation of a straight line is in the form $y=mx+b$, then m represents the slope.

$y=2x+3$ is in the form $y=mx+b$ with $m=2$; therefore, the slope is 2.

The slope is **2**.

5. If two lines are parallel, they must have the same slope.

If the equation of a line is in the form $y=mx+b$, then m represents the slope. The given equation, $y=3x-4$, is in the $y=mx+b$ form with $m=3$. Its slope is 3.

Examine each choice in turn to see which one also has a slope of 3:

(1) $y=4x-3$ is in the $y=mx+b$ form with $m=4$.
Therefore, it is *not* parallel to $y=3x-4$.

(2) $y=3x+4$ is in the $y=mx+b$ form with $m=3$.
Therefore, it *is* parallel to $y=3x-4$.

(3) $y=-3x+4$ is in the $y=mx+b$ form with $m=-3$.
Therefore, it is *not* parallel to $y=3x-4$.

(4) $y=3$ can be written in the $y=mx+b$ form as $y=0x+3$. From this form, $m=0$. Therefore, $y=3$ is *not* parallel to $y=3x-4$.

The correct solution is **(2)**.

6. An equation of a line whose slope is m and whose y-intercept is b can be written in the form $y=mx+b$.

In this case, $m=2$ and $b=-2$. Hence, an equation of the line is $y=2x-2$.

The correct choice is **(4)**.

10. QUESTIONS ON INEQUALITIES

1. Which inequality is represented by the accompanying graph?

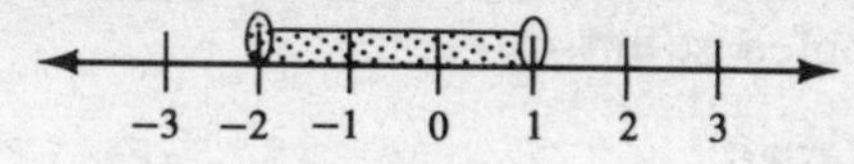

(1) $-2 < x < 1$ (2) $-2 < x \leq 1$ (3) $-2 \leq x < 1$
(4) $-2 \leq x \leq 1$

2. Which graph shows the solution to $(x < 3) \vee (x \geq 5)$?

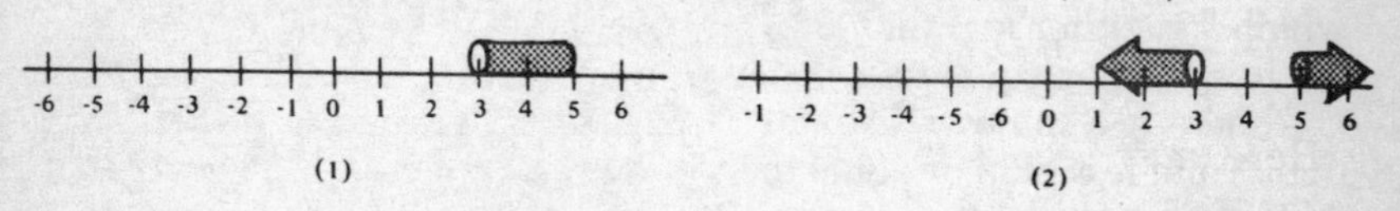

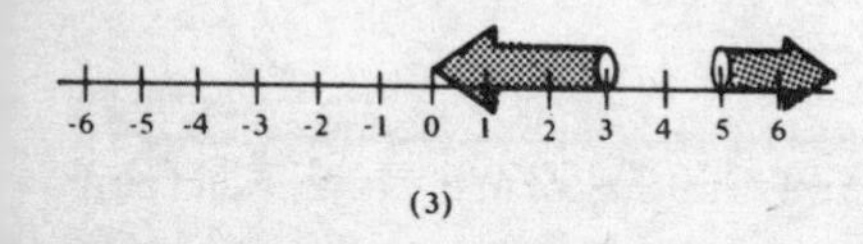

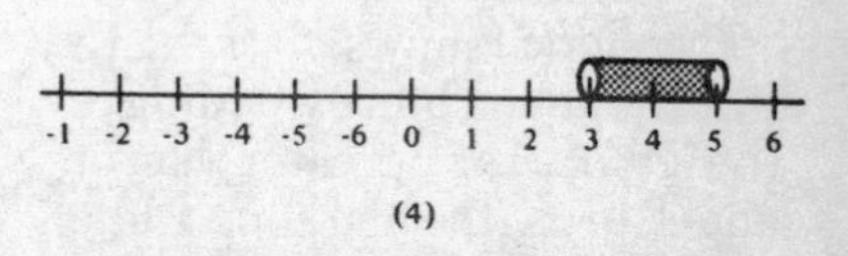

3. The inequality $2x > x + 3$ is equivalent to
(1) $x > 1$ (2) $x > \frac{3}{2}$ (3) $x = 3$ (4) $x > 3$

4. What are the numbers in the solution set of $4 \leq x < 7$ if x is an integer?
(1) 5,6 (2) 5,6,7 (3) 4,5,6 (4) 4,5,6,7

5. The inequality $2x + 5 > x + 3$ is equivalent to
(1) $x > -2$ (2) $x > 2$ (3) $x > 8$ (4) $x > 4$

6. Given the replacement set {5,6,7,8}. Which member of the replacement set will make the statement $(x < 6) \vee (x < 8)$ false?

Solutions to Questions on Inequalities

1. The graph shows an inequality extending from -2 to $+1$. The heavy, shaded circle at -2 indicates that -2 is included, that is, that the inequality represents some numbers greater than or equal to -2. The unshaded, open circle

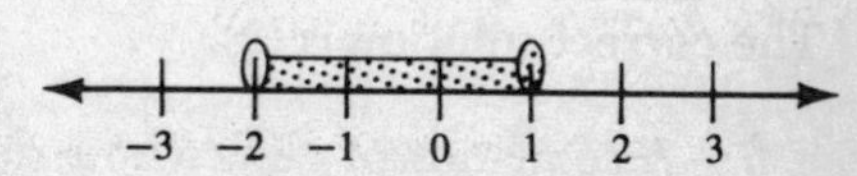

at $+1$ indicates that $+1$ is not included, that is, that the inequality represents some numbers less than $+1$, but not including $+1$. The inequality is therefore $-2 \leq x < 1$.

The correct choice is (**3**).

2. $(x < 3) \vee (x \geq 5)$ stands for the set of values of x for which x is either less than 3 or greater than or equal to 5.

Choice (1) shows x greater than 3 and less than or equal to 5. Therefore, this is incorrect.

Choice (2) shows x less than 3 (the darkened line extending to the left of 3) or x greater than or equal to 5 (the darkened line extending to the right of 5 and including 5). This is the correct choice.

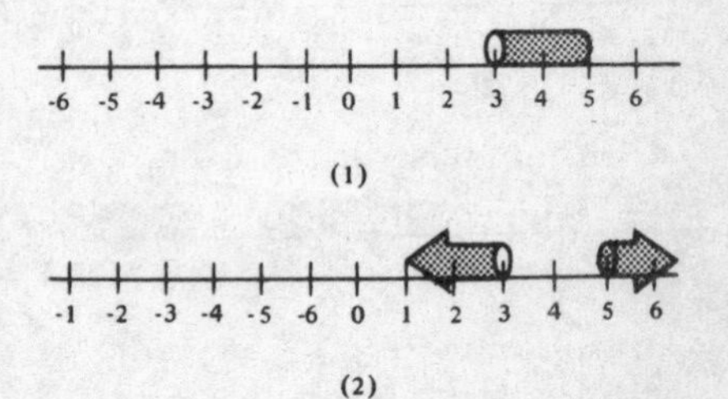

Choice (3) shows x less than 3 or x greater than 5. The open circle at $x = 5$ denotes the fact that $x = 5$ is *not* included. Therefore, this choice is incorrect.

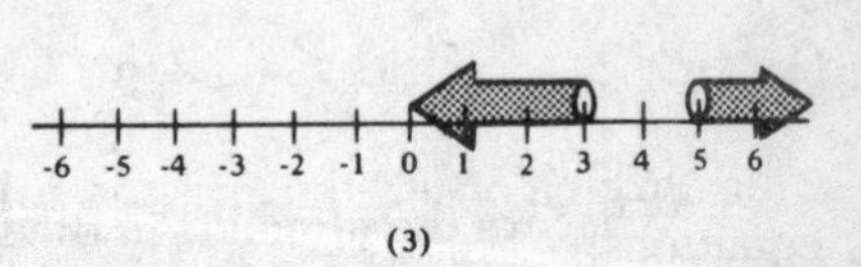

Choice (4) shows x greater than 3 and less than 5. Therefore, this choice is incorrect.

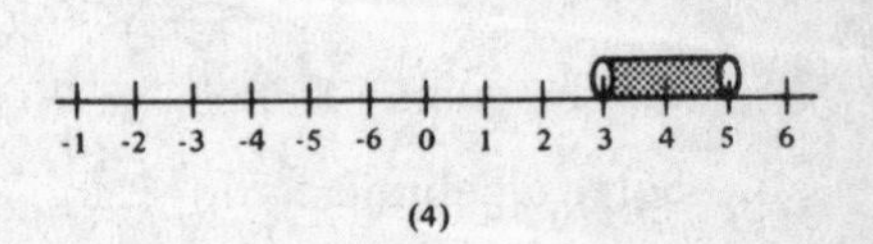

The correct choice is (**2**).

3.

Add $-x$ (the additive inverse of x) to both sides of the inequality:

$$\begin{array}{r} 2x > x + 3 \\ \underline{-x = -x} \\ x > \quad 3 \end{array}$$

The correct choice is (**4**).

4. $4 \leq x < 7$ $\qquad$ x is an integer

The given relationship means that x is greater than or equal to 4 but less than 7.

The only set of integers that fits these requirements is {4,5,6}. The correct choice is (3).

5.

$$2x + 5 > x + 3$$

Add -5 (the additive inverse of 5) and also add $-x$ (the additive inverse of x) to both sides of the inequality:

$$\begin{array}{rcl} -x - 5 & = & -x - 5 \\ \hline x & > & -2 \end{array}$$

The correct choice is (1).

6. $(x < 6) \vee (x < 8)$ is the *disjunction* of $(x < 6)$ and $(x < 8)$. The disjunction is true if either $(x < 6)$ or $(x < 8)$ or both are true. Hence, it is false if neither $(x < 6)$ nor $(x < 8)$ is true.

Consider the elements in the replacement set {5,6,7,8}. $(x < 6)$ is false if $x = 6$, 7, or 8. $(x < 8)$ is false if $x = 8$. Hence, both $(x < 6)$ and $(x < 8)$ are false if $x = 8$. Thus, $(x < 6) \vee (x < 8)$ is false if $x = 8$.

8 will make it false.

11. QUESTIONS ON SYSTEMS OF EQUATIONS AND INEQUALITIES (ALGEBRAIC AND GRAPHICAL SOLUTIONS)

1. Solve for x:

$$\begin{aligned} x + y &= 7 \\ 2x - y &= 2 \end{aligned}$$

2. Which ordered pair is the solution to the following system of equations?

$$\begin{aligned} 3x + 2y &= 4 \\ -2x + 2y &= 24 \end{aligned}$$

(1) (2,−1) (2) (−4,8) (3) (−4,−8) (4) (2,−5)

3. Solve algebraically for x and y and check:

$$\begin{aligned} 2x + y &= 6 \\ x &= 3y + 10 \end{aligned}$$

4. Solve graphically and check:

$$\begin{aligned} y &= x + 3 \\ 2x + y &= 3 \end{aligned}$$

5. a. On the same set of coordinate axes, graph the following system of inequalities:

$$\begin{aligned} y &> x + 4 \\ x + y &\leq 2 \end{aligned}$$

b. Which point is in the solution set of the graph drawn in answer to part a?
(1) (2,3) (2) (−5,2) (3) (0,6)
(4) (−1,0)

6. On the same set of coordinate axes, graph the following system of inequalities and label the solution set A:

$$2y \geq x - 4$$
$$y < 3x$$

Solutions to Questions on Systems of Equations and Inequalities (algebraic and graphical solutions)

1.

$$x + y = 7$$

Adding the two equations together will eliminate y:

$$2x - y = 2$$
$$3x \quad = 9$$

Multiply both sides by $\frac{1}{3}$ (the multiplicative inverse of 3):

$$\frac{1}{3}(3x) = \frac{1}{3}(9)$$
$$x = 3$$

The solution for x is 3.

2.

$$3x + 2y = 4$$
$$-2x + 2y = 24$$

Subtracting the second equation from the first will eliminate y.

To subtract, change the signs of each term in the equation being subtracted and proceed as in addition:

$$3x + 2y = 4$$
$$\overset{+}{\ominus} 2x \overset{-}{\oplus} 2y = \overset{-}{\oplus} 24$$
$$5x \quad = -20$$

Multiply both sides by $\frac{1}{5}$ (the multiplicative inverse of 5):

$$\frac{1}{5}(5x) \quad = \frac{1}{5}(-20)$$
$$x = -4$$

Substitute -4 for x in the first equation:

$$3x + 2y = 4$$
$$3(-4) + 2y = 4$$
$$-12 + 2y = 4$$

Add 12 (the additive inverse of -12) to both sides of the equation:

$$12 \quad = 12$$
$$2y = 16$$

Multiply both sides by $\frac{1}{2}$ (the multiplicative inverse of 2):

$$\frac{1}{2}(2y) = \frac{1}{2}(16)$$
$$y = 8$$

The solution to the system is $x = -4$, $y = 8$ or $(-4,8)$.
The correct choice is **(2)**.

ALTERNATIVE SOLUTION: The correct ordered pair may be found by trying each of the four choices in turn. The values of x and y from each choice would have to be substituted in *both* equations; the correct ordered pair must satisfy *both* equations. Note that this procedure might require 8 testings (the 4 choices tried in each of the two equations).

3.

$$2x + y = 6$$
$$x = 3y + 10$$

Rearrange the second equation by adding $-3y$ (the additive inverse of $3y$) to both sides:

$$\begin{aligned} x &= 3y + 10 \\ -3y &= -3y \\ \hline x - 3y &= 10 \end{aligned}$$

Multiply each term in the first equation by 3: $6x + 3y = 18$

Add the new form of the second equation, thus eliminating y:

$$\begin{aligned} x - 3y &= 10 \\ \hline 7x \quad &= 28 \end{aligned}$$

Multiply both sides by $\frac{1}{7}$ (the multiplicative inverse of 7):

$$\frac{1}{7}(7x) = \frac{1}{7}(28)$$
$$x = 4$$

Substitute 4 for x in the first equation:

$$2(4) + y = 6$$
$$8 + y = 6$$

Add -8 (the additive inverse of 8) to both sides:

$$\begin{aligned} -8 \quad &= -8 \\ \hline y &= -2 \end{aligned}$$

The solution is $\boldsymbol{x = 4, y = -2}$.

CHECK: Substitute 4 for x and -2 for y in both original equations to see if they are satisfied:

$$\begin{aligned} 2x + y &= 6 \\ 2(4) - 2 &\stackrel{?}{=} 6 \\ 8 - 2 &\stackrel{?}{=} 6 \\ 6 &= 6 \checkmark \end{aligned} \qquad \begin{aligned} x &= 3y + 10 \\ 4 &\stackrel{?}{=} 3(-2) + 10 \\ 4 &\stackrel{?}{=} -6 + 10 \\ 4 &= 4 \checkmark \end{aligned}$$

4.
$$y = x + 3$$
$$2x + y = 3$$

To solve graphically, the graphs of both equations are drawn on the same set of axes. The coordinates of the point of intersection of the two graphs represent the solution to the system.

STEP 1: Draw the graph of $y = x + 3$. Set up a table of values by choosing 3 convenient values for x and substituting them in the equation to find the corresponding values of y:

x	$x + 3$	$= y$
0	$0 + 3$	$= 3$
3	$3 + 3$	$= 6$
-3	$-3 + 3$	$= 0$

Plot the points (0,3), (3,6), and (−3,0). They should lie on a straight line. Draw this line; it is the graph of $y = x + 3$.

STEP 2: Draw the graph of $2x + y = 3$. To do this, it is advisable to first rearrange the equation so that it is in a form in which it is solved for y in terms of x:

Add $-2x$ (the additive inverse of $2x$) to both sides:

$$\begin{array}{rcr} 2x + y & = & 3 \\ \underline{-2x \qquad} & \underline{=} & \underline{-2x} \\ y & = & 3 - 2x \end{array}$$

Set up a table of values by choosing 3 convenient values for x and substituting them in the equation to find the corresponding values of y:

x	$3 - 2x$	$= y$
0	$3 - 2(0) = 3 - 0$	$= 3$
3	$3 - 2(3) = 3 - 6$	$= -3$
-3	$3 - 2(-3) = 3 + 6$	$= 9$

Plot the points (0,3), (3,−3), and (−3,9). They should lie in a straight line. Draw a line through them; this line is the graph of $2x + y = 3$.

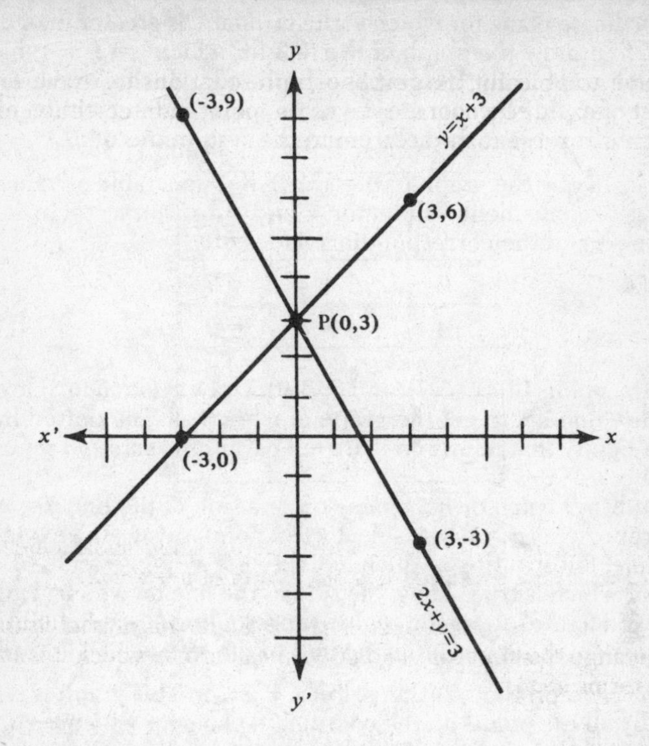

STEP 3: The common solution is represented by P, the point of intersection of the two graphs. The coordinates of P are (0,3) or $x = 0$ and $y = 3$.

The solution to the system is $\boldsymbol{x = 0, y = 3}$ **or {(0,3)}**.

CHECK: The solution is checked by substituting 0 for x and 3 for y in *both* of the two *original* equations to see if they are satisfied:

$y = x + 3$	$2x + y = 3$
$3 \stackrel{?}{=} 0 + 3$	$2(0) + 3 \stackrel{?}{=} 3$
$3 = 3\checkmark$	$0 + 3 \stackrel{?}{=} 3$
	$3 = 3\checkmark$

5. a. $y > x + 4$
$x + y \leq 2$

STEP 1: Graph the solution set of the inequality $y > x + 4$. The graph of the inequality $y > x + 4$ is represented by all the points on the coordinate plane for which y, the ordinate, is greater than $x + 4$. Hence, first draw the graph of the line for which $y = x + 4$; having this graph will permit locating the region for which $y > x + 4$.

Select any three convenient values for x and substitute in the equation $y = x + 4$ to find the corresponding values of y:

x	$x + 4$	$= y$
0	$0 + 4$	$= 4$
3	$3 + 4$	$= 7$
-4	$-4 + 4$	$= 0$

Plot the points (0,4), (3,7), and (−4,0). Draw a *dotted line* through these three points to get the graph of $y = x + 4$. The dotted line is used to signify that points on it are *not* part of the solution set of the inequality $y > x + 4$.

To find the *region* or *half-plane* on one side of the line $y = x + 4$ which represents $y > x + 4$, select a test point, say (1,8), on one side of the line. Substituting in the inequality $y > x + 4$ gives $8 > 1 + 4$, or $8 > 5$, which is true. Thus, the side of the line on which (1,8) lies (above and to the left) is the region representing $y > x + 4$. Shade it with cross-hatching extending up and to the left.

STEP 2: Graph the solution set of $x + y \leq 2$. This graph is represented by all the points on the coordinate plane for which $x + y < 2$ in addition to those points on the line for which $x + y = 2$. Hence, the line $x + y = 2$ is first graphed. To make it convenient to find points on the line, solve for y in terms of x:

Add $-x$ (the additive inverse of x) to both sides of the equation:

$$\begin{array}{r} x + y = 2 \\ \underline{-x \qquad = -x} \\ y = 2 - x \end{array}$$

Set up a table by selecting any three convenient values for x and substituting in the equation $y = 2 - x$ to find the corresponding values of y:

x	$2 - x$	$= y$
0	$2 - 0$	$= 2$
3	$2 - 3$	$= -1$
5	$2 - 5$	$= -3$

Plot the points (0,2), (3,−1), and (5,−3). Draw a *solid line* through these three points to get the graph of $x + y = 2$. The solid line is used to signify that points on it are part of the solution set of $x + y \leq 2$.

To find the *region* or *half-plane* on one side of the line $x + y = 2$ for which $x + y < 2$, select a test point, say (−2,−1) on one side of the line. Substituting (−2,−1) in the inequality $x + y < 2$ results in $-2 - 1 < 2$, or $-3 < 2$, which is true. Thus, the side of the line where (−2,−1) is located (below and to the left) is the region representing $x + y < 2$. Shade this region with cross-hatching extending down and to the left.

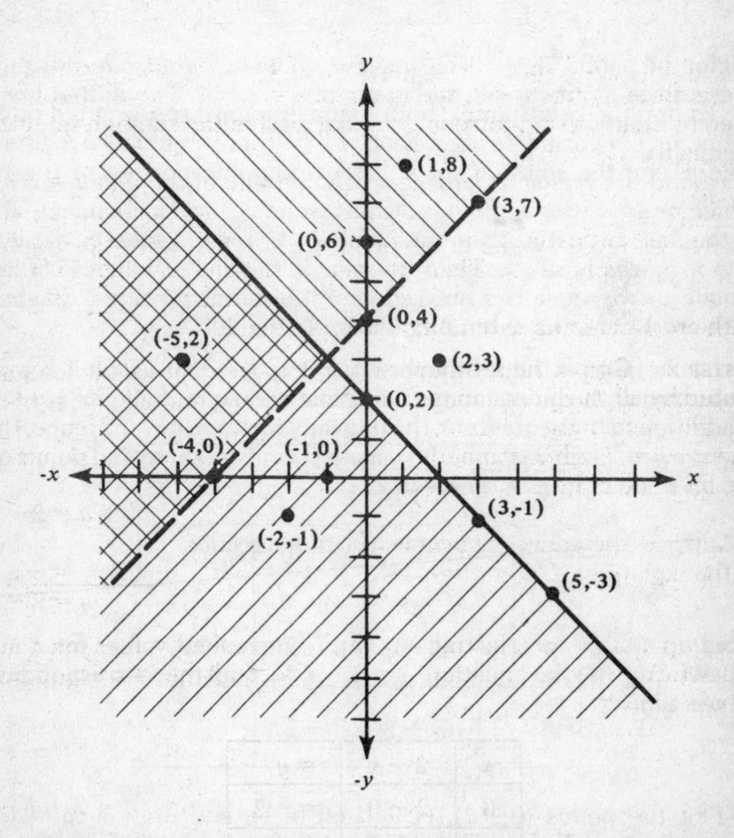

b. If a point is in the solution set, it will lie in the region in which the graph shows *both* sets of cross-hatching; points in this region satisfy *both* inequalities. Test each point in turn by locating it on the graph:

(1) (2,3) does not lie in the cross-hatched area at all.
(2) (−5,2) lies in the region with both sets of cross-hatching. Therefore, it is in the solution set of the graph.
(3) (0,6) lies in a region cross-hatched only for $y > x + 4$; it satisfies this inequality but not the other.
(4) (−1,0) lies in a region cross-hatched only for $x + y \leq 2$; it satisfies this inequality but not the other one.

The only point in the solution set of the graph is **(−5,2)**.

6. $2y \geq x - 4$
$y < 3x$

To find the solution set, both inequalities are represented on the same set of coordinate axes. Each will be represented by a region on the plane; the solution set is the overlapping portion of the two regions.

STEP 1: To draw the graph of $2y \geq x - 4$, first draw the graph of the equation $2y = x - 4$.

Solve the equation for y by dividing by 2: $y = \frac{1}{2}x - 2$.

Set up a table of values by choosing 3 convenient values for x and substituting in the equation to find the corresponding values of y:

x	$\frac{1}{2}x - 2$	$= y$
0	$\frac{1}{2}(0) - 2 = 0 - 2$	$= -2$
4	$\frac{1}{2}(4) - 2 = 2 - 2$	$= 0$
6	$\frac{1}{2}(6) - 2 = 3 - 2$	$= 1$

Plot the points (0,−2), (4,0), and (6,1), and draw a *solid* line through them. The solid line indicates that points on it are part of the solution set of $2y \geq x - 4$; in fact, they constitute the part represented by $2y = x - 4$.

Now the points on the graph represented by $2y > x - 4$ must be indicated. Points that satisfy the inequality occupy an entire region on one side of the line $2y = x - 4$. Such a region is called a *half-plane*. To find which side of the line represents the half-plane $2y > x - 4$, select a test point, say (2,2). Substituting in the inequality gives:

$$2(2) \stackrel{?}{>} 2 - 4$$
$$4 > -2 \text{ which is true.}$$

Therefore, (2,2) lies within the region for which $2y > x - 4$. This region is shaded with cross-hatching extending up and to the right.

STEP 2: To draw the graph of $y < 3x$, first draw the graph of the equation $y = 3x$.

Set up a table of values by choosing 3 convenient values for x and substituting in the equation to find the corresponding values of y:

x	$3x$	$= y$
0	3(0)	= 0
2	3(2)	= 6
3	3(3)	= 9

Plot the points (0,0), (2,6), and (3,9). Draw a *dotted* line through these points; the dotted line indicates that points on it are *not* part of the graph of $y < 3x$ (the line is actually the graph of $y = 3x$). Now it must be determined on which side of the line $y = 3x$ the points lie which represent the graph of $y < 3x$. Select a test point on one side, say the point (2,2). Substituting in the inequality $y < 3x$ gives:

$$2 \stackrel{?}{<} 3(2)$$
$$2 < 6 \text{ which is true.}$$

Therefore, (2,2) lies within the region for which $y < 3x$. This region is shaded with cross-hatching extending down and to the right.

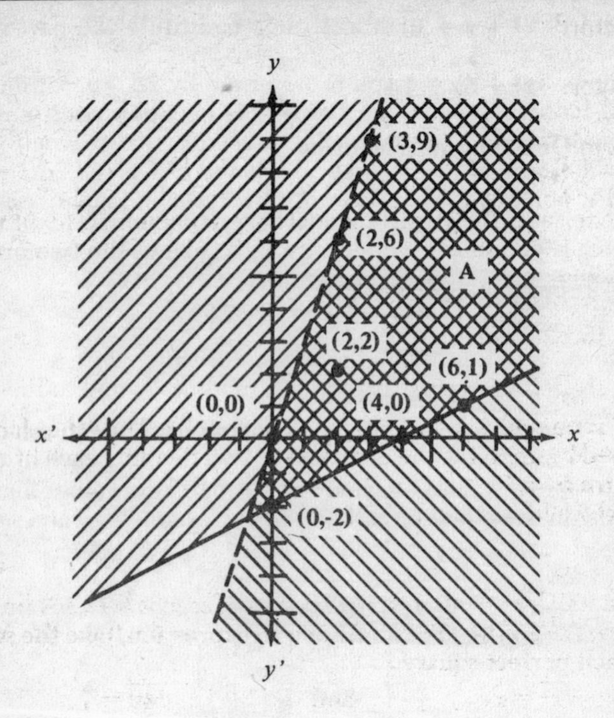

STEP 3: The solution set of the system of inequalities is the region covered by both sets of cross-hatching, including the solid line along one boundary; this region is labeled **A**.

12. QUESTIONS ON FACTORING

1. Factor: $x^2 - 7x$

2. Factor: $x^2 - 5x$

3. Factor: $x^2 - 49$

4. Factor: $x^2 - 36$

5. Factor: $x^2 + x - 30$

6. Factor: $x^2 + 5x - 14$

Solutions to Questions on Factoring

1. $x^2 - 7x$

The two terms in the given binomial have a *common factor* of x. The other factor is obtained by dividing x into each of the two original terms, yielding $x - 7$.

Therefore: $x^2 - 7x = x(x - 7)$

The factored form is $\mathbf{x(x - 7)}$.

2. $x^2 - 5x$

x^2 and $5x$ contain a *highest common factor* of x. The other factor is determined by applying the distributive law, dividing each of x^2 and $-5x$ in turn by x: $x(x - 5)$

The factored form is $\mathbf{x(x - 5)}$.

3. $x^2 - 49$

The binomial, $x^2 - 49$, represents the *difference between two perfect squares,* x^2 and 49. To factor such an expression, take the square root of each perfect square:

$\sqrt{x^2} = x$ and $\sqrt{49} = 7$

One factor will be the sum of the respective square roots and the other factor will be the difference of the square roots:

$x^2 - 49 = (x + 7)(x - 7)$

The correct answer is $\mathbf{(x + 7)(x - 7)}$.

4. $x^2 - 36$

The binomial, $x^2 - 36$, represents the *difference between two perfect squares,* x^2 and 36. Take the square root of each perfect square:

$\sqrt{x^2} = x$ and $\sqrt{36} = 6$

One factor is the sum of the respective square roots and the other factor is the difference of the square roots: $(x + 6)(x - 6)$

The correct answer is $\mathbf{(x + 6)(x - 6)}$.

5. The given expression is a *quadratic trinomial:* $x^2 + x - 30$

The factors of a quadratic trinomial are two binomials.

The factors of the first term, x^2, are x and x, and they become the first terms of the binomials: $(x \quad)(x \quad)$

The factors of the last term, -30, become the second terms of the binomials but they must be chosen in such a way that the product of the inner terms and the product of the outer terms add up to the middle term, $+x$, of the original trinomial. Try $+6$ and -5 as the factors of -30:

$+6x =$ inner product
$(x + 6)(x - 5)$
$-5x =$ outer product

Since $+6x$ and $-5x$ add up to $+x$, these are the correct factors: $(x + 6)(x - 5)$

The factored form is $\mathbf{(x + 6)(x - 5)}$.

6. The given expression is a *quadratic trinomial:* $x^2 + 5x - 14$

The factors of a quadratic trinomial are two binomials. The first terms of the binomials are the factors of the first term, x^2, of the trinomial: $(x \quad)(x \quad)$

The second terms of the binomials are the factors of the last term, -14, of the trinomial. These factors must be chosen in such a way that the sum of the product of the inner terms and the product of the outer terms is equal to the middle term, $+5x$, of the original trinomial. -14 has factors of 14 and -1, -14 and 1, $+7$ and -2, and -7 and $+2$. Try $+7$ and -2:

$+7x =$ inner product
$(x + 7)(x - 2)$
$-2x =$ outer product

Since $(+7x) + (-2x) = +5x$, these are the correct factors: $(x + 7)(x - 2)$

The factored form is $\mathbf{(x + 7)(x - 2)}$.

13. QUESTIONS ON QUADRATIC EQUATIONS

1. The solution set of the equation $x^2 - 3x = 0$ is
(1) $\{3\}$ (2) $\{-3\}$ (3) $\{3,-3\}$ (4) $\{0,3\}$

2. The solution set of $x^2 - x - 6 = 0$ is
(1) $\{1,-6\}$ (2) $\{-3,2\}$ (3) $\{3,-2\}$ (4) $\{5,1\}$

3. The solution set of the equation $x^2 - 7x + 10 = 0$ is
(1) $\{2,5\}$ (2) $\{2,-5\}$ (3) $\{-2,5\}$ (4) $\{-2,-5\}$

4. The solution set of the equation $x^2 - 2x - 3 = 0$ is
(1) $\{-1,-2\}$ (2) $\{-2,-3\}$ (3) $\{-1,3\}$ (4) $\{1,-3\}$

5. What is the solution set for the equation $x^2 + 2x - 15 = 0$?
(1) $\{3,-5\}$ (2) $\{3,5\}$ (3) $\{-3,-5\}$ (4) $\{-3,5\}$

6. Which is a root of the equation $x^2 = 3x + 10$?
(1) -1 (2) 2 (3) 5 (4) 4

Solutions to Questions on Quadratic Equations

1. The solution set consists of those values of x which satisfy the equation. To find the solution set, solve the equation: $x^2 - 3x = 0$
Factor the left side: $x(x - 3) = 0$
When the product of two factors is 0, either one, or both, must be 0: $x = 0$ *or* $x - 3 = 0$
Solve the right equation by adding 3 (the additive inverse of -3) to both sides: $\underline{3 = 3}$
$x = 3$
The solution set is $\{0,3\}$.
The correct choice is (4).

2. $x^2 - x - 6 = 0$
The solution set is found by solving the equation. This is a *quadratic equation* which can be solved by factoring. The left side is a *quadratic trinomial* which can be factored into two binomials. $x^2 - x - 6 = 0$

STEP 1: The factors of the first term, x^2, are x and x, and they constitute the first terms of each binomial factor: $(x \quad)(x \quad) = 0$

STEP 2: The last term, -6, must be factored into two factors which will be the second terms of the binomials. The factors must be chosen in such a way that the product of the two inner terms added to the product of the two outer terms equals the middle term of the original trinomial. Try $-6 = (-3)(+2)$:

$-3x$ = inner product

$$(x - 3)(x + 2) = 0$$

$+2x$ = outer product

The sum of the inner and outer products is $-3x + 2x$ or $-x$, which is equal to the middle term of the original trinomial; thus, the correct factors have been chosen and the equation can be written as:

$$(x - 3)(x + 2) = 0$$

Since the product of two factors is zero, either factor may be equal to zero:

$$x - 3 = 0 \text{ or } x + 2 = 0$$

Add the appropriate additive inverse to both sides, $+3$ in the case of the left equation and -2 in the case of the right one:

$$\underline{3 = 3} \qquad \underline{-2 = -2}$$
$$x = 3 \qquad x = -2$$

Thus, the solution set is $\{3, -2\}$.

The correct choice is (3).

3. The given equation is a *quadratic equation:*

$$x^2 - 7x + 10 = 0$$

The left side is a *quadratic trinomial* which can be factored into the product of two binomials. The factors of the first term, x^2, are x and x, and they become the first terms of the binomials:

$$(x \quad)(x \quad) = 0$$

A pair of factors of the last term, $+10$, become the second terms of the binomials. $+10$ has several pairs of factors; the pair chosen must be such that the sum of the product of the inner terms and the product of the outer terms equals the middle term, $-7x$, of the original trinomial. Try -5 and -2 as the factors of $+10$:

$-5x$ = inner product

$$(x - 5)(x - 2) = 0$$

$-2x$ = outer product

Since $(-5x) + (-2x) = -7x$, these are the correct factors:

$$(x - 5)(x - 2) = 0$$

If the product of two factors is 0, either factor may equal 0:

$$x - 5 = 0 \text{ or } x - 2 = 0$$

Add the appropriate additive inverse to both sides of the equation, 5 in the case of the left equation and 2 in the case of the right equation:

$$\begin{array}{r} 5=5 \\ \hline x \quad =5 \end{array} \qquad \begin{array}{r} 2=2 \\ \hline x \quad =2 \end{array}$$

The solution set of the equation is {2,5}.
The correct choice is (1).

ALTERNATIVE SOLUTION: The question may also be solved by substituting each pair of numbers from the four choices to see which has two roots both of which satisfy the equation.

4. The given equation is a *quadratic equation:*

$$x^2 - 2x - 3 = 0$$

Factor the *quadratic trinomial* on the left side. It factors into the product of two binomials. In choosing the factors check to see that the product of the inner terms of the binomials added to the product of the outer terms equals the middle term, $-2x$, of the original trinomial:

$$-3x = \text{inner product}$$
$$(x - 3)(x + 1) = 0$$
$$+x = \text{outer product}$$

Since $(-3x) + (+x)$ add up to $-2x$, these are the correct factors:

$$(x - 3)(x + 1) = 0$$

If the product of two factors equals 0, either factor may equal 0:

$$x - 3 = 0 \text{ } or \text{ } x + 1 = 0$$

Add the appropriate additive inverse to both sides of the equation, 3 in the case of the left equation and -1 in the case of the right equation:

$$\begin{array}{r} 3=3 \\ \hline x \quad =3 \end{array} \qquad \begin{array}{r} -1=-1 \\ \hline x \quad =-1 \end{array}$$

The solution set is {-1,3}.
The correct choice is (3).

5. The given equation is a *quadratic equation:*

$$x^2 + 2x - 15 = 0$$

The *quadratic trinomial* on the left side can be factored into the product of two binomials. In choosing the factors, check to see that the sum of the product of the inner terms of the binomials and the product of the outer terms equals the middle term, $+2x$, of the original trinomial:

$$+5x = \text{inner product}$$
$$(x + 5)(x - 3) = 0$$
$$-3x = \text{outer product}$$

Since $(+5x) + (-3x) = 2x$, these are the correct factors:

$$(x + 5)(x - 3) = 0$$

If the product of two factors is zero, either factor may equal zero:

$$x + 5 = 0 \text{ or } x - 3 = 0$$

Add the appropriate additive inverse to both sides of the equation, -5 in the case of the left equation and 3 in the case of the right equation:

$$\begin{array}{ll} \underline{\quad -5 = -5} & \underline{\quad 3 = 3} \\ x \quad = -5 & x \quad = 3 \end{array}$$

The solution set is $\{-5,3\}$.
The correct choice is **(1)**.

6. $x^2 = 3x + 10$

If a number is a root of an equation it must satisfy the equation when substituted for x. Try each of the choices in turn:

(1) -1: $(-1)^2 \stackrel{?}{=} 3(-1) + 10$
$1 \stackrel{?}{=} -3 + 10$
$1 \neq 7$
-1 is not a root.

(2) 2: $(2)^2 \stackrel{?}{=} 3(2) + 10$
$4 \stackrel{?}{=} 6 + 10$
$4 \neq 16$
2 is not a root.

(3) 5: $(5)^2 \stackrel{?}{=} 3(5) + 10$
$25 \stackrel{?}{=} 15 + 10$
$25 = 25\checkmark$
5 is a root.

(4) 4: $(4)^2 \stackrel{?}{=} 3(4) + 10$
$16 \stackrel{?}{=} 12 + 10$
$16 \neq 22$
4 is not a root.

The correct choice is (3).

ALTERNATIVE SOLUTION: Since $x^2 = 3x + 10$ is a *quadratic equation* it may be solved by factoring. The equation will first have to be rearranged into the form $x^2 - 3x - 10 = 0$ with all terms on one side equal to 0 on the other side. The left side can then be factored. A solution by factoring will yield two roots, one of which should be among the four choices.

14. QUESTIONS ON VERBAL PROBLEMS

1. If 19 is subtracted from three times a certain number, the difference is 110. What is the number?

2. The sum of the squares of two positive consecutive odd integers is 74. What are the integers? [*Only an algebraic solution will be accepted.*]

3. Find three consecutive positive odd integers such that the square of the smallest exceeds twice the largest by 7. [*Only an algebraic solution will be accepted.*]

4. The length of a rectangle is 1 centimeter less than twice the width. If the perimeter of the rectangle is 76 centimeters, find the number of centimeters in *each* dimension of the rectangle. [*Only an algebraic solution will be accepted.*]

5. In triangle ABC, angle A is 30° more than angle B. Angle C equals the sum of angle A and angle B. Find the measures of *each* of the three angles. [*Only an algebraic solution will be accepted.*]

6. The measure of the base of a parallelogram is 4 meters greater than the measure of the altitude to that base. If the area of the parallelogram is 32 square meters, find the number of meters in the measures of the base and altitude.

Solutions to Questions on Verbal Problems

1. Let x = the number.

Three times a certain number minus 19 is 110.

$$\underbrace{\text{Three times a certain number}}_{3x}\ \underbrace{\text{minus}}_{-}\ \underbrace{19}_{19}\ \underbrace{\text{is}}_{=}\ \underbrace{110}_{110}$$

The equation to be used is: $3x - 19 = 110$

Add 19 (the additive inverse of -19) to both sides of the equation: $\quad 19 = 19$

$$3x = 129$$

Multiply both sides by $\frac{1}{3}$ (the multiplicative inverse of 3): $\frac{1}{3}(3x) = \frac{1}{3}(129)$

$$x = 43$$

The number is **43.**

2. Let x = the first positive odd integer.

Then $x + 2$ = the next consecutive odd integer.

The square of one plus the square of the other is 74.

$$\underbrace{\text{The square of one}}_{x^2}\ \underbrace{\text{plus}}_{+}\ \underbrace{\text{the square of the other}}_{(x+2)^2}\ \underbrace{\text{is}}_{=}\ \underbrace{74}_{74}$$

The equation to be used is: $x^2 + (x + 2)^2 = 74$

Multiply out $(x + 2)^2$:

$$\begin{array}{r} x + 2 \\ \underline{x + 2} \\ x^2 + 2x \\ \underline{\quad 2x + 4} \\ x^2 + 4x + 4 \end{array}$$

Replace $(x + 2)^2$ in the equation by its expanded value: $x^2 + x^2 + 4x + 4 = 74$

Combine like terms: $2x^2 + 4x + 4 = 74$

Add -74 (the additive inverse of 74) to both sides:

$$\begin{array}{r} -74 = -74 \\ \hline 2x^2 + 4x - 70 = 0 \end{array}$$

Divide each term on both sides by 2: $x^2 + 2x - 35 = 0$

The left side is a *quadratic trinomial* which can be factored into 2 binomials. The factors of the first term, x^2, are x and x, and they represent the first terms of each binomial factor: $(x \quad)(x \quad) = 0$

The factors of the last term, -35, must be chosen in such a way that the product of the 2 outer terms of the binomials added to the product of the 2 inner terms equals the middle term, $2x$, of the trinomial. Try $-35 = (+7)(-5)$:

$-5x$ = outer product

$(x + 7)(x - 5) = 0$

$7x$ = inner product

This factoring is correct since $(-5x) + (7x) = 2x$.

Since the product of two factors is zero, either one or both of the factors must be zero: $x + 7 = 0$ *or* $x - 5 = 0$

Add the appropriate additive inverse to each side, -7 in the case of the left equation, and $+5$ in the case of the right:

$$\begin{array}{rr} \underline{-7 = -7} & \quad \underline{5 = 5} \\ x = -7 & \quad x = 5 \end{array}$$

Reject -7 since the question calls for a *positive* number: $x = 5$

$x + 2 = 7$

The integers are **5** and **7**.

3. An example of consecutive odd integers is 5, 7, 9, ... Each consecutive odd integer can be obtained by adding 2 to the previous odd integer.

Let n = first odd integer.
Then $n + 2$ = second consecutive odd integer.
And $n + 4$ = third consecutive odd integer.

The square of the smallest equals twice the largest plus 7.

$$n^2 \quad = \quad 2(n+4) \quad + \quad 7$$

The equation to be used is: $n^2 = 2(n + 4) + 7$

Remove parentheses by using the distributive property: $n^2 = 2n + 8 + 7$

Combine like terms: $n^2 = 2n + 15$

Add $-2n$ (the additive inverse of $2n$) and also add -15 (the additive inverse of 15) to both sides:

$$\begin{array}{r} -2n - 15 = -2n - 15 \\ \hline n^2 - 2n - 15 = 0 \end{array}$$

This is a *quadratic equation* which can be solved by factoring. The left side is a quadratic trinomial. Its factors are two binomials. The factors of the first term, n^2, are n and n, and they become the first terms of each binomial factor: $(n \quad)(n \quad) = 0$

The factors of the last term, -15, become the other terms of the two binomials. The factors of -15 must be selected in such a way that the inner product of the binomial terms added to the outer product equals the middle term, $-2n$, of the trinomial. Try $-15 = (-5)(+3)$:

$-5n$ = inner product
$$(n - 5)(n + 3) = 0$$
$+3n$ = outer product

Since $(-5n) + (+3n) = -2n$, these are the correct factors.

Since the product of two factors is zero, either factor may be equal to zero: $n - 5 = 0$ *or* $n + 3 = 0$

Add the appropriate additive inverse to both sides, $+5$ in the case of the left equation and -3 in the case of the right:

$$\begin{array}{r} 5 = 5 \\ \hline n = 5 \end{array} \qquad \begin{array}{r} -3 = -3 \\ \hline n = -3 \end{array}$$

-3 must be rejected since the problem requires *positive* integers. If $n = 5$, then $n + 2 = 7$, and $n + 4 = 9$.

The three consecutive odd integers are **5, 7, and 9.**

4. Let $x =$ the width of the rectangle in centimeters.
Then $2x - 1 =$ the length of the rectangle in centimeters.
The perimeter of a rectangle is equal to the sum of the four sides. Since the opposite sides of a rectangle are equal, the perimeter is twice the width plus twice the length, or $P = 2W + 2L$
In this case, $P = 76$, $W = x$, and $L = 2x - 1$. Therefore, the equation to be used is:

$$2(x) + 2(2x - 1) = 76$$

Remove parentheses by applying the distributive law:

$$2x + 4x - 2 = 76$$

Combine like terms:

$$6x - 2 = 76$$

Add 2 (the additive inverse of -2) to both sides:

$$\frac{2 = 2}{6x = 78}$$

Multiply both sides by $\frac{1}{6}$ (the multiplicative inverse of 6):

$$\frac{1}{6}(6x) = \frac{1}{6}(78)$$

$$x = 13$$

$$2x - 1 = 25$$

The width is **13** centimeters; the length is **25** centimeters.

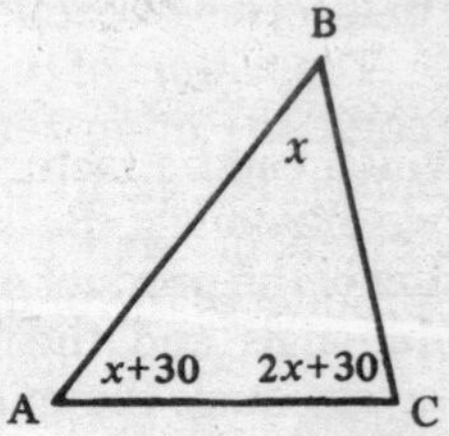

5. Let $x =$ the measure in degrees of $\measuredangle B$.
Then $x + 30 =$ the measure in degrees of $\measuredangle A$.
And $2x + 30 =$ the measure in degrees of $\measuredangle C$.

The sum of the measure of the three angles of a triangle is 180°. Therefore, the equation to be used is:

$$x + x + 30 + 2x + 30 = 180$$

Combine like terms:

$$4x + 60 = 180$$

Add -60 (the additive inverse of 60) to both sides:

$$\frac{-60 = -60}{4x = 120}$$

Multiply both sides by $\frac{1}{4}$ (the multiplicative inverse of 4):

$$\frac{1}{4}(4x) = \frac{1}{4}(120)$$

$$x = 30\ (\measuredangle B)$$

$$x + 30 = 60\ (\measuredangle A)$$

$$2x + 30 = 90\ (\measuredangle C)$$

$\measuredangle A = 60°$, $\measuredangle B = 30°$, $\measuredangle C = 90°$.

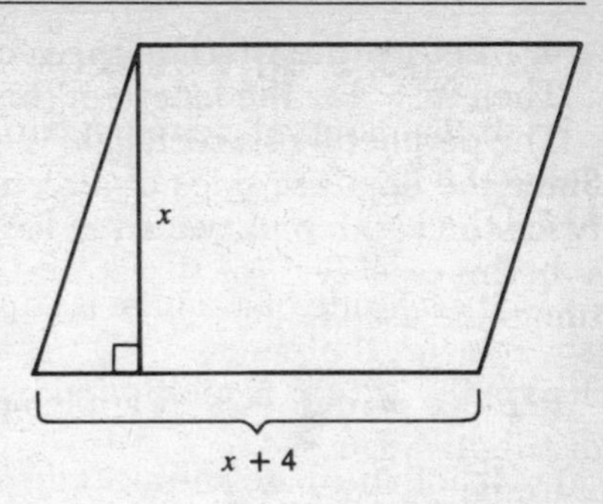

6. Let $x =$ the measure of the altitude.

Then $x + 4 =$ the measure of the base.

The area of a parallelogram is equal to the product of the measure of the base and the measure of the altitude to that base:

$$x(x + 4) = 32$$

Remove parentheses by applying the distributive law of multiplication over addition:

$$x^2 + 4x = 32$$

The equation is a *quadratic equation.* Rearrange it so that all terms are on one side equal to zero by adding -32 (the additive inverse of 32) to both sides:

$$\begin{array}{r} -32 = -32 \\ \hline x^2 + 4x - 32 = 0 \end{array}$$

The left side is a *quadratic trinomial* which can be factored into the product of two binomials. Be sure to check that the sum of the product of the inner terms of the binomials and the product of the outer terms equals the middle term, $+4x$, of the original trinomial:

$$+8x = \text{inner product}$$
$$(x + 8)(x - 4) = 0$$
$$-4x = \text{outer product}$$

Since $(+8x) + (-4x) = +4x$ these are the correct factors:

$$(x + 8)(x - 4) = 0$$

If the product of two factors equals zero, either factor may equal zero:

$$x + 8 = 0 \text{ or } x - 4 = 0$$

Add the appropriate additive inverse to both sides of the equation, -8 in the case of the left equation and 4 in the case of the right equation:

$$\begin{array}{r} -8 = -8 \\ \hline x = -8 \end{array} \qquad \begin{array}{r} 4 = 4 \\ \hline x = 4 \end{array}$$

Reject the negative value as meaningless for a length:

$$x = 4$$
$$x + 4 = 8$$

The measure of the altitude is 4 meters and the measure of the base is 8 meters.

15. QUESTIONS ON VARIATION

1. If the radius of a circle is tripled, then the area of the circle is multiplied by
(1) 27 (2) 9 (3) 3 (4) 6

2. If each side of a square is tripled, the perimeter of the square
(1) remains the same (2) is increased by 3 (3) is multiplied by 3 (4) is multiplied by 9

3. If both the base and altitude of a triangle are doubled, the area of the triangle will be multiplied by
(1) $\frac{1}{2}$ (2) 2 (3) $\frac{1}{4}$ (4) 4

4. The area of a triangle is $\frac{1}{2}bh$. If the base of the triangle is doubled and the height of the triangle is multiplied by 3, then the area is
(1) multiplied by 6 (2) multiplied by 3 (3) increased by 6 (4) increased by 3

5. If each side of a square is doubled, then its area
(1) is doubled (2) is multiplied by 4 (3) is halved
(4) stays the same

Solutions to Questions on Variation

1. The area, A, of a circle is given by the formula $A = \pi r^2$, where r is the length of the radius.

If the radius is tripled, then r is replaced by $3r$. Calling the new area A' (to distinguish it from the old area, A):

$$A' = \pi(3r)^2$$
$$A' = 9\pi r^2$$

The new area, A', is thus 9 times as large as the old area, A.
The correct choice is **(2)**.

ALTERNATIVE SOLUTION: Choose a convenient value for the length of the radius, say $r = 2$, and use it to compute the corresponding area of the circle:

$$A = \pi r^2$$
$$A = \pi(2)^2$$
$$A = 4\pi$$

If the radius of 2 is tripled, it becomes $3 \times 2 = 6$. Now compute the area of this larger circle:

$$A' = \pi r^2$$
$$A' = \pi(6)^2$$
$$A' = 36\pi$$

Comparing the old area of 4π with the new area of 36π, it is seen that the area has been multiplied by 9.

Again, the correct choice is (**2**).

2. Since all four sides of a square are equal in measure, the perimeter, p, of a square equals 4 times the length of one side, s: $p = 4s$

If the side of length s is tripled, the length becomes $3s$. The perimeter, p', of the enlarged square will then be given by the formula: $p' = 4(3s)$

Remove the parentheses: $p' = 12s$

$12s$ is 3 times $4s$, that is, the new perimeter, p', is 3 times as large as the original perimeter, p.

The correct choice is (**3**).

ALTERNATIVE SOLUTION: Choose a convenient value for the measure of the side of the square, say 5. Then the perimeter $= 4(5)$ or 20.

If the side is tripled, its measure becomes 15. The new perimeter is 4(15) or 60. 60 is 3 times as large as the old perimeter, 20.

3. The area, A, of a triangle is equal to one-half the product of its base, b, and altitude, h: $A = \frac{1}{2} bh$

If the base and altitude are both doubled, the new base can be represented by $2b$ and the new altitude by $2h$. If A' represents the area of the enlarged triangle, then: $A' = \frac{1}{2}(2b)(2h)$

Remove the parentheses: $A' = 2bh$

$2bh$ is 4 times $\frac{1}{2} bh$, that is, the enlarged area, A', is 4 times the original area, A.

The correct choice is (**4**).

ALTERNATIVE SOLUTION: The question may be solved by choosing arbitrary values for the base and altitude, say $b = 3$ and $h = 2$. Then, since $A = \frac{1}{2}bh$:

$$A = \frac{1}{2}(3)(2)$$

$$A = \frac{1}{2}(6)$$

$$A = 3$$

Doubling the base and altitude will make them 6 and 4 respectively. The new area, A', will be:

$$A' = \frac{1}{2}(6)(4)$$

$$A' = \frac{1}{2}(24)$$

$$A' = 12$$

Since 12 is 4 times the old area, 3, the area of the triangle has been multiplied by 4.

4. Represent the area of the original triangle by A:

$$A = \frac{1}{2}bh$$

If the base is doubled, the new base is $2b$. If the height is multiplied by 3, the new height is $3h$. If A' represents the new area, then:

$$A' = \frac{1}{2}(2b)(3h)$$

$$A' = \frac{1}{2}(6bh)$$

$$A' = 3bh$$

$3bh$ is 6 times as large as $\frac{1}{2}bh$. Therefore, the area has been multiplied by 6.

The correct choice is (1).

ALTERNATIVE SOLUTION: The question may be solved by choosing arbitrary values for the base and height of the triangle, say $b = 2$ and $h = 3$. Then the original area, A, is:

$$A = \frac{1}{2}(2)(3)$$

$$A = 3$$

If the base is doubled, the new base is 4. If the height is multiplied by 3, the new height is 9. Then the area, A', of the enlarged triangle is:

$$A' = \frac{1}{2}(4)(9)$$

$$A' = \frac{1}{2}(36)$$

$$A' = 18$$

Since the new area, 18, is 6 times the old area, 3, the area has been multiplied by 6.

5. The area, A, of a square whose side has length s is: $A = s^2$

If the side is doubled, its length becomes $2s$. Then the area, A', of the enlarged square is: $A' = (2s)^2$

Remove the parentheses: $A' = 4s^2$

Since $4s^2$ is 4 times s^2, the new area, A', is 4 times as large as the original area, A.

The correct choice is **(2)**.

ALTERNATIVE SOLUTION: Choose an arbitrary value for the length of the side of the square, say $s = 5$:

$$A = 5^2$$
$$A = 25$$

If the side is doubled, it becomes 10. The new area, A', is then:

$$A' = 10^2$$
$$A' = 100$$

Since the new area, 100, is 4 times the original area, 25, the area has been multiplied by 4.

16. QUESTIONS ON LITERAL EQUATIONS; EXPRESSING RELATIONS ALGEBRAICALLY

1. Solve for x in terms of a, b, and c: $ax - b = c$

2. Solve for B in terms of V and h: $V = \frac{1}{2}Bh$

3. The length of a rectangle is 5 more than its width. If the width is represented by w, which expression represents the area of the rectangle?
(1) $w^2 + 5w$ (2) $w^2 + 5$ (3) $5w^2$ (4) $4w + 10$

4. The perimeter of a square is represented by $12x - 4$. Express the length of one side of the square in terms of x.

5. The length and width of a rectangle are represented by $(x + 7)$ and $(x - 3)$. If the area of the rectangle is 24, which equation can be used to find x?
(1) $(x + 7) + (x - 3) = 24$ (2) $2(x + 7) + 2(x - 3) = 24$
(3) $(x + 7)^2 + (x - 3)^2 = 24$ (4) $(x + 7)(x - 3) = 24$

6. As shown in the accompanying diagram, a square with side s is inscribed in a circle with radius r. Which expression represents the area of the shaded region?

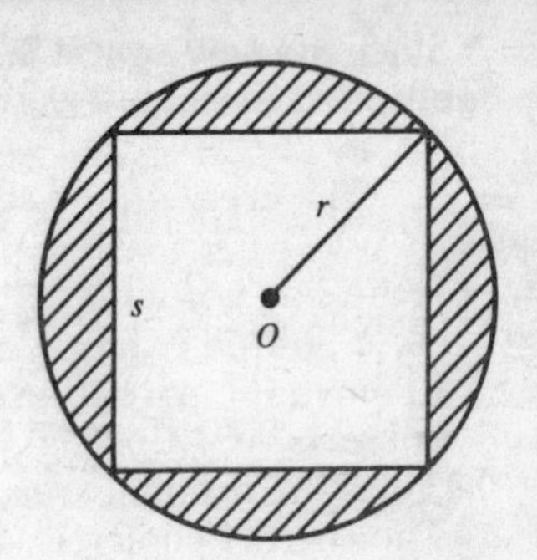

(1) $s^2 - \pi r^2$ (2) $\pi r^2 - s^2$ (3) $\pi r^2 - 4s$ (4) $4s - \pi r^2$

Solutions to Questions on Literal Equations; Expressing Relations Algebraically

1.

$$ax - b = c$$

Add $+b$ (the additive inverse of $-b$) to both sides of the equation:

$$\begin{aligned} +b &= +b \\ \hline ax &= b + c \end{aligned}$$

Multiply both sides of the equation by $\frac{1}{a}$ (the multiplicative inverse of a):

$$\frac{1}{a}(ax) = \frac{1}{a}(b + c)$$

$$x = \frac{b + c}{a}$$

The correct answer is $\frac{b+c}{a}$.

2.

$$V = \frac{1}{3}Bh$$

Clear fractions by multiplying both sides of the equation by 3:

$$3(V) = 3\left(\frac{1}{3}Bh\right)$$

$$3V = Bh$$

Divide both sides of the equation by h:

$$\frac{3V}{h} = \frac{Bh}{h}$$

$$\frac{3V}{h} = B$$

$B = \frac{3V}{h}$.

3. Since the length is 5 more than the width, w, the length is represented by $w + 5$.

The area of a rectangle equals its length times its width. Therefore, the area is given by:

$$w(w + 5)$$

Remove parentheses by applying the distributive law of multiplication: $w^2 + 5w$

The correct choice is **(1)**.

4. Let s = the length of one side of the square. The perimeter of a square is the sum of the lengths of all 4 sides. Since the 4 sides of a square are all equal:

$$4s = 12x - 4$$

Multiply both sides by $\frac{1}{4}$ (the multiplicative inverse of 4):

$$\frac{1}{4}(4s) = \frac{1}{4}(12x - 4)$$

$$s = 3x - 1$$

The length of one side of the square is **$3x - 1$**.

5. The area of a rectangle is equal to the product of its length and width: $(x + 7)(x - 3) = 24$

The correct choice is **(4)**.

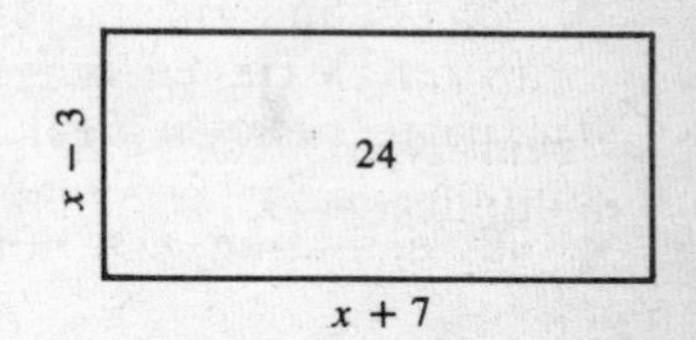

6. The area, A, of a circle whose radius is r is given by the formula: $A = \pi r^2$

The area, A, of a square whose side is s is given by the formula: $A = s^2$

The shaded area is the area of the circle minus the area of the square: $\pi r^2 - s^2$

The correct choice is **(2)**.

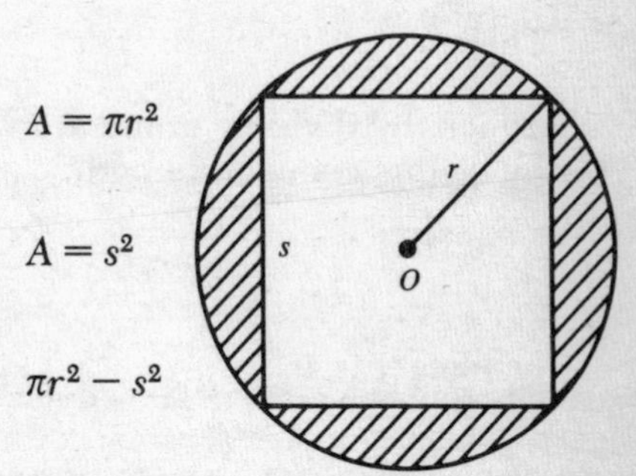

17. QUESTIONS ON FACTORIAL *n*

1. The symbol for "factorial 4" is 4! What is the value of 4!?
(1) 24 (2) 16 (3) 8 (4) 4

2. An expression equivalent to 3! is
(1) 3(2)(1) (2) 3(3) (3) 3(2)(1)(0) (4) 3(3)(3)

3. The expression 5! is equivalent to
(1) 5 (2) 15 (3) 20 (4) 120

4. What integer does $\frac{4!}{3!}$ equal?

Solutions to Questions on Factorial *n*

1. By definition, factorial n, or $n!$, equals $n(n-1)(n-2) \ldots (3)(2)(1)$. Hence, $n! = 4(3)(2)(1) = 24$.
The correct choice is **(1)**.

2. Factorial n is the product of n and each of the integers less than n down to and including 1. Hence, $3! = 3(2)(1)$.
The correct choice is **(1)**.

3. By definition, factorial n, or $n!$, equals $n(n-1)(n-2) \ldots (3)(2)(1)$. Hence, $5! = 5(4)(3)(2)(1) = 20(6)(1) = 120$.
The correct choice is **(4)**.

4. By definition, $n! = n(n-1)(n-2) \ldots (3)(2)(1)$

Therefore: $$\frac{4!}{3!} = \frac{4(3)(2)(1)}{3(2)(1)}$$

Reduce, by dividing numerator and denominator by factors common to both: $$\frac{4!}{3!} = \frac{4(\overset{1}{\cancel{3}})(\overset{1}{\cancel{2}})(1)}{\underset{1}{\cancel{3}}(\underset{1}{\cancel{2}})(1)}$$

$$\frac{4!}{3!} = 4$$

The integer is **4**.

18. QUESTIONS ON AREAS, PERIMETERS, CIRCUMFERENCES OF COMMON FIGURES

1. The perimeter of a square is 36. What is the length of one side of the square?

2. Find the area of the triangle whose vertices are (0,0), (0,4), and (5,0).

3. What is the area of a circle whose radius is 5?
(1) 100π (2) 25π (3) 10π (4) 5π

4. The area of a triangle is 24 square centimeters and the base measures 6 centimeters. Find the number of centimeters in the measure of the altitude to that base.

5. In the accompanying diagram, $ABCD$ is a rectangle. Diameter $\overline{MN}$ of circle O is perpendicular to $\overline{BC}$ at M and to $\overline{AD}$ at N, $AD = 8$, and $CD = 6$. (Answers may be left in terms of π.)

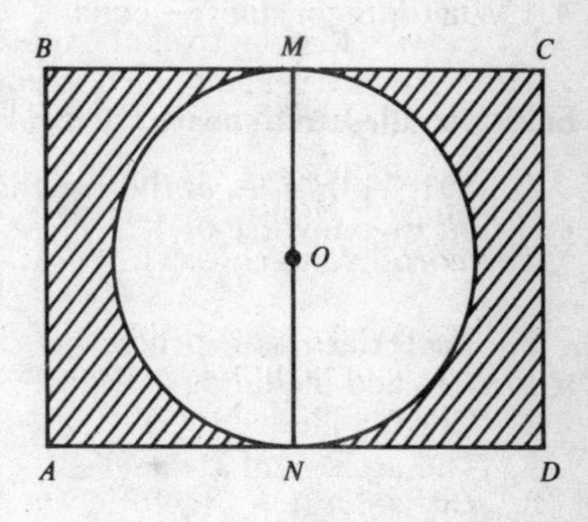

a. What is the perimeter of rectangle $ABCD$?

b. What is the circumference of circle O?

c. What is the area of rectangle $ABCD$?

d. What is the area of circle O?

e. What is the area of the shaded region of the diagram?

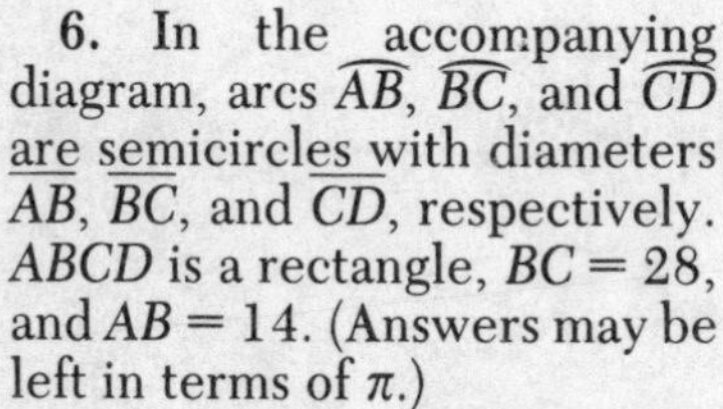

6. In the accompanying diagram, arcs $\overset{\frown}{AB}$, $\overset{\frown}{BC}$, and $\overset{\frown}{CD}$ are semicircles with diameters $\overline{AB}$, $\overline{BC}$, and $\overline{CD}$, respectively. $ABCD$ is a rectangle, $BC = 28$, and $AB = 14$. (Answers may be left in terms of π.)

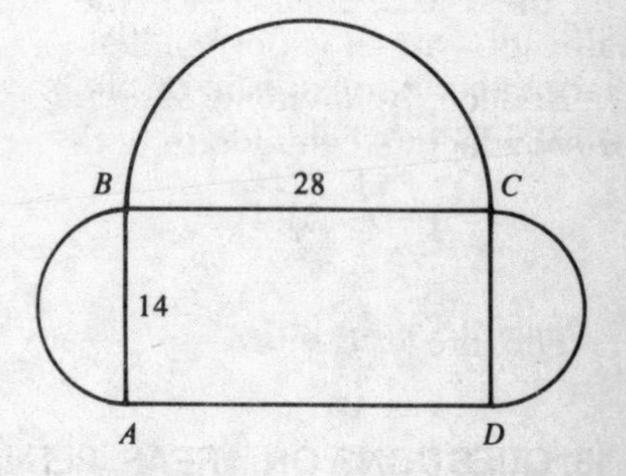

a. Find the area of $ABCD$.

b. Find the area of the region enclosed by diameter $\overline{BC}$ and semicircle $\overset{\frown}{BC}$.

c. Find the area of the region enclosed by diameter $\overline{AB}$ and semicircle $\overset{\frown}{AB}$.

d. Find the area of the entire region.

Solutions to Questions on Areas, Perimeters, Circumferences of Common Figures

1. Let x = the length of one side of the square.
The perimeter of any polygon equals the sum of all the sides; in a square, all 4 sides are equal: $4x = 36$

Multiply both sides of the equation by $\frac{1}{4}$ (the multiplicative inverse of 4): $\frac{1}{4}(4x) = \frac{1}{4}(36)$

$x = 9$

The length of one side is **9.**

2. The area, A, of a triangle is given by $A = \frac{1}{2}bh$ where b is the length of the base and h is the length of the altitude to that base.

If the base is considered to lie on the x-axis, then $b = 5$; the altitude (which is perpendicular to the base) will lie on the y-axis, with $h = 4$.

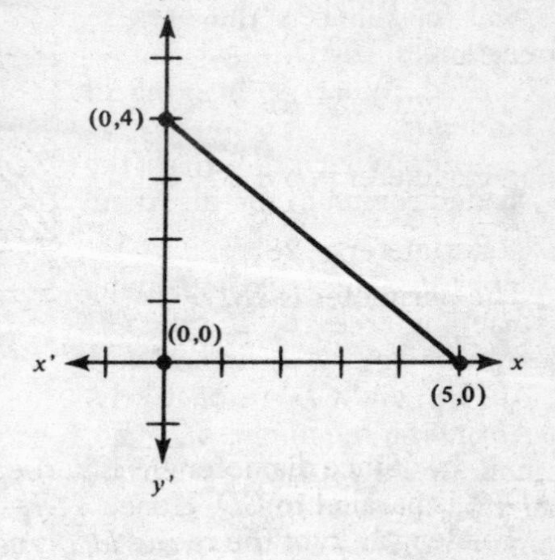

$A = \frac{1}{2}(5)(4)$

$A = \frac{1}{2}(20)$

$A = 10$

The area is **10.**

3. The area, A, of a circle is given by the formula $A = \pi r^2$, where r is the length of the radius.
Here, $r = 5$: $A = \pi(5)^2$

$A = 25\pi$

The correct choice is **(2).**

4. The area, A, of a triangle is given by the formula $A = \frac{1}{2}bh$ where b is the measure of the base and h is the measure of the altitude to that base. Since it is given that $A = 24$ and $b = 6$, substitute these values in the formula:

$$24 = \frac{1}{2}(6)h$$

Perform the indicated multiplication:

$$24 = 3h$$

Divide both sides of the equation by 3:

$$\frac{24}{3} = \frac{3h}{3}$$

$$8 = h$$

The measure of the altitude is **8**.

5. **a.** The opposite sides of a rectangle are equal:
$BC = AD = 8$; $AB = CD = 6$
The perimeter of a rectangle equals the sum of lengths of all four sides:

$$\text{Perimeter} = AB + BC + CD + AD$$
$$\text{Perimeter} = 6 + 8 + 6 + 8$$
$$\text{Perimeter} = 28$$

The perimeter is **28**.

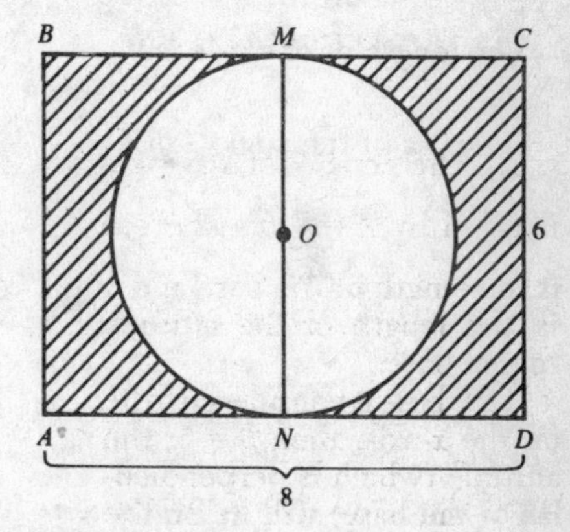

b. Since diameter $\overline{MN}$ is perpendicular to $\overline{BC}$ and $\overline{AD}$, it is parallel to $\overline{CD}$. Hence, $MN = CD = 6$.
The length, r, of the radius $\overline{MO}$ is one-half the diameter:

$$r = 3$$

The circumference, c, of a circle is given by the formula $c = 2\pi r$ where r is the length of the radius:

$$c = 2\pi(3)$$
$$c = 6\pi$$

The circumference is $\mathbf{6\pi}$.

c. The area of a rectangle is equal to the product of its length and width:

$$\text{Area of } ABCD = (8)(6)$$
$$\text{Area of } ABCD = 48$$

The area of $ABCD$ is **48**.

d. The area, A, of a circle is given by the formula $A = \pi r^2$ where r is the radius:

$$A = \pi(3)^2$$
$$A = 9\pi$$

The area of the circle is **9π.**

e. The area of the shaded region is equal to the area of the rectangle minus the area of the circle: $48 - 9\pi$

The area of the shaded region is **48 − 9π.**

6. a. The area of a rectangle is equal to the product of its length and width:

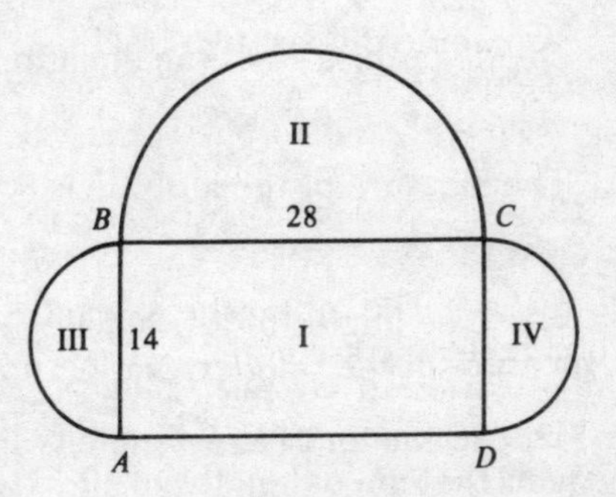

$$\begin{array}{r} 28 \\ \times 14 \\ \hline 112 \\ 28 \\ \hline 392 \end{array}$$

Area of $ABCD = 14(28)$
Area of $ABCD = 392$

The area of rectangle $ABCD$ is **392.**

b. The area, A, of a circle is given by the formula $A = \pi r^2$ where r is the radius. Since $\overline{BC}$ is a diameter, the radius of region II $= \frac{1}{2}(BC) = \frac{1}{2}(28) = 14$. Since region II is half a circle, its area is $\frac{1}{2}A$ or $\frac{1}{2}\pi r^2 = \frac{1}{2}\pi(14)^2$.

$$\begin{array}{r} 14 \\ \times 14 \\ \hline 56 \\ 14 \\ \hline 196 \end{array}$$

$$\frac{1}{2}\pi(14)^2 = \frac{1}{2}\pi(196) = 98\pi$$

The area of region II is **98π.**

c. Since $\overline{AB}$ is a diameter, the radius of region III $= \frac{1}{2}(AB) = \frac{1}{2}(14) = 7$. Since region III is half a circle, its area $= \frac{1}{2}\pi r^2 = \frac{1}{2}\pi(7)^2 = \frac{1}{2}\pi(49) = \frac{49\pi}{2}$.

The area of region III is $\frac{49\pi}{2}$.

d. The area of region IV is the same as the area of region III or $\frac{49\pi}{2}$.

The area of the entire region = area of I + area of II + area of III + area of IV: Total area $= 392 + 98\pi + \frac{49\pi}{2} + \frac{49\pi}{2}$

Combine the last two fractions: Total area $= 392 + 98\pi + 49\pi$

Combine like terms: Total area $= 392 + 147\pi$

The area of the entire region is $\mathbf{392 + 147\pi}$.

19. QUESTIONS ON GEOMETRY ($\cong$, $\parallel$ LINES, COMPLS., SUPPLS.)

1. The angles of a triangle are in the ratio of $1:2:3$. Find the measure of the *smallest* angle.

2. If two angles of a triangle measure 30° and 70°, what is the number of degrees of the third angle of the triangle?

3. As shown in the accompanying figure, $\overleftrightarrow{AB}$ is parallel to $\overleftrightarrow{CD}$, and $\overleftrightarrow{AB}$ and $\overleftrightarrow{CD}$ are cut by transversal $\overleftrightarrow{EF}$ at E and F, respectively. If the measure of $\measuredangle AEF$ equals $x + 30°$, and the measure of $\measuredangle DFE$ equals 70°, find x.

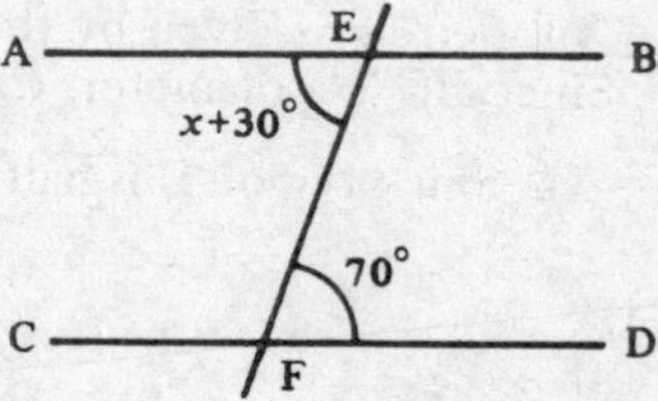

4. In the accompanying diagram, $\overleftrightarrow{AB} \parallel \overleftrightarrow{CD}$, $\overline{EF}$ intersects $\overleftrightarrow{AB}$ at E and $\overleftrightarrow{CD}$ at F, and $\overline{GH}$ intersects $\overleftrightarrow{AB}$ at G and $\overline{EF}$ at H. If $m\angle EGH = 40°$, $m\angle GHE = 80°$, and $m\angle EFD = x$, what is the value of x?

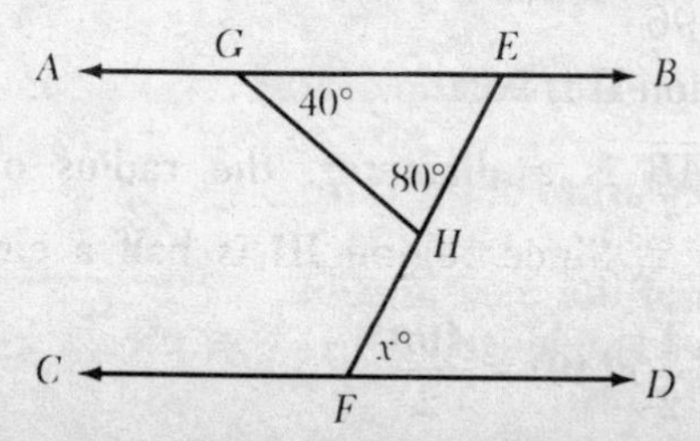

5. In the accompanying figure, ℓ, m, and n are lines with $\ell \perp m$. Which angles are complementary?

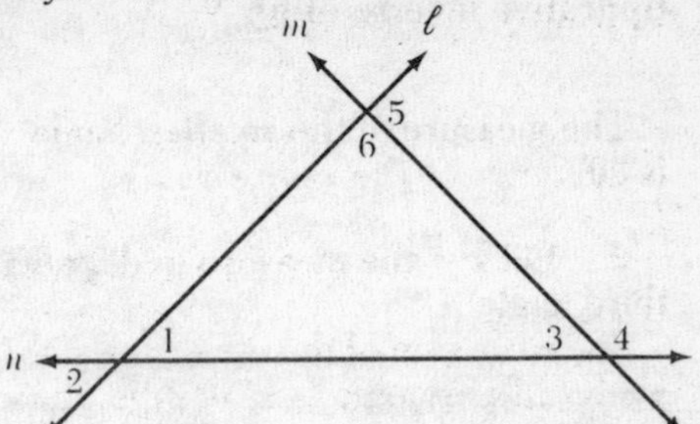

(1) 1 and 3
(2) 1 and 2
(3) 3 and 4
(4) 3 and 5

6. If two angles of a triangle are complementary, find the number of degrees in the third angle of the triangle.

7. Two complementary angles are in the ratio of 1 : 4. Find the measure of the *smaller* angle.

8. In the accompanying diagram, $\overleftrightarrow{AB}$ intersects $\overleftrightarrow{PQ}$ and $\overleftrightarrow{RS}$ at C and D, respectively. If $\overleftrightarrow{PQ} \parallel \overleftrightarrow{RS}$, $m\angle RDB = 2x - 10$, and $m\angle QCA = 3x - 65$, find x.

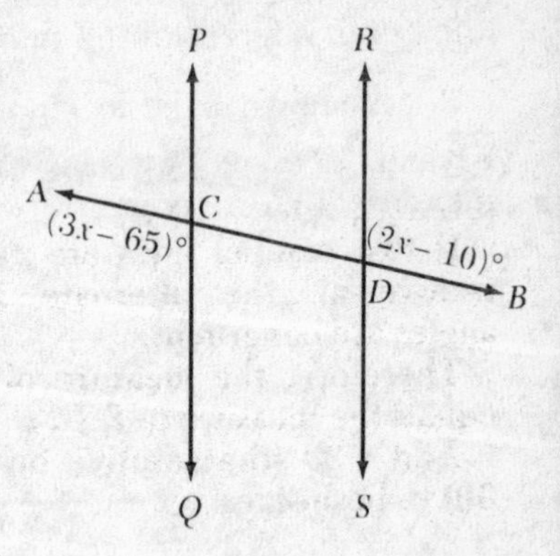

Solutions to Questions on Geometry (≅, ∥ lines, compls., suppls.)

1. Let x = the measure of the *smallest* angle.
Then $2x$ = the measure of the next angle.
And $3x$ = the measure of the third angle.
Since the sum of the measures of the 3 angles of a triangle is 180°:

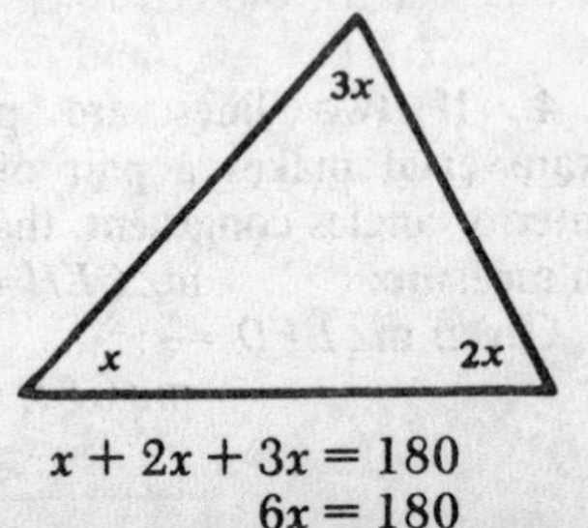

$$x + 2x + 3x = 180$$

Combine like terms:

$$6x = 180$$

Multiply both sides by $\frac{1}{6}$ (the multiplicative inverse of 6):

$$\frac{1}{6}(6x) = \frac{1}{6}(180)$$
$$x = 30$$

The measure of the *smallest* angle is **30°**.

2. Let x = the measure in degrees of the third angle.

Since the sum of the measures of the three angles of a triangle is 180°, the equation to be used is:

$$30 + 70 + x = 180$$

Combine like terms:

$$100 + x = 180$$

Add -100 (the additive inverse of 100) to both sides of the equation:

$$\begin{aligned} -100 &= -100 \\ \hline x &= 80 \end{aligned}$$

The number of degrees in the third angle is **80.**

3. $\measuredangle AEF$ and $\measuredangle DFE$ are alternate interior angles.

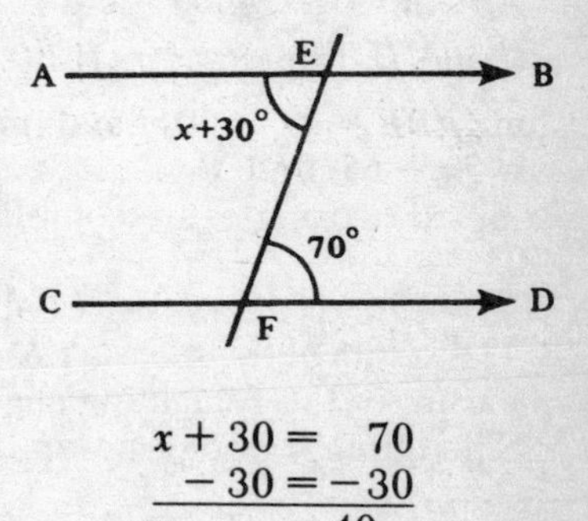

If two parallel lines are cut by a transversal, the alternate interior angles are congruent.

Therefore, the measure of $\measuredangle AEF$ equals the measure of $\measuredangle DFE$:

$$x + 30 = 70$$

Add -30 (the additive inverse of 30) to both sides:

$$\begin{aligned} -30 &= -30 \\ \hline x &= 40 \end{aligned}$$

The value of x is **40.**

4. If two lines are parallel, a transversal makes a pair of alternate interior angles congruent, that is, equal in measure: $m\angle GEH = m\angle EFD$

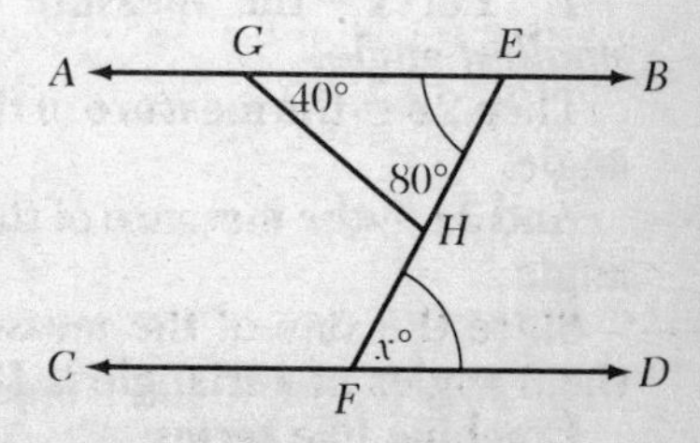

Given $m\angle EFD = x$:

$$m\angle GEH = x$$

The sum of the measures of the three angles of a triangle is 180: $m\angle EGH + m\angle GHE + m\angle GEH = 180$

$m\angle EGH = 40$; $m\angle GHE = 80$: $40 + 80 + x = 180$

Combine like terms: $120 + x = 180$

Add -120 (the additive inverse of 120) to both sides of the equation: $-120 \quad = -120$

$x = 60$

The value of x is **60**.

5. If $\ell \perp m$, $\angle 6$ must be a right angle. The triangle formed by lines ℓ, m, and n is a right triangle. The acute angles of a right triangle are complementary; hence, $\angle 1$ and $\angle 3$ are complementary.

The correct choice is **(1)**.

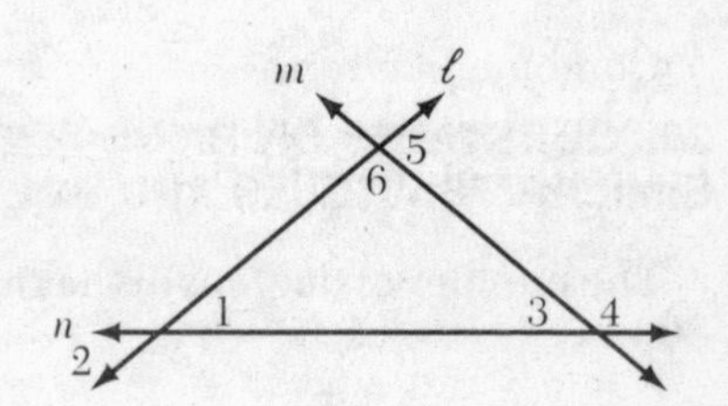

6. If two angles are complementary, the sum of their measures is 90°.

The sum of the measures of all 3 angles of a triangle is 180°.

If the sum of the measures of 2 angles is 90°, then 90° is left as the measure of the remaining angle.

The number of degrees in the third angle is **90**.

7. Let x = the measure of the smaller angle.

Then $4x$ = the measure of the larger angle.

The sum of the measures of two complementary angles is 90°: $x + 4x = 90$

Combine like terms: $5x = 90$

Divide both sides of the equation by 5: $\frac{5x}{5} = \frac{90}{5}$

$x = 18$

The *smaller* angle has a measure of **18°**.

8. Vertical angles are congruent, that is, equal in measure: $m\angle PCD = m\angle QCA$

Since $m\angle QCA = 3x - 65$:

$m\angle PCD = 3x - 65$

If two lines are parallel, a transversal makes a pair of corresponding angles congruent, that is, equal in measure:

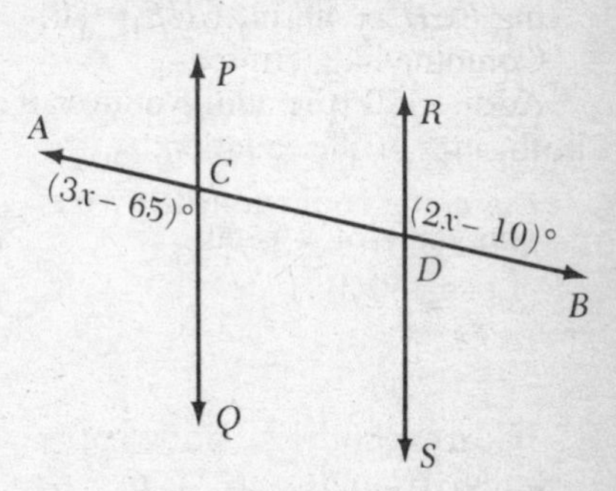

$m\angle PCD = m\angle RDB$

$3x - 65 = 2x - 10$

$3x - 2x = -10 + 65$

$x = 55$

$x = 55$.

20. QUESTIONS ON RATIO AND PROPORTION (INCLUDING SIMILAR TRIANGLES AND POLYGONS)

1. In the accompanying diagrams of triangles RST and UVW, $\angle R \cong \angle U$ and $\angle S \cong \angle V$. If $RS = 4$, $ST = 6$, and $UV = 8$, find VW.

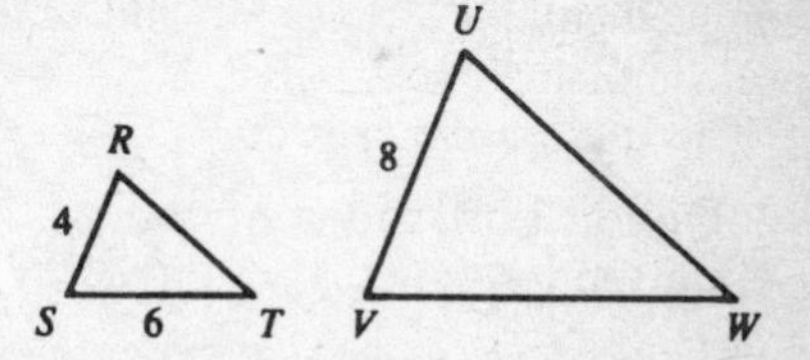

2. Right triangles ABC and DEF are similar. In $\triangle ABC$, the lengths of the legs are 3 and 4. In $\triangle DEF$, the length of the longer of the two legs is 12. What is the length of the shorter leg of $\triangle DEF$?

3. The lengths of the sides of a triangle are 24, 20, and 12. If the longest side of a similar triangle is 6, what is the length of its *shortest* side?

4. A person 5 feet tall casts a shadow of 12 feet at the same time that a tree casts a shadow of 60 feet. Find the number of feet in the height of the tree.

5. The lengths of corresponding sides of two similar polygons are in the ratio 2 : 5. If the perimeter of the larger polygon is 100, what is the perimeter of the smaller polygon?

6. If a car can travel 51 kilometers on 3 liters of gasoline, how many kilometers can it travel under the same conditions on 7 liters?

Solutions to Questions on Ratio and Proportion (including similar triangles and polygons)

1. Since it is given that $\angle R \cong \angle U$ and $\angle S \cong \angle V$, $\triangle RST \sim \triangle UVW$ (two triangles are similar if two angles of one are congruent to two angles of the other).

Let $x = VW$.

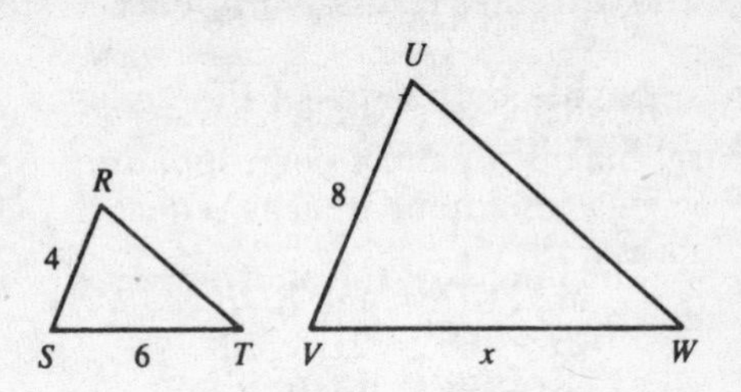

Corresponding sides of similar triangles are in proportion:

$$\frac{x}{6} = \frac{8}{4}$$

In a proportion, the product of the means equals the product of the extremes (cross-multiply):

$$4x = 6(8)$$
$$4x = 48$$

Divide both sides of the equation by 4:

$$\frac{4x}{4} = \frac{48}{4}$$
$$x = 12$$

$VW = 12$.

2. Since the triangles are similar, the longer leg, 12, of $\triangle DEF$ will correspond to the longer leg, 4, of $\triangle ABC$.

Let x = the length of the shorter leg of $\triangle DEF$.

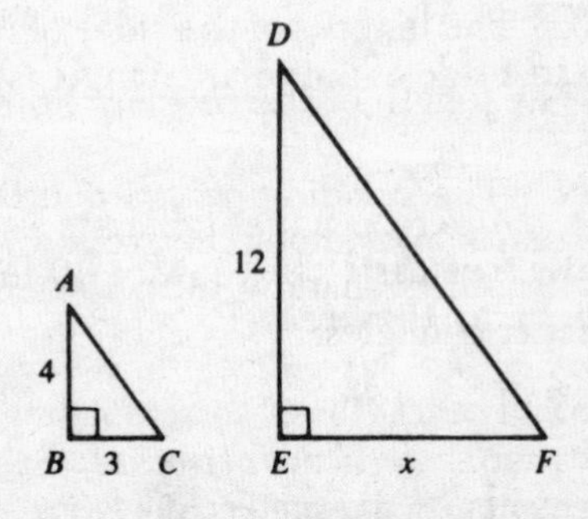

If two triangles are similar, the lengths of their corresponding sides are in proportion:

$$\frac{x}{3} = \frac{12}{4}$$

In a proportion, the product of the means equals the product of the extremes (cross-multiply):

$$4x = 3(12)$$
$$4x = 36$$

Divide both sides of the equation by 4:

$$\frac{4x}{4} = \frac{36}{4}$$
$$x = 9$$

The length of the shorter leg is 9.

3. Let x = the length of the *shortest* side.

The measures of the corresponding sides of similar triangles are in proportion. Note that 6 corresponds to 24 (the *longest* side of the first triangle) and x corresponds to 12 (the *shortest* side of the first triangle):

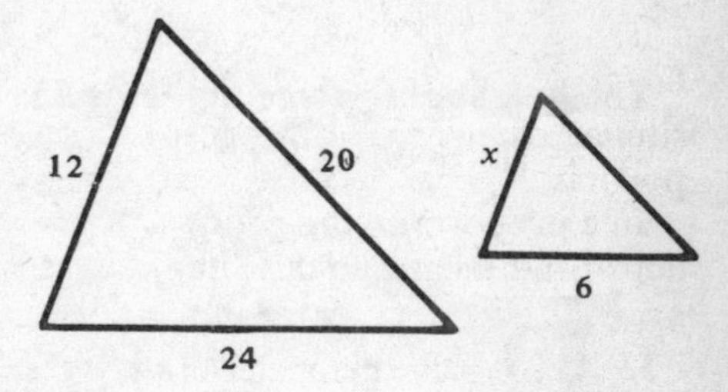

$$\frac{6}{24} = \frac{x}{12}$$

Reduce the fraction, $\frac{6}{24}$, by dividing the numerator and denominator by 6:

$$\frac{1}{4} = \frac{x}{12}$$

In a proportion, the product of the means equals the product of the extremes (cross-multiply):

$$4x = 12$$

Multiply both sides by $\frac{1}{4}$ (the multiplicative inverse of 4):

$$\frac{1}{4}(4x) = \frac{1}{4}(12)$$

The length of the *shortest* side is 3.

$$x = 3$$

4. The standing person and the tree are vertical; their shadows are horizontal (along the ground). The person and his shadow and the tree and its shadow form the vertical and horizontal legs of two similar triangles:

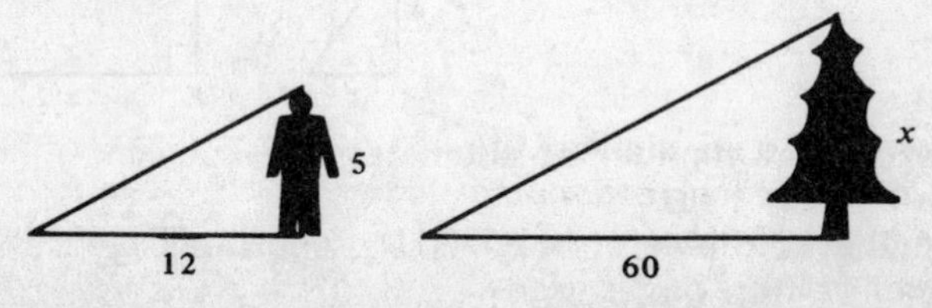

Let x = the number of feet in the height of the tree.

The corresponding sides of similar triangles are in proportion. That is, the ratio of a pair of corresponding sides is the same as the ratio of another pair of corresponding sides:

$$\frac{x}{5} = \frac{60}{12}$$

In a proportion, the product of the means equals the product of the extremes (cross-multiply):

$$12x = 300$$

Multiply each side by $\frac{1}{12}$ (the multiplicative inverse of 12):

$$\frac{1}{12}(12x) = \frac{1}{12}(300)$$

$$x = 25$$

The height of the tree in feet is **25**.

5. Let x = the perimeter of the smaller polygon.

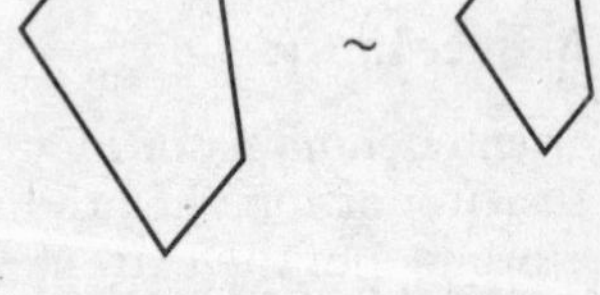

Perimeter = 100 Perimeter = x

The perimeters of two similar polygons have the same ratio as any two corresponding sides. Since the sides are in the ratio 2 : 5:

$$\frac{x}{100} = \frac{2}{5}$$

In a proportion, the product of the means equals the product of the extremes (cross-multiply):

$$5x = 2(100)$$
$$5x = 200$$
$$x = 40$$

The perimeter of the smaller polygon is **40**.

6. Let x = the number of kilometers the car can travel on 7 liters of gas.

The number of kilometers traveled is proportional to the number of liters of gas used:

$$\frac{x}{7} = \frac{51}{3}$$

In a proportion, the product of the means is equal to the product of the extremes (cross-multiply): $3x = 7(51)$

$$3x = 357$$

Divide both sides of the equation by 3: $\frac{3x}{3} = \frac{357}{3}$

$$x = 119$$

The car can travel 119 kilometers on 7 liters of gasoline.

21. QUESTIONS ON THE PYTHAGOREAN THEOREM

1. The lengths of the legs of a right triangle are 2 and 5. Find, in radical form, the length of the hypotenuse.

2. The length of the hypotenuse of a right triangle is 8 and the length of one leg is 5. The length of the other leg is
(1) $\sqrt{39}$ (2) $\sqrt{89}$ (3) 3 (4) 13

3. In the accompanying diagram of rectangle $ABCD$, $AB = 5$ and $BC = 12$. What is the length of $\overline{AC}$?

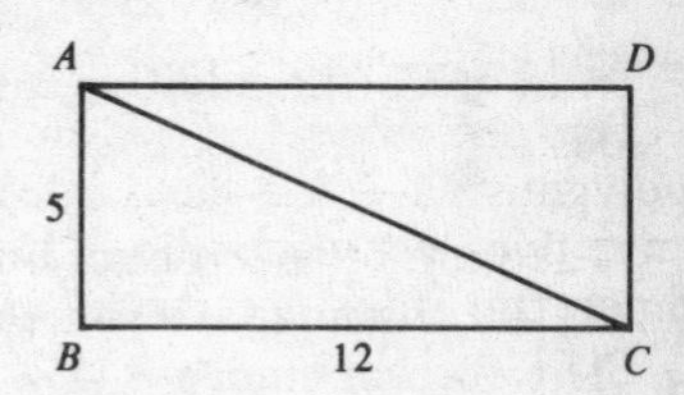

4. What is the length of a diagonal of a rectangle whose dimensions are 5 by 7?
(1) 5 (2) 8 (3) $\sqrt{24}$ (4) $\sqrt{74}$

5. Express in radical form the length of the diagonal of a square whose sides are each 2.

Solutions to Questions on the Pythagorean Theorem

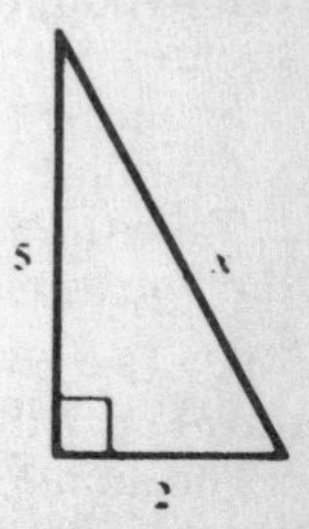

1. Let x = the length of the hypotenuse. By the Pythagorean Theorem, in a right triangle, the square of the length of the hypotenuse equals the sum of the squares of the lengths of the two legs: $x^2 = 5^2 + 2^2$

$x^2 = 25 + 4$

$x^2 = 29$

Take the square root of both sides: $x = \pm\sqrt{29}$

Reject the negative value as meaningless: $x = \sqrt{29}$

The length of the hypotenuse is $\sqrt{29}$.

2.

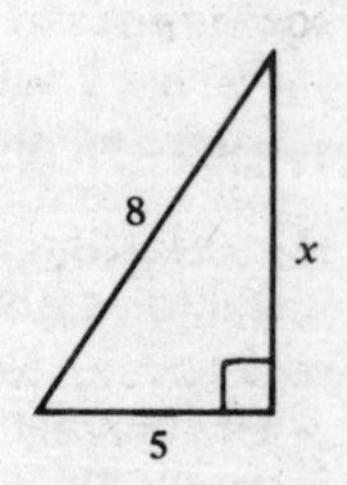

Let x = the length of the other leg.

By the Pythagorean Theorem, the sum of the squares of the measures of the two legs of a right triangle = the square of the measure of the hypotenuse: $x^2 + 5^2 = 8^2$

$x^2 + 25 = 64$

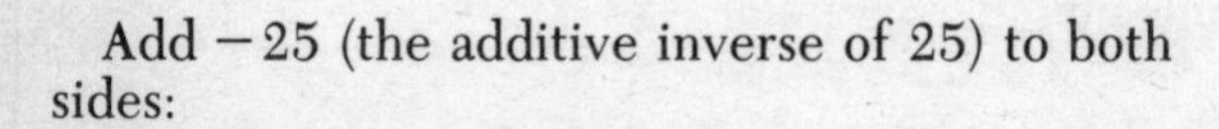

Add -25 (the additive inverse of 25) to both sides:

$$\begin{aligned} -25 &= -25 \\ \hline x^2 &= 39 \end{aligned}$$

Take the square root of both sides: $x = \pm\sqrt{39}$

Reject the negative value as meaningless: $x = \sqrt{39}$

The correct choice is **(1)**.

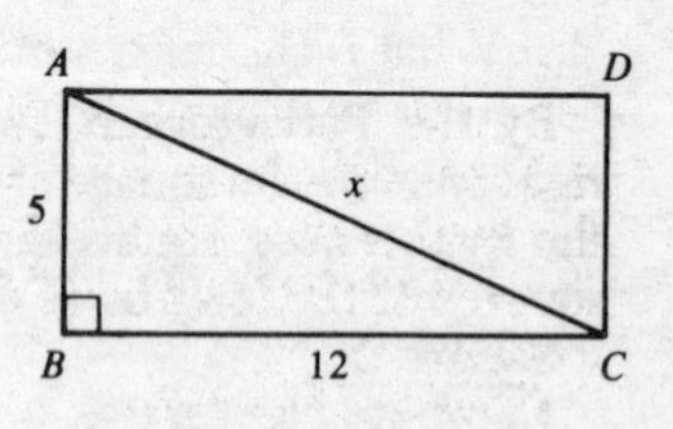

3. Since all angles of a rectangle are right angles, $\angle B$ is a right angle and thus $\triangle ABC$ is a right triangle.

Let x = the length of $\overline{AC}$.

By the Pythagorean Theorem, in a right triangle the square of the length of the hypotenuse equals the sum of the squares of the lengths of the legs: $x^2 = 5^2 + 12^2$

Square 5 and square 12: $x^2 = 25 + 144$

Combine like terms: $x^2 = 169$

Take the square root of both sides of the equation: $x = \pm\sqrt{169}$

$x = \pm 13$

Reject the negative value as meaningless for a length: $x = 13$

The length of $\overline{AC}$ is **13**.

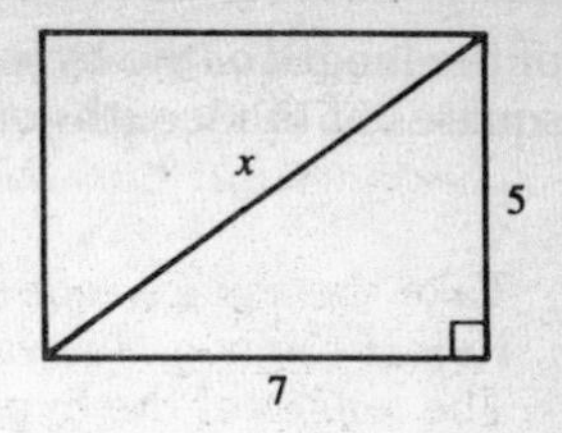

4. Each angle of a rectangle is a right angle. Therefore, the diagonal forms a right triangle with two of the sides.

Let x = the length of the diagonal.

By the Pythagorean Theorem, in a right triangle the square of the length of the hypotenuse equals the sum of the squares of the lengths of the legs: $x^2 = 7^2 + 5^2$

Square 7 and square 5: $x^2 = 49 + 25$

Combine like terms: $x^2 = 74$

Take the square root of both sides of the equation: $x = \pm\sqrt{74}$

Reject the negative value as meaningless for a length: $x = \sqrt{74}$

The correct choice is (**4**).

5. All angles of a square are right angles. Therefore, a diagonal forms a right triangle with two sides of the square.

Let x = the length of the diagonal.

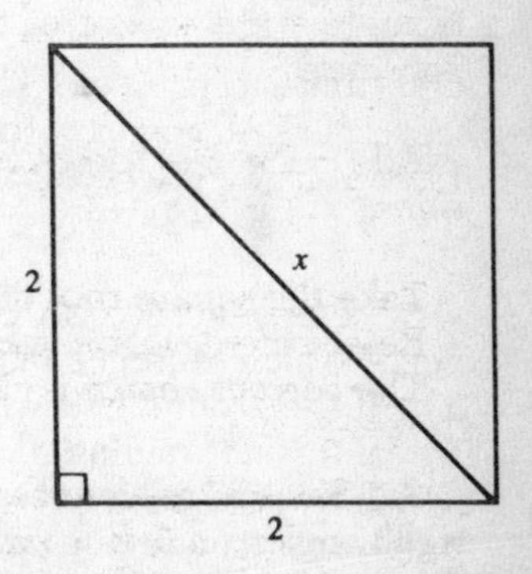

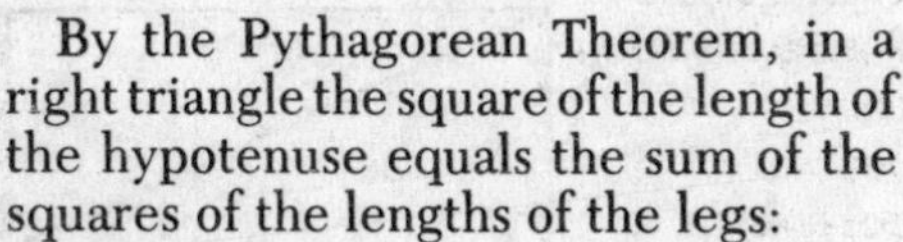

By the Pythagorean Theorem, in a right triangle the square of the length of the hypotenuse equals the sum of the squares of the lengths of the legs: $x^2 = 2^2 + 2^2$

Square 2: $x^2 = 4 + 4$

Combine like terms: $x^2 = 8$

Take the square root of both sides of the equation: $x = \pm\sqrt{8}$

Reject the negative value as meaningless for a length: $x = \sqrt{8}$

This answer may be written in simpler form by factoring out the perfect square factor from the radicand: $x = \sqrt{4(2)}$

Remove the perfect square factor from under the radical sign by taking its square root and writing it outside as a coefficient of the radical: $x = 2\sqrt{2}$

The length of the diagonal is $\sqrt{8}$ or $2\sqrt{2}$.

22. QUESTIONS ON LOGIC (SYMBOLIC REP., LOGICAL FORMS, TRUTH TABLES)

1. P represents "It is cold" and Q represents "I will go skiing." Using P and Q, write in symbolic form: "If it is cold, then I will *not* go skiing."

2. If p represents "He is tall," and q represents "He is handsome," write in symbolic form using p and q: "He is *not* tall, and he is handsome."

3. Let p represent "The polygon has exactly 3 sides," and let q represent "All angles of the polygon are right angles." Which is true if the polygon is a rectangle?
(1) $p \wedge q$ (2) $p \vee q$ (3) p (4) $\sim q$

4. This sentence is true: "If it is raining, then the ground gets wet." Which sentence must also be true?
(1) If the ground gets wet, then it is raining.
(2) If it is not raining, then the ground does not get wet.
(3) If the ground gets wet, then it is not raining.
(4) If the ground does not get wet, then it is not raining.

5. What is the converse of $\sim p \rightarrow q$?
(1) $p \rightarrow q$ (2) $p \rightarrow \sim q$ (3) $\sim q \rightarrow p$ (4) $q \rightarrow \sim p$

6. Which is the inverse of $\sim p \rightarrow q$?
(1) $p \rightarrow \sim q$ (3) $\sim p \rightarrow \sim q$
(2) $q \rightarrow \sim p$ (4) $\sim q \rightarrow \sim p$

7. What is the inverse of the statement, "If n is an odd integer, then $n + 2$ is an odd integer"?
(1) If n is an odd integer, then $n + 2$ is not an odd integer.
(2) If n is not an odd integer, then $n + 2$ is not an odd integer.
(3) If $n + 2$ is an odd integer, then n is an odd integer.
(4) If $n + 2$ is not an odd integer, then n is not an odd integer.

8. Which is the contrapositive of the statement, "If today is Monday, then tomorrow will be Tuesday"?
(1) Tomorrow is Tuesday if today is Monday.
(2) If tomorrow is Tuesday, then today is Monday.
(3) If today is not Monday, then tomorrow is not Tuesday.
(4) If tomorow is not Tuesday, then today is not Monday.

9. a. On your answer paper, copy and complete the truth table for the statement

$$(\sim p \rightarrow q) \leftrightarrow (p \vee q).$$

p	q	$\sim p$	$\sim p \rightarrow q$	$p \vee q$	$(\sim p \rightarrow q) \leftrightarrow (p \vee q)$

b. Is $(\sim p \rightarrow q) \leftrightarrow (p \vee q)$ a tautology?
c. Let p represent: "I do my homework."
Let q represent: "I get into trouble."
Which sentence is equivalent to $(p \vee q)$?
(1) If I do not do my homework, I will get into trouble.
(2) I do my homework or I do not get into trouble.
(3) If I do my homework, I get into trouble.
(4) I do my homework and I get into trouble.

10. a. Copy and complete the truth table for the statement $(p \rightarrow q) \leftrightarrow (q \vee \sim p)$.

p	q	$p \rightarrow q$	$\sim p$	$q \vee \sim p$	$(p \rightarrow q) \leftrightarrow (q \vee \sim p)$

b. Why is $(p \rightarrow q) \leftrightarrow (q \vee \sim p)$ a tautology?
c. In $(p \rightarrow q) \leftrightarrow (q \vee \sim p)$, let p represent "We pollute the water," and let q represent, "The fish will die."
Which statement is logically equivalent to "If we pollute the water, then the fish will die"?
(1) The fish will die or we do not pollute the water.
(2) We pollute the water and the fish will die.
(3) If we do not pollute the water, then the fish will not die.
(4) If the fish die, then we pollute the water.

Solutions to Questions on Logic (symbolic rep., logical forms, truth tables)

1. P represents "It is cold."
Q represents "I will go skiing."
"I will *not* go skiing" is the negation of Q, which is represented by $\sim Q$.
"If it is cold, then I will *not* go skiing" is the implication, P implies the negation of Q, which is represented by $P \rightarrow \sim Q$.
The symbolic form is $\mathbf{P \rightarrow \sim Q}$.

2. p represents "He is tall" and q represents "He is handsome"
"He is *not* tall" is the *negation* of p; the negation of p is represented by $\sim p$.
"He is *not* tall, and he is handsome" is the *conjunction* of $\sim p$ and q; the conjunction is represented by $\wedge$.
"He is *not* tall, and he is handsome" is represented by $\sim p \wedge q$.
The symbolic representation is $\mathbf{\sim p \wedge q}$.

3. p represents "The polygon has exactly 3 sides."
q represents "All angles of the polygon are right angles."
If the polygon is a rectangle, it has 4 sides; therefore, p is false. The angles of a rectangle are all right angles; therefore, q is true.
Consider each choice in turn:
(1) $p \wedge q$ is the *conjunction* of p and q. If it is true, it asserts that p and q are *both* true. It has been shown above that p is false; hence, choice (1) is *not* correct.
(2) $p \vee q$ is the *disjunction* of p and q. If it is true, either p or q, or both, are true. This agrees with the discussion above in which it has been shown that p is false but q is true. Hence, choice (2) is correct.
(3) If this choice is accepted as true, it asserts that p is true. This contradicts the reasoning above that p is false. Hence, choice (3) is *not* correct.
(4) $\sim q$ is the *negation* of q. If $\sim q$ is true, it states that "not all angles of a rectangle are right angles." This contradicts the facts concerning a rectangle. Hence, choice (4) is *not* correct.
The correct choice is **(2)**.

4. The given true statement is "If it is raining, then the ground gets wet."
Consider each choice in turn:
(1) "If the ground gets wet, then it is raining" is the *converse* of the given statement since it has been formed by interchanging the antecedent (hypothesis) and consequent (conclusion) of the given statement. If a statement is true, its converse may or may not be true.

(2) "If it is not raining, then the ground does not get wet" is the *inverse* of the given statement since it has been formed by negating the antecedent and consequent of the given statement. If a statement is true, its inverse may or may not be true.
(3) "If the ground gets wet, then it is not raining" has been formed by interchanging the negation of the original antecedent with the original consequent. This statement is false; its *contrapositive* (to which it is logically equivalent) would be "If it is raining, then the ground does not get wet" which contradicts the given statement.
(4) "If the ground does not get wet, then it is not raining" is the *contrapositive* of the given statement since it has been formed by negating both the original antecedent and the original consequent and then interchanging them. A statement and its contrapositive are logically equivalent, that is, if one is true, the other is also true. Thus, the statement of choice (4) is always true.
The correct choice is **(4)**.

5. $\sim p \rightarrow q$ is an implication which states that the negation of p implies q. $\sim p$ is the antecedent of the implication and q is the consequent.
The *converse* of an implication is formed by interchanging the antecedent and the consequent of the given implication. In the converse, q becomes the antecedent and $\sim p$ becomes the consequent:
The converse of $\sim p \rightarrow q$ is $q \rightarrow \sim p$.
The correct choice is **(4)**.

6.

$$\sim p \rightarrow q$$

The inverse of a statement is formed by negating both its antecedent (hypothesis or "if clause") and its consequent (conclusion or "then clause").
The negation of $\sim p$ is p.
The negation of q is $\sim q$.
Therefore, the inverse of $\sim p \rightarrow q$ is $p \rightarrow \sim q$.
The correct choice is **(1)**.

7. The *inverse* of a statement is formed by changing the antecedent (hypothesis or "if clause") to its negation, and also changing the consequent (conclusion or "then clause") to its negation.
The negation of "n is an odd integer" is "n is not an odd integer."
The negation of "$n + 2$ is an odd integer" is "$n + 2$ is not an odd integer."
Therefore, the inverse of "If n is an odd integer, then $n + 2$ is an odd integer" is "If n is not an odd integer, then $n + 2$ is not an odd integer."
The correct choice is **(2)**.

8. The given statement is "If today is Monday, then tomorrow will be Tuesday."

The *contrapositive* of a statement is formed by negating its antecedent (hypothesis or "if clause"), negating its consequent (conclusion or "then clause"), and then interchanging the two resulting statements.

The negation of "Today is Monday" is "Today is not Monday."

The negation of "Tomorrow will be Tuesday" is "Tomorrow is not Tuesday."

Thus, the contrapositive of "If today is Monday, then tomorrow will be Tuesday" is "If tomorrow is not Tuesday, then today is not Monday."

The correct choice is **(4)**.

9. **a.** STEP 1: Let T represent "true" and F represent "false." Fill in the columns for p and q with all possible combinations of T and F; this will require 4 lines.

STEP 2: For each line, fill in the column for $\sim p$ with the opposite of the entry for p, since $\sim p$ is the negation of p.

STEP 3: Use the 2nd and 3rd columns to determine the appropriate entry, T or F, for the column headed $\sim p \rightarrow q$, which means "the negation of p implies q."

STEP 4: Use the entries for p and q to determine the appropriate entry, T or F, for each line of the column headed $p \vee q$. This column represents the *disjunction* of p and q, that is, it states that either p or q or both are true.

STEP 5: Determine the appropriate entry, T or F, for each line of the final column, $(\sim p \rightarrow q) \leftrightarrow (p \vee q)$ by using the entries in the two preceding columns. The final column states that the two preceding columns are equivalent, that is, that both are true or both are false.

p	q	$\sim p$	$\sim p \rightarrow q$	$p \vee q$	$(\sim p \rightarrow q) \leftrightarrow (p \vee q)$
T	T	F	T	T	T
T	F	F	T	T	T
F	T	T	T	T	T
F	F	T	F	F	T

b. A *tautology* is a statement formed by combining other propositions or statements ($p, q, r, \ldots$) which is true regardless of the truth or falsity of $p, q, r, \ldots$. Since all entries in the column for $(\sim p \rightarrow q) \leftrightarrow (p \vee q)$ are T, this is a tautology.

The statement **is a tautology.**

c. p represents "I do my homework."

q represents "I get into trouble."

$(p \vee q)$ stands for the *disjunction* of p and q, that is, "I do my homework or I get into trouble." Either p and q are both true, or at least one of them is true.

(1) "If I do not do my homework, I will get into trouble" is equivalent to saying if p is false, q must be true. This is equivalent to the disjunction $(p \vee q)$.

(2) "I do my homework or I do not get into trouble" says that either p is true or q is false. This is *not* equivalent to $(p \vee q)$.

(3) "If I do my homework, I get into trouble" says that if p is true, q is also true. This is *not* equivalent to $(p \vee q)$.

(4) "I do my homework and I get into trouble" says that p is true and q is true. It is not an implication, and is *not* equivalent to $(p \vee q)$.

The correct choice is **(1)**.

10. a. STEP 1: Using "T" to represent true and "F" to represent false, fill in the columns for p and q with all possible combinations of T and F; this will require 4 lines.

STEP 2: The third column, $p \rightarrow q$, stands for the *implication*, p implies q. For each line under this column, determine the appropriate entry, "T" or "F", according to the values shown in the columns for p and q.

STEP 3: The fourth column, $\sim p$, represents the *negation* of p. The entry for each line in this column should be the opposite of the entry in the column for p.

STEP 4: The column, $q \vee \sim p$, represents the *disjunction* of q and the negation of p. It asserts that either q or not p, or both, are true. Use the entries in the columns for q and $\sim p$ to determine the appropriate entry for each line under $q \vee \sim p$.

STEP 5: The last column, $(p \rightarrow q) \leftrightarrow (q \vee \sim p)$, stands for the *equivalence* of $p \rightarrow q$ and $q \vee \sim p$. Its value will be "T" if the values of $p \rightarrow q$ and $q \vee \sim p$ are the same as each other; its value will be "F" if their values are different from each other.

p	q	$p \rightarrow q$	$\sim p$	$q \vee \sim p$	$(p \rightarrow q) \leftrightarrow (q \vee \sim p)$
T	T	T	F	T	T
T	F	F	F	F	T
F	T	T	T	T	T
F	F	T	T	T	T

b. A *tautology* is a statement formed by combining other propositions or statements, $p, q, r, \ldots$, which is true regardless of the truth or falsity of $p, q, r, \ldots$.

In the table above, the truth values for $(p \rightarrow q) \leftrightarrow (q \vee \sim p)$ are all "T" no matter what the truth values are for p and q on the same line of the table. Hence, $(p \rightarrow q) \leftrightarrow (q \vee \sim p)$ is a tautology.

c. p represents "We pollute the water."
q represents "The fish will die."

The given statement, "If we pollute the water, then the fish will die" is represented symbolically by the *implication*, $p \rightarrow q$.

In parts a and b above, it has been shown that $(p \rightarrow q) \leftrightarrow (q \vee \sim p)$ is a tautology, that is, that $p \rightarrow q$ is *logically equivalent* to $q \vee \sim p$.

For what statement is $q \vee \sim p$ the symbolic representation? $\sim p$ is the *negation* of p and therefore stands for "We do not pollute the water." $q \vee \sim p$ is the *disjunction* of q and $\sim p$, and therefore stands for "The fish will die or we do not pollute the water." This is choice (1).

The correct choice is **(1)**.

23. QUESTIONS ON PROBABILITY (INCL. TREE DIAGRAMS, SAMPLE SPACES)

1. The mail consists of 3 bills, 2 advertisements, and 1 letter. If the mail is opened randomly, what is the probability that an advertisement is opened first?

2. If the probability of an event happening is $\frac{2}{5}$, what is the probability of the event *not* happening?

3. Two fair dice are tossed. Each die has six faces numbered 1 to 6. What is the probability that each die shows a 5?

4. A box contains 4 nickels, 3 dimes, and 2 quarters. One coin is drawn, put aside, and then another coin is drawn. What is the probability that the two coins total 9¢?

5. The probability of the Bears beating the Eagles is $\frac{1}{2}$. The probability of the Bears beating the Cubs is $\frac{3}{5}$. What is the probability of the Bears winning both games?

6. From an ordinary deck of 52 cards, one card is drawn. What is the probability that the card drawn is either a king or a seven?

(1) $\frac{26}{52}$ (2) $\frac{13}{52}$ (3) $\frac{8}{52}$ (4) $\frac{2}{52}$

7. A die and a coin are tossed simultaneously. The die is fair and has six faces.

a. Draw a tree diagram or list the sample space of all possible pairs of outcomes.

b. What is the probability of obtaining a 6 on the die and a head on the coin?

c. What is the probability of obtaining an odd number on the die and a tail on the coin?

d. What is the probability of obtaining a head on the coin?

8. The first step of an experiment is to pick one number from the set {1,2,3}. The second step of the experiment is to pick one number from the set {1,4,9}.

a. Draw a tree diagram or list the sample space of all possible pairs of outcomes.

b. Determine the probability that:

(1) both numbers are the same
(2) the second number is the square of the first
(3) both numbers are odd

Solutions to Questions on Probability (incl. tree diagrams, sample spaces)

1. Probability of an event $= \frac{\text{number of successful outcomes}}{\text{total number of possible outcomes}}$.

There are 2 advertisements in the mail; these represent the 2 favorable cases for opening an advertisement first.

Three bills, 2 advertisements, and 1 letter constitute 6 pieces of mail, any one of which could be the first to be opened. Thus, the total number of cases is 6.

Probability of opening an advertisement first $= \frac{2}{6}$.

The probability is $\frac{2}{6}$.

2. Probability of an event $= \dfrac{\text{number of successful outcomes}}{\text{total number of possible outcomes}}$.

Since the probability of the event happening is $\frac{2}{5}$, the event has 2 ways to happen (successful outcomes) out of a total of 5 possible cases.

The number of outcomes in which an event will *not* happen = the total number of outcomes minus the number of outcomes in which it will happen. Here, the number of cases in which the event will *not* happen $= 5 - 2 = 3$.

Probability of an event *not* happening =

$$\frac{\text{number of outcomes in which event will not happen}}{\text{total number of possible outcomes}} = \frac{3}{5}.$$

The probability of the event not happening is $\mathbf{\frac{3}{5}}$.

ALTERNATE SOLUTION: An event happening or *not* happening covers all possible outcomes. Therefore, the probability of an event happening added to the probability of it *not* happening equals certainty, which is represented by 1. Hence, the probability of an event *not* happening = 1 − the probability of it happening. In this case, the probability of the event *not* happening $= 1 - \frac{2}{5} = \frac{3}{5}$.

3. Probability of an event $= \dfrac{\text{number of successful outcomes}}{\text{total number of possible outcomes}}$.

Since there is only one 5 on each of the die faces, there is only 1 way in which both dice can show 5. Each die by itself can show any of 6 possible faces. Any one of the 6 possibilities for the first die may be matched with any one of the 6 possibilities for the second die, making a total of $6 \times 6 = 36$ ways in which the two dice may be tossed.

Probability of two 5's $= \frac{1}{36}$.

The probability that each die will show a 5 is $\mathbf{\frac{1}{36}}$.

4. Probability of an event $= \dfrac{\text{number of successful outcomes}}{\text{total number of possible outcomes}}$.

There is *no* way to obtain a total of 9¢ by using nickels, dimes, and quarters. Hence, the number of successful outcomes for such an event is 0. There is no need to compute the total number of outcomes, t, as long as it is realized that $t \neq 0$.

Probability of a total of 9¢ $= \frac{0}{t} = 0$.

Note: The reason it is important to realize that $t \neq 0$ is so that it is possible to divide 0 by t; division by 0 is undefined.

The probability is **0.**

5. The probability of $\frac{1}{2}$ means that the Bears will beat the Eagles in $\frac{1}{2}$ of the games they play against the Eagles. Similarly, the probability of $\frac{3}{5}$ means that the Bears will beat the Cubs in $\frac{3}{5}$ of the games they play against the Cubs. If a Bears-Eagles game is followed by a Bears-Cubs game, then the $\frac{1}{2}$ of the times the Bears win the first game will be followed $\frac{3}{5}$ of the time by their winning the second game. Thus, the probability that the Bears will win both games is $\frac{1}{2} \times \frac{3}{5} = \frac{3}{10}$.

The probability is $\mathbf{\frac{3}{10}}$.

6. Probability of an event $= \frac{\text{number of successful outcomes}}{\text{total number of possible outcomes}}$.

In an ordinary deck of cards, there are 4 kings and 4 sevens, giving 8 successful outcomes for the drawing of either a king or a seven. Since any one of the 52 cards in the deck may be drawn, the total number of possible outcomes is 52.

Thus, the probability of drawing either a king or a seven $= \frac{8}{52}$.

The correct choice is **(3)**.

7. a. A fair die toss has an equal chance of turning up a 1, 2, 3, 4, 5, or 6. To represent this, the die toss is shown by 6 branches on the tree diagram.

Each die toss may be matched with either a head (H) or a tail (T) resulting from the coin toss. Thus, the tree diagram represents the coin toss by two branches above each of the die toss branches, one for H and one for T.

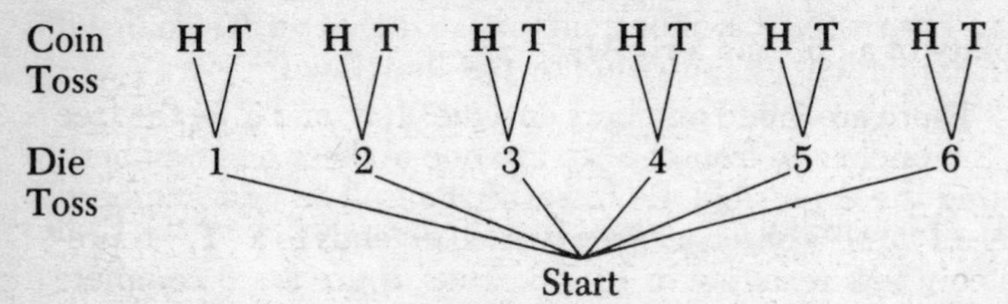

The results of a simultaneous toss of a die and a coin may also be listed in a sample space. The die toss result is shown in the left column; the coin toss result in the right. Each of the 6 possible outcomes for the die toss is paired with the 2 possible outcomes (HEADS or TAILS) for the coin toss. There are thus 12 lines, each representing a different possible pair, in the sample space:

Die Toss	Coin Toss
1	H
1	T
2	H
2	T
3	H
3	T
4	H
4	T
5	H
5	T
6	H
6	T

b. In the tree diagram, there is only one branch from the Start that leads to a "6" for the die toss. From the "6" there is only one branch that leads to "H" for the coin toss. Thus, there is only one path that produces a die showing "6" and a coin showing HEADS. There are 12 possible paths in all. If the sample space is used to determine the answer, only 1 line of the 12 contains both a "6" and an "H."

$$\text{Probability of an event} = \frac{\text{number of successful outcomes}}{\text{total number of possible outcomes}} = \frac{1}{12}.$$

Probability of a "6" and a HEAD $= \mathbf{\frac{1}{12}}$.

c. There are 3 odd numbers on a die: 1, 3, and 5. In the tree diagram, 3 branches go from the Start to one of these odd numbers, representing the 3 possible die tosses that result in odd numbers. From each of these 3 branches, one branch extends to a "T," representing a coin toss resulting in TAILS. Thus, there are 3 complete paths satisfying the two requirements of the question. There is a total of 12 possible paths. If the sample space is used, it will be noted that 3 of the 12 lines contain an odd number and a "T."

$$\text{Probability of an event} = \frac{\text{number of successful outcomes}}{\text{total number of possible outcomes}} = \frac{3}{12}.$$

Probability of an odd number and TAILS $= \mathbf{\frac{3}{12}}$.

d. All 6 branches from the Start to the die toss outcome can be used as the first step toward ultimately reaching a branch leading to "H" for the coin toss. For each of these 6 lower branches, there is only one branch leading to a coin toss of HEADS. Thus, there are 6 paths from the Start leading to possible HEADS. The total number of all possible paths is 12. If the sample space is used to solve, note that there are 6 lines out of the 12 containing an "H"; it does not matter what is shown in the "Die Toss" column.

$$\text{Probability of an event} = \frac{\text{number of successful outcomes}}{\text{total number of possible outcomes}} = \frac{6}{12}.$$

Probability of obtaining a HEAD $= \mathbf{\frac{6}{12}}$.

8. a. The tree diagram is shown first. The tree diagram contains 3 branches leading from the Start to represent the first step since either 1, 2, or 3 may be picked on the first step. For each of the "first step" branches, 3 "second step" branches lead to 1, 4, or 9 as the possible second number to be picked:

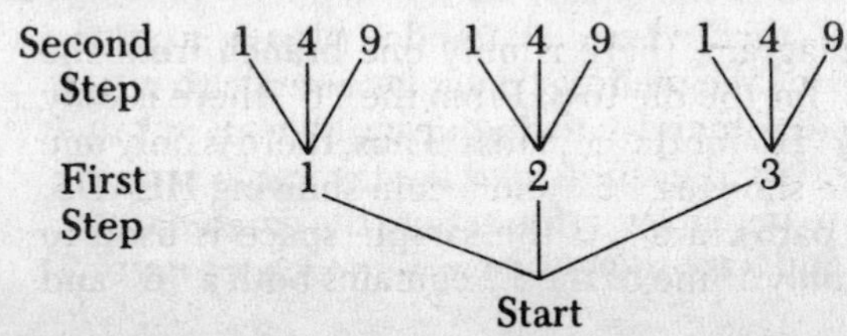

The sample space contains two columns, one for each of the two steps in the experiment. The column for the first step shows the possible numbers, 1, 2, and 3, for that step. Each number in the first column is paired with 1, 4, and 9, in the second column; 1, 4, and 9 are the possible numbers that may be picked in the second step following the first pick of 1, 2, or 3:

First Step	Second Step
1	1
1	4
1	9
2	1
2	4
2	9
3	1
3	4
3	9

b. Probability of an event = $\frac{\text{number of successful outcomes}}{\text{total number of possible outcomes}}$.

(1) The only way in which both numbers can be the same is if "1" is chosen both times. Using the tree diagram, there is only one path leading to both selections of "1"—the leftmost path. There are 9 possible paths in all. If the sample space is used, there is only one line in which both columns contain a "1"—the first line. There are 9 lines in all. Therefore, the probability that both numbers are the same $= \frac{1}{9}$.

The probability is $\frac{1}{9}$.

(2) If the second number is the square of the first, the possible selections may be 1 and 1^2 (that is, 1 and 1), 2 and 2^2 (that is, 2 and 4), or 3 and 3^2 (that is, 3 and 9). Using the tree diagram, one path leads to 1 and 1, a second path leads to 2 and 4, and a third path leads to 3 and 9. Thus, 3 paths out of the total possible of 9 are favorable cases. If the sample space is used to obtain the information, the combinations of 1 and 1, 2 and 4, and 3 and 9 are shown on one line each, a total of 3

lines (favorable cases) out of the 9 possible lines. Therefore, the probability that the second number is the square of the first $= \frac{3}{9}$.

The probability is $\frac{3}{9}$.

(3) Using the tree diagram, the paths leading to both odd numbers are the paths 1 – 1, 1 – 9, 3 – 1, and 3 – 9. Thus, there are 4 possible paths out of the total of 9 which represent successful outcomes. Using the sample space, the lines containing 1 and 1, 1 and 9, 3 and 1, and 3 and 9 are the successful outcomes of lines containing both odd numbers. There are 9 lines in all. Therefore, the probability that both numbers are odd is $\frac{4}{9}$.

The probability is $\frac{4}{9}$.

24. QUESTIONS ON COMBINATIONS (ARRANGEMENTS, PERMUTATIONS)

1. If a boy has 5 shirts and 3 pairs of pants, how many possible outfits consisting of one shirt and one pair of pants can be chosen?

2. A school cafeteria offers 6 kinds of sandwiches and 3 kinds of beverages. If a lunch consists of a sandwich and a beverage, how many different lunches can a student choose?

3. A boy has 3 shirts, 2 pairs of slacks, and 3 pairs of shoes. Find the total number of possible outfits he can wear consisting of a shirt, a pair of slacks, and a pair of shoes.

4. How many four-digit numbers can be formed from the digits 3, 4, 5, and 6 if no repetition is allowed?

5. In how many different ways can 4 students be arranged in a row?

6. What is the total number of possible 5-letter arrangements of the letters *D*, *I*, *S*, *C*, and *O*, if each letter is used only once in each arrangement?

Solutions to Questions on Combinations (arrangements, permutations)

1. The boy has 5 choices for shirts. For each of these 5 choices, he has 3 choices of pairs of pants. The total number of possible outfits is therefore $5 \times 3 = 15$.
The number of possible outfits is **15.**

2. A lunch consists of a sandwich and a beverage. There are 6 different sandwiches from which to choose. Each of these 6 ways to choose a sandwich may be matched with each of the 3 kinds of beverages.
Therefore, the number of different lunches $= 6 \times 3 = 18$.
The number of different lunches is **18.**

3. With each of the 3 shirts, either one of the 2 pairs of slacks may be worn. Thus, 3×2 or 6 shirt-slacks outfits are possible.
With each of the 6 shirt-slacks outfits, there are 3 possible choices for pairs of shoes. Thus, 6×3 or 18 possible shirt-slacks-shoes outfits can be worn.
There are **18** possibilities.

4. There is a choice of 4 digits (3, 4, 5, or 6) for filling the first place of the four-digit number. For filling the second place, there is a choice of 3 digits (since one has been used for the first place and may not be repeated). Similarly, there is a choice of 2 digits for the third place. The one remaining digit must be used to fill the fourth place. Thus, there are $4 \times 3 \times 2 \times 1$ or 24 different four-digit numbers that can be formed.
24 numbers can be formed.

5. An arrangement of 4 students in a row is made by selecting one of the 4 students for the first place, one of the remaining 3 for the second place, one of the 2 remaining for the third place, and then putting the one that is left in the last place.
There are thus 4 choices for the first place; and for each of these, there are 3 choices for the second place; for each of these, there are 2 choices for the third place; finally, there is only 1 choice for the last place. In other words, there is a total of $4 \times 3 \times 2 \times 1 = 24$ different arrangements in all.
There are **24** different arrangements.

6. Each arrangement of the letters *D*, *I*, *S*, *C*, and *O* forms a 5-letter "word." Forming such a word may be considered to be the successive acts of choosing the first letter in it, then the second letter, etc., until finally the fifth letter is filled in.

The first letter may be chosen in 5 different ways. Since the letter chosen for the first place cannot be used again, there are 4 choices left for the second place. Then, 3 letters remain as choices for the third place, 2 for the fourth, and 1 for the fifth. Thus, the total number of possible arrangements $= 5 \times 4 \times 3 \times 2 \times 1 = 120$.

The number of possible 5-letter arrangements is **120.**

25. QUESTIONS ON STATISTICS

1. If student heights are 176 cm, 172 cm, 160 cm, and 160 cm, what is the mean height of these students?

2. What is the mode of the following data?

2.6, 2.8, 2.8, 2.7, 2.9, 2.4

3. Five girls in a club reported on the number of boxes of cookies that they sold: 20, 20, 40, 50, and 70. Which is true?

(1) The median is 20.
(2) The mean is 20.
(3) The median is equal to the mean.
(4) The median is equal to the mode.

4. Given the following table:

Interval	Frequency
91–100	2
81–90	2
71–80	3
61–70	4

Which interval contains the median?

(1) 91–100 (2) 81–90 (3) 71–80 (4) 61–70

5. The following data are heights (in centimeters) of a group of 15 students: 165, 160, 173, 150, 188, 150, 173, 155, 163, 152, 175, 183, 151, 163, 178.

a. On your answer paper, copy and complete the table below.

Interval	Number (frequency)
180–189	
170–179	
160–169	
150–159	

b. On graph paper, construct a frequency histogram based on the grouped data.

c. In what interval is the median for the grouped data?

6. The table below gives the distribution of test scores for a class of 20 students.

Test Score Interval	Number of Students (frequency)
91–100	1
81–90	3
71–80	3
61–70	7
51–60	6

a Draw a *frequency* histogram for the given data.

b Which interval contains the median?

c Which interval contains the lower quartile?

7. The Play-Craft Company, maker of table tennis paddles, listed the following salaries in its annual report:

Employee	*Position*	*Salary*
Pat Thomas	Manager	$100,000
Don Pierce	Assistant manager	$50,000
Linda Jones	Foreman	$25,000
Jim Jeffrey	Skilled worker	$16,000
Donna Love	Skilled worker	$15,000

John Hanna	Skilled worker	\$14,000
Jill Walker	Skilled worker	\$14,000
Ben Black	Skilled worker	\$13,000
Tony Burch	Secretary	\$11,750
John Slack	Custodian	\$10,000

a. Using all ten salaries,
(1) find the mean,
(2) find the median,
(3) find the mode.

b. How much more than the mean is the manager's salary?

c. If an employee is selected at random, what is the probability that this person's salary is greater than the mean?

Solutions to Questions on Statistics

1. The *mean* is obtained by adding all the items and dividing their sum by the number of items (in this case, by 4):

$$\begin{array}{r} 176 \\ 172 \\ 160 \\ \underline{160} \\ 4\overline{)668} \\ 167 \end{array}$$

The mean height is **167** cm.

2. The mode is the item of data occurring most frequently. In 2.6, 2.8, 2.8, 2.7, 2.9, 2.4, the item 2.8 occurs twice while all others appear only once. Hence, 2.8 is the mode.
The mode is **2.8.**

3. 20, 20, 40, 50, 70
The *median* is the middle item when items are arranged in order of size. Here, the third item, which is 40, is the median. This rules out choice (1) which asserts that 20 is the median.
The *mean* is the sum of all the items divided by the number of items. In this case, mean $= \frac{20 + 20 + 40 + 50 + 70}{5} = \frac{200}{5} = 40.$

This rules out choice (2) which asserts that the mean is 20.
The preceding calculations have shown that the median is 40 and also that the mean is 40. Therefore, choice (3) is true since it asserts that the median is equal to the mean.

The *mode* is the item which appears most frequently. In this case, 20 is the mode since it appears twice while each of the other items appears only once. The median has already been calculated to equal 40; hence, choice (4) is false since it asserts that the median is equal to the mode.

The correct choice is (3).

4. The median is the middle item when a set of data is arranged in order of size.

First add the frequencies: $2 + 2 + 3 + 4 = 11$. Therefore, the middle item is the 6th; there are 5 items above it and 5 items below it. Counting down from the top, the 91 – 100 interval contains 2 items and the 81 – 90 interval contains 2, making a total of 4. 2 more are needed to reach the 6th item. Since the 71 – 80 interval contains the next 3 items, the 6th is contained within the 71 – 80 interval.

The correct choice is (3).

5. a.

Interval	Number (frequency)
180 – 189	2
170 – 179	4
160 – 169	4
150 – 159	5

b.

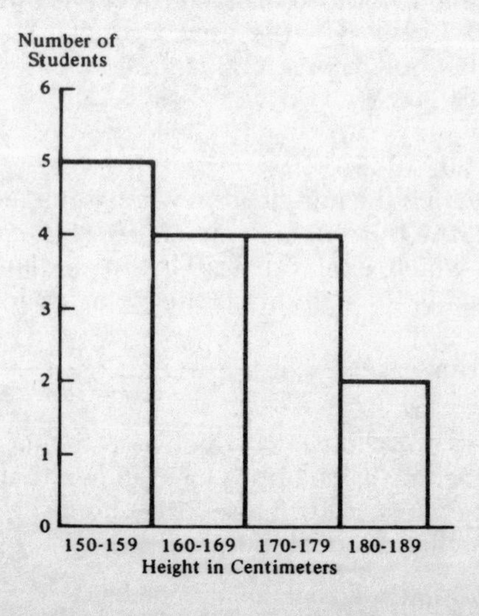

c. The *median* is the middle item when items are arranged in order of size. Since there are 15 heights given, the eighth one would represent the median; there are 7 items on each side of it.

Counting down from the top of the table shown in part a, the first two intervals, 180–189 and 170–179, account for 2 + 4 = 6 items. Thus, 2 more items are needed to reach the eighth. Both the next two would lie in the 160–169 interval since it contains 4 items.

The median lies in the **160–169** interval.

6.

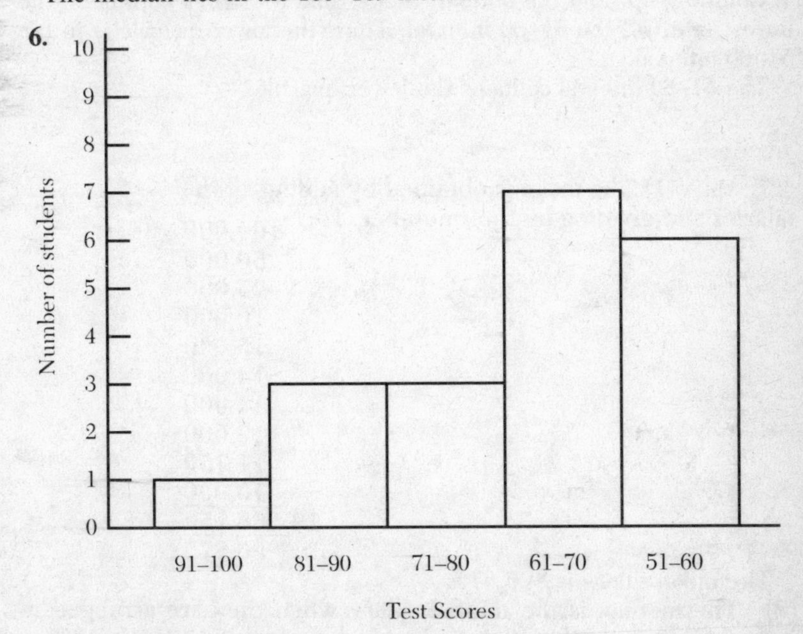

b. Find the total number of students by adding all the frequencies: 1 + 3 + 3 + 7 + 6 = 20

If the scores are arranged in order of size, the *median* is the middle score if the number of scores is odd; or, if there is an even number of scores (as is the case here), the median is the score midway between the two middle scores. Since there are 20 scores, the median score will lie midway between the 10th and 11th scores.

Counting up from the bottom, there are 6 scores in the 51–60 interval. To count up to the 10th score, 4 more are needed from the next interval (the 61–70 interval); 5 more are needed from the same interval to reach the 11th score. Both the 10th and 11th scores are in the 61–70 interval. The score midway between them (the median) must also lie in this interval.

The **61–70** interval contains the median.

c. The *lower quartile* is the score separating the lowest one-quarter of all scores from the remaining three-quarters.

Since there are 20 scores in all and $\frac{1}{4}(20) = 5$, the lowest 5 scores comprise the lowest one-quarter; the remaining 15 scores comprise the upper three-quarters. The lower quartile is the score midway between the 5th and 6th scores.

Counting up from the bottom, the 5th and 6th scores both lie in the lowest interval, the 51–60 interval. Thus, the lower quartile is in the 51–60 interval.

The **51–60** interval contains the lower quartile.

7. a. (1) The mean is obtained by adding all the salaries and dividing by their number, 10:

$$\begin{array}{r} 100{,}000 \\ 50{,}000 \\ 25{,}000 \\ 16{,}000 \\ 15{,}000 \\ 14{,}000 \\ 14{,}000 \\ 13{,}000 \\ 11{,}750 \\ \underline{10{,}000} \\ 10\overline{)268{,}750} \\ 26{,}875 \end{array}$$

The mean salary is **$26,875.**

(2) The median is the middle salary when they are arranged in order of size. Since there is an even number of salaries, 10, there is no single middle salary. In such a case, the median is taken to be midway between the two middle salaries (the 5th and 6th); half the salaries will then be above the median and half will be below it. Counting down from the top, the 5th salary is $15,000 and the 6th is $14,000. Thus, the median (midway between them) is $14,500.

The median salary is **$14,500.**

(3) The mode is the salary which occurs most often. $14,000 occurs twice; all other salaries occur once each.

The mode is **$14,000.**

b.

Manager's salary	=	$100,000
Mean	=	26,875
Difference	=	$ 73,125

The manager's salary is **$73,125** more than the mean.

c. Probability of an event = $\dfrac{\text{number of successful outcomes}}{\text{total possible number of outcomes}}$

There are only two employees (the manager and the assistant manager) with salaries greater than the mean. The number of successful outcomes for selecting an employee with a salary greater than the mean is thus 2.

The total number of cases for the random selection of an employee is 10.

The probability of selecting an employee with a salary greater than the mean is $\frac{2}{10}$.

The probability is $\frac{2}{10}$.

26. QUESTIONS ON PROPERTIES OF TRIANGLES AND POLYGONS

1. If the lengths of two sides of a triangle are 6 and 8, the length of the third side may be
   (1) 7
   (2) 2
   (3) 14
   (4) 15

2. A regular hexagon has a perimeter of $12x - 30$. Express the length of one side of the hexagon as a binomial in terms of x.

3. In rectangle $ABCD$, $AC = 2x + 15$ and $BD = 4x - 5$. Find x.

4. If the diagonals of a parallelogram are perpendicular and not congruent, then the parallelogram is
   (1) a rectangle
   (2) a rhombus
   (3) a square
   (4) an isosceles trapezoid

5. A set contains five quadrilaterals: a parallelogram, a rectangle, a rhombus, a square, and a trapezoid. If one quadrilateral is selected at random from the set, what is the probability that the figure selected will have congruent opposite angles?

6. The diagonals of a rhombus have lengths of 8 centimeters and 6 centimeters. The perimeter of the rhombus is

(1) 20 cm (3) 5 cm
(2) 24 cm (4) 14 cm

Solutions to Questions on Properties of Triangles and Polygons

1. The sum of the lengths of two sides of a triangle must be greater than the length of the third side. Use this relationship to check each choice as a possible length for the third side of the triangle whose other two sides have lengths of 6 and 8:

(1) 7: $6 + 8 \overset{?}{>} 7$ $6 + 7 \overset{?}{>} 8$ $8 + 7 \overset{?}{>} 6$
$14 > 7$ ✓ $13 > 8$ ✓ $15 > 6$ ✓ 7 is possible.

(2) 2: $6 + 8 \overset{?}{>} 2$ $6 + 2 \overset{?}{>} 8$
$14 > 2$ ✓ $8 \not> 8$ The third side cannot be 2.

(3) 14: $6 + 8 \overset{?}{>} 14$
$14 \not> 14$ The third side cannot be 14.

(4) 15: $6 + 8 \overset{?}{>} 15$
$14 \not> 15$ The third side cannot be 15.

The correct choice is **(1)**.

2. A regular hexagon is a polygon with 6 congruent sides and 6 congruent angles. Let $y =$ the length of one side of the hexagon. The perimeter of the hexagon equals the sum of the lengths of the 6 sides:

$$6y = 12x - 30$$

Divide each term on both sides of the equation by 6:

$$\frac{6y}{6} = \frac{12x}{6} - \frac{30}{6}$$
$$y = 2x - 5$$

The length of one side is $\mathbf{2x - 5}$.

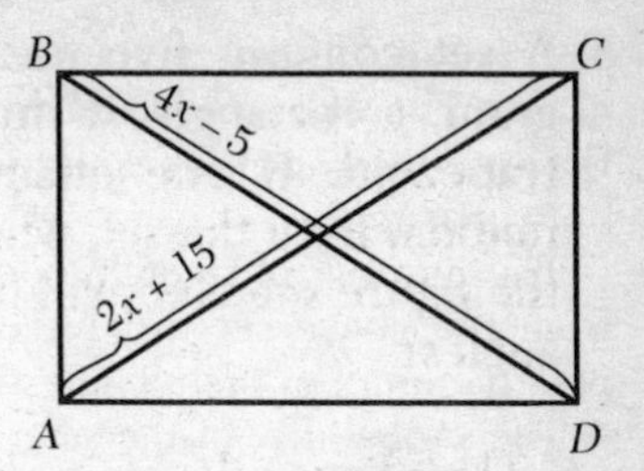

3. In rectangle $ABCD$, $\overline{AC}$ and $\overline{BD}$ are diagonals. The diagonals of a rectangle are congruent:

$$\begin{aligned} 2x + 15 &= 4x - 5 \\ 15 + 5 &= 4x - 2x \\ 20 &= 2x \\ 10 &= x \end{aligned}$$

x = **10.**

4. Consider each choice:

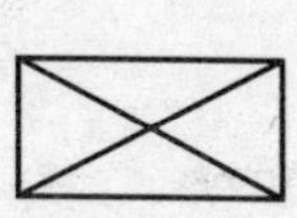
(1) Rectangle

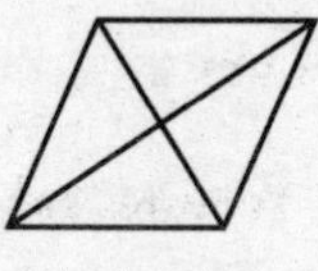
(2) Rhombus

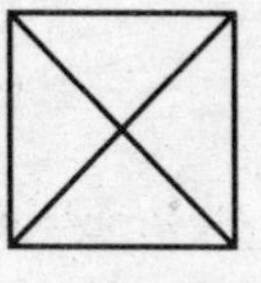
(3) Square

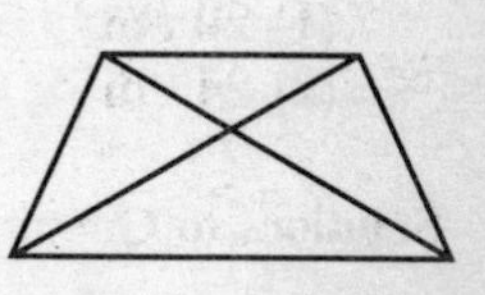
(4) Isosceles Trapezoid

Choice (4) is immediately ruled out because an isosceles trapezoid is not a parallelogram; the legs of a trapezoid are not parallel.

The fact that the diagonals of the parallelogram are perpendicular rules out choice (1), a rectangle; the diagonals of a rectangle are perpendicular only in the special case that it is a square.

The fact that the diagonals are not congruent rules out both choice (1), a rectangle, and choice (3), a square.

Choice (2), a rhombus, is the only remaining choice. Actually, that this choice is correct can be proved from the information given. Since the diagonals of any parallelogram bisect each other, if they are also perpendicular to each other they form four congruent right triangles. The four sides of the parallelogram are corresponding parts of these triangles and hence are all congruent, making the parallelogram equilateral, and hence a rhombus.

The correct choice is **(2)**.

5. Probability of

$$\text{an event occurring} = \frac{\text{number of successful outcomes}}{\text{total possible number of outcomes}}.$$

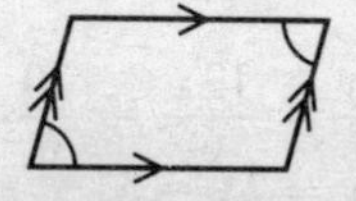
Parallelogram

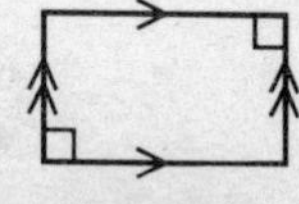
Rectangle

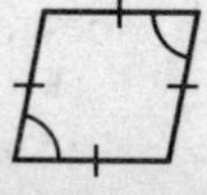
Rhombus

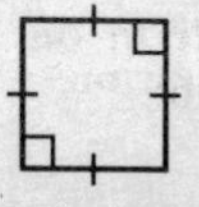
Square

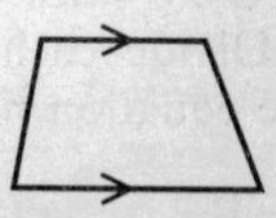
Trapezoid

A parallelogram, rectangle, rhombus, and square are all parallelograms. Therefore, they all have a property common to all parallelograms: the opposite angles are congruent. In a trapezoid, the opposite angles are *not* congruent.

The number of favorable outcomes for a random selection of a figure from the above set to have opposite angles congruent is 4. The total possible number of outcomes is 5 since there are 5 figures in all.

The probability of the selection of a figure with opposite angles congruent is $\frac{4}{5}$.

The probability is $\mathbf{\frac{4}{5}}$

6. The diagonals of a rhombus bisect each other since a rhombus is a parallelogram. The diagonals of a rhombus are also perpendicular to each other. Thus, the diagonals divide the rhombus into four congruent right triangles, each having legs whose lengths are 3 cm and 4 cm. The hypotenuse of each right triangle is a side of the rhombus. From the 3-4-5 right triangle (or by use of the Pythagorean Theorem), each side of the rhombus has a length of 5 cm.

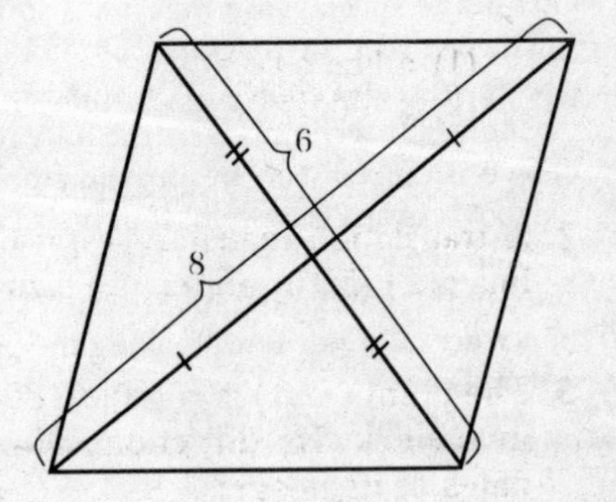

The perimeter of the rhombus is the sum of the lengths of all four sides. The perimeter = 4(5) or 20 cm.

The correct choice is **(1)**.

27. QUESTIONS ON TRANSFORMATIONS (REFLECTIONS, TRANSLATIONS, ROTATIONS, DILATIONS).

1. In the diagram below, $\triangle ABC \cong \triangle A'B'C'$. Under which type of transformation is $\triangle A'B'C'$ the image of $\triangle ABC$?

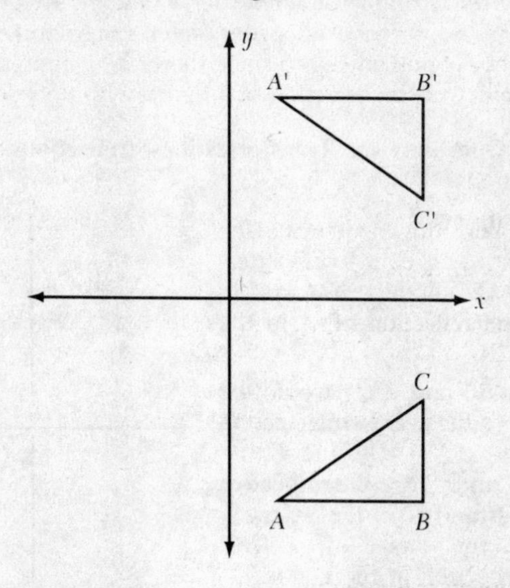

(1) a line reflection
(2) a rotation
(3) a translation
(4) a dilation

2. A translation maps (1,4) onto (7,–3). Write the image of (5,10) under the same translation.

3. The point (–2,1) is rotated 180° about the origin in a clockwise direction. What are the coordinates of its image?

4. What are the coordinates of P', the image of the point $P(2,3)$, under the transformation $r_{x=4} \circ r_{x\text{-axis}}$?

5. Find the image of $(3, -2)$ under the dilation D_2.

6. When point $A(-2,5)$ is reflected in the line $x = 1$, the image is

(1) (5,2) (3) (4,5)
(2) (−2,−3) (4) (0,5)

Solutions to Questions on Transformations (reflections, translations, rotations, dilations)

1. If $\overline{AA'}$ were drawn, it would be perpendicular to the x-axis, intersecting it at M, so that $AM = MA'$. Thus, A' is the reflection of A in the x-axis.

Similarly, $\overline{BB'}$ and $\overline{CC'}$ would be perpendicular to the x-axis, intersecting it at N so that $BN = NB'$ and $CN = NC'$. Thus, B' and C' are the respective reflections of B and C in the x-axis.

$\triangle A'B'C'$ is the image of $\triangle ABC$ under a line reflection in the x-axis.

The correct choice is **(1)**.

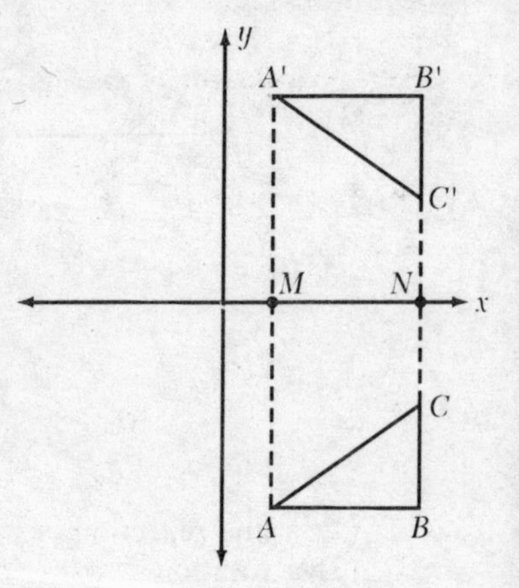

2. Let T, defined by $(x,y) \rightarrow (x + a, y + b)$, be the translation that maps a point (x,y) onto its image $(x + a, y + b)$. Since T maps (1,4) onto (7,−3), let $x = 1$, $y = 4$ and $x + a = 7$, $y + b = -3$.

Substitute 1 for x and 4 for y in the expression for the coordinates of the image:

$$1 + a = 7 \qquad 4 + b = -3$$
$$a = 7 - 1 \qquad b = -3 - 4$$
$$a = 6 \qquad b = -7$$

Therefore, T is $(x,y) \rightarrow (x + 6, y - 7)$.

Apply this definition of T to find the image of (5,10):

$$(5,10) \rightarrow (5 + 6, 10 - 7)$$
$$(5,10) \rightarrow (11,3)$$

The image is (11,3).

3. A rotation of 180° about the origin maps a point $A(x, y)$ onto its image $A'(-x, -y)$; that is, x is replaced by $-x$ and y is replaced by $-y$.

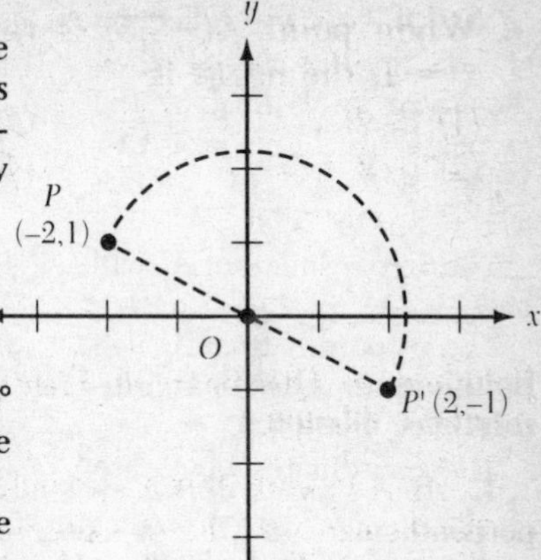

If point $P(-2, 1)$ is rotated 180° about the origin, its image will be $P'(2, -1)$.

The coordinates of the image are **(2, −1)**.

4.

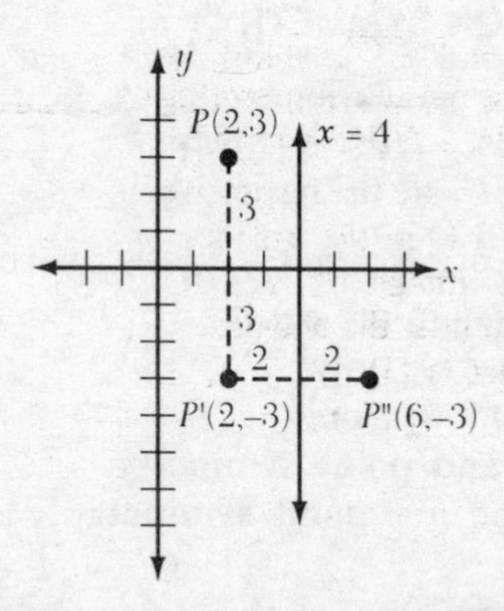

The transformation, $r_{x=4} \circ r_{x\text{-}axis}$, means that a reflection in the *x-axis* is followed by a reflection in the line $x = 4$.

The image, P' of $P(2,3)$ under the transformation $r_{x\text{-axis}}$ is the image formed by the reflection of P in the x-axis; P' is therefore the point (2,3).

P'', the image of P' under the transformation $r_{x=4}$ is the image formed by the reflection of P' in the line $x = 4$; P'' is therefore the point $(6, -3)$.

Thus, $P(2,3)$ has the image $P''(6, -3)$ under the transformation $r_{x=4} \circ r_{x\text{-}axis}$.

The coordinates of the image are $(6, -3)$.

5. The dilation D_2 maps a point (x, y) onto its image $(2x, 2y)$.

Thus, point $(3, -2)$ under the dilation D_2 is mapped onto its image $(2[3], 2[-2])$ or $(6, -4)$.

The image is **(6, −4)**.

6. If A' is the reflection of A in the line $x = 1$, then the line $x = 1$ must be the perpendicular bisector of $\overline{AA'}$. Thus, $AD = DA'$. Since $AD = 3$, $DA' = 3$.

The x coordinate of A' will be $1 + DA'$ or $1 + 3$ or 4.

The y coordinate of A' will be the same as the y coordinate of A, that is, 5.

The coordinates of A' are (4, 5).

The correct choice is **(3)**.

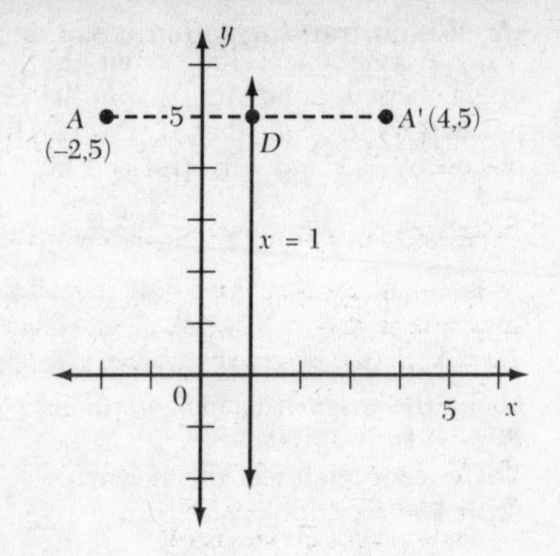

28. QUESTIONS ON SYMMETRY

1. Which letter has point symmetry?

(1) **A** (2) **C** (3) **S** (4) **R**

2. Which kind of symmetry does a rectangle have?

(1) line symmetry, only
(2) point symmetry, only
(3) both line and point symmetry
(4) neither line nor point symmetry

3. Which symbol has 90° rotational symmetry?

(1) **X** (3) **S**
(2) **H** (4) **8**

4. Which letter has rotational symmetry?

(1) **E** (3) **G**
(2) **T** (4) **H**

5. Which geometric figure has 72° rotational symmetry?

(1) square (3) rhombus
(2) regular pentagon (4) regular hexagon

6. Which transformation is *not* an isometry?

(1) $T_{(5,3)}$ (3) $r_{x\text{-axis}}$

(2) D_2 (4) $\text{Rot}_{(0,90°)}$

Solutions to Questions on Symmetry

1. A figure has point symmetry if there exists a point X, such that if any line is drawn through point X, and intersects the figure in a point P, it will also intersect the figure in another point P' with $PX = XP'$.

The letter **S** has point symmetry as shown in the diagram. For letters **A**, **C**, and **R**, there is no point X with the required property.

The correct choice is (**3**).

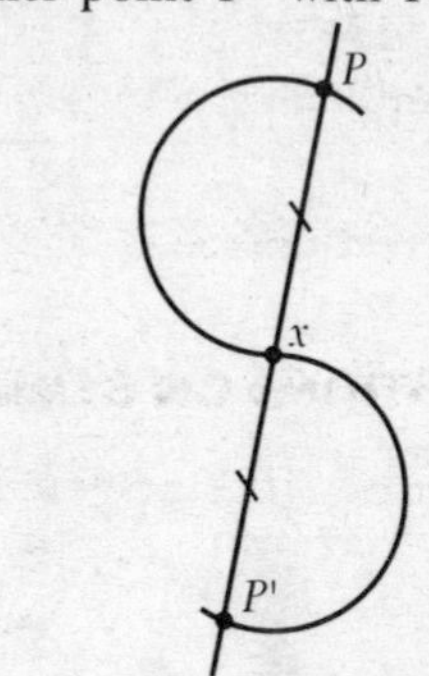

2. A rectangle has line symmetry in two ways.

Draw $\overline{AB}$ parallel to the longer sides and midway between them. If the rectangle is folded over along $\overline{AB}$, every point, P, of the rectangle above the line $\overline{AB}$ will be matched with a point, P', of the rectangle below $\overline{AB}$. Therefore, the rectangle has line symmetry with respect to $\overline{AB}$.

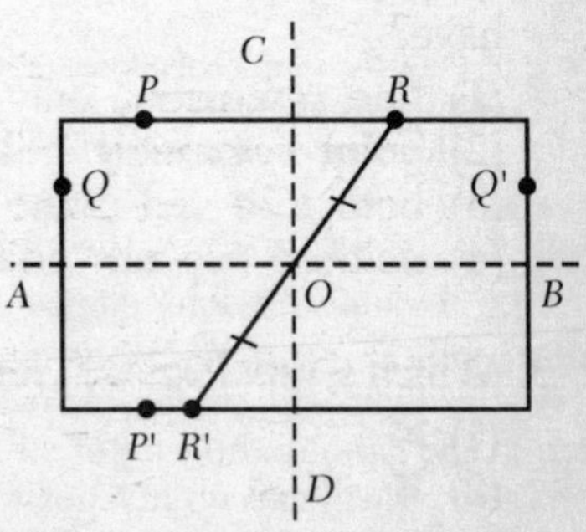

Similarly, if $\overline{CD}$ is drawn parallel to the shorter sides of the rectangle and midway between them, the rectangle may be folded over along $\overline{CD}$ with the result that any point, Q, of the rectangle to the left of $\overline{CD}$ will be matched with a point, Q', of the rectangle to the right of $\overline{CD}$. Therefore, the rectangle has line symmetry with respect to $\overline{CD}$.

The rectangle also has point symmetry. Consider the point O at the intersection of $\overline{AB}$ and $\overline{CD}$. Any line drawn through O will intersect the rectangle in two points, R and R', such that $RO = OR'$. Thus, O is the point center of symmetry for the rectangle.

The correct choice is (**3**).

3. For a figure to have 90° rotational symmetry, every point on its image after a 90° rotation must coincide with a point on its original position.

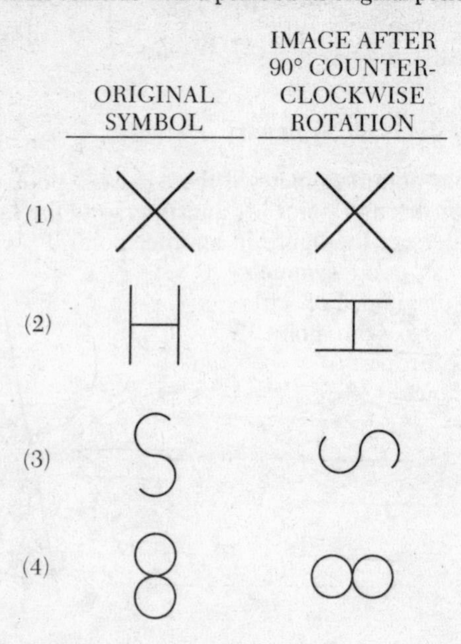

Only the symbol of choice (1) has every point of its image coinciding with a point of its original position.

The correct choice is **(1)**.

4. Examine each choice in turn:

If the figure in choice (4) is rotated 180° about the center of its crossbar, it will end up in a position in which every point of it will be matched with a point in its original position. Thus, choice (4) has rotational symmetry.

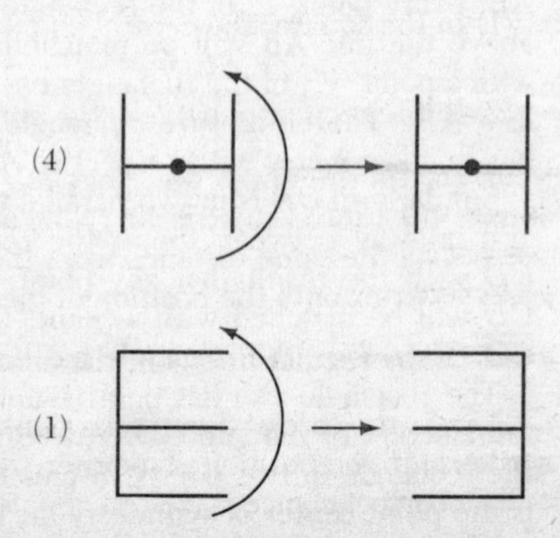

If the figure in choice (1) is rotated, except for a rotation of 360°, it cannot end up in a position where it would appear similar to its original orientation.

If the figure in choice (2) is rotated, again except for the special case of a 360° rotation, its final position will present a different appearance from its original.

Any rotation of the figure in choice (3), except for 360°, will result in a similar situation to that for choices (1) and (2).

The correct choice is **(4)**.

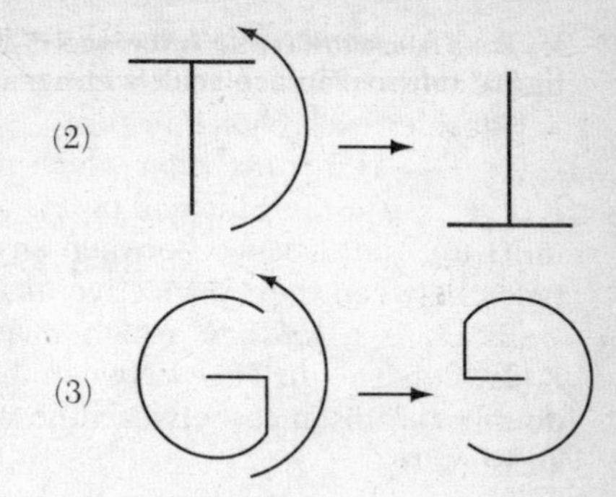

5.

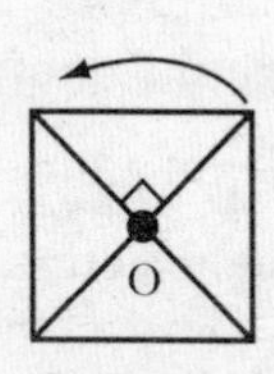

(1) Square

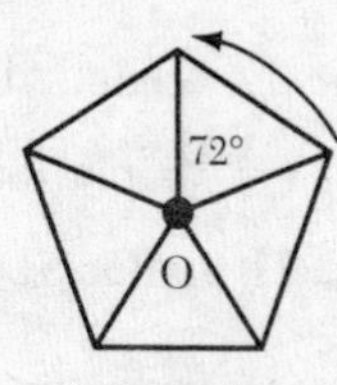

(2) Regular Pentagon

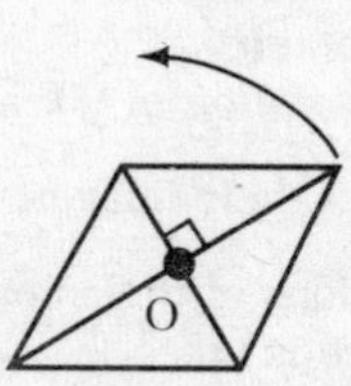

(3) Rhombus

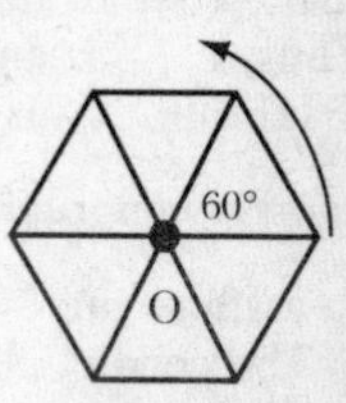

(4) Regular Hexagon

The "center" of each figure is labeled as O in the diagram.

A figure has 72° rotational symmetry if a rotation of 72° results in each of its vertices being carried onto the previous position of another vertex.

In order for a rotation of 72° to carry one vertex onto another, the central angle formed by the radii to two vertices must be 72°:

(1) In the square, the central angle $= \frac{360°}{4} = 90°$.

(2) In the regular pentagon, the central angle $= \frac{360°}{5} = 72°$.

(3) In the rhombus, the diagonals are perpendicular to each other; hence, the central angle is 90°. Note also that in the rhombus the vertices are not all the same distance from the "center," so a rotation would not carry a vertex onto the position of the next adjacent vertex.

(4) In the regular hexagon, the central angle $= \frac{360°}{6} = 60°$.

Only in the case of the regular pentagon will a rotation of 72° carry one vertex onto the position of another.

The correct choice is **(2)**.

6. An *isometry* is a transformation that results in the mapping of a figure onto an image that is congruent to the original figure.

Consider each choice in turn:

(1) $T_{(5,3)}$ is a *translation* which maps a point $P(x, y)$ onto its image, $P'(x + 5, y + 3)$. Since all points are moved 5 units to the right and 3 units up, the distance between any two points is the same as the distance between their respective images. Thus, $T_{(5,3)}$ is an isometry.

(2) D_2 is a *dilation* which maps a point $P(x, y)$ onto its image, $P'(2x, 2y)$. The distance between the images of any two points becomes double the distance between the original two points. Thus, D_2 is *not* an isometry.

(3) $r_{x\text{-axis}}$ is a *reflection* in the x-axis which maps a point $P(x, y)$ onto its image, $P'(x, -y)$. The result is equivalent to "flipping" figures over the x-axis; all mirror images remain congruent to the original figures. Thus, $r_{x\text{-axis}}$ is an isometry.

(4) $\text{Rot}_{(0,90°)}$ is a *rotation* of 90° about the origin. All original figures remain congruent to their images; they are simply rotated $\frac{1}{4}$ of a turn to assume the positions of their images. Thus, $\text{Rot}_{(0,90°)}$ is an isometry.

The correct choice is **(2)**.

29. QUESTIONS ON AREAS BY COORDINATE GEOMETRY

1. What is the area of rectangle $ABCD$ whose vertices are $A(3,2)$, $B(3,-2)$, $C(-3,-2)$, and $D(-3,2)$?

2. What is the area of a parallelogram if the coordinates of its vertices are $(0, -2)$, $(3,2)$, $(8,2)$, and $(5, -2)$?

3. What is the area of the triangle whose vertices are $(0,0)$, $(3,0)$, and $(0,4)$?

4. Given: $\triangle ABC$ with vertices $A(2,1)$, $B(10,7)$, and $C(4,10)$.

   a Find the area of triangle ABC.

5. Triangle ABC has vertices $A(-3,-4)$, $B(-1,7)$, and $C(3,5)$. Find the area of $\triangle ABC$.

6. Find the area of quadrilateral $ABCD$ with vertices $A(5,2)$, $B(0,5)$, $C(-2,-2)$, and $D(0,0)$.

Solutions to Questions on Areas by Coordinate Geometry

1. The area, A, of a rectangle is equal to the product of its length and width:

$$A = (CB)(AB)$$
$$A = (6)(4)$$
$$A = 24$$

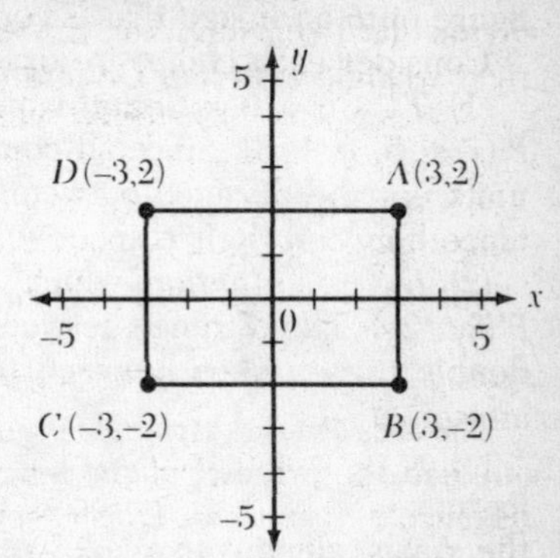

The area is **24**.

2. Drop altitude $\overline{BE}$ perpendicular to $\overline{AD}$. The coordinates of E are (3, −2).

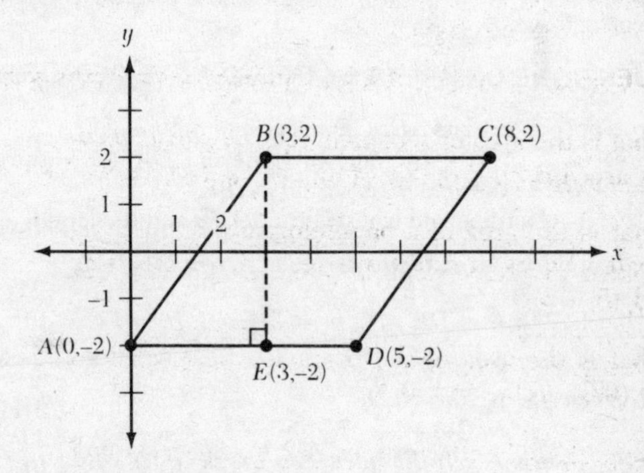

The area of a parallelogram equals the product of the length of its base and its altitude:

$AD = 5 - 0 = 5$
$BE = 2 - (-2) = 2 + 2 = 4$

Area of $\square ABCD = AD \times BE$

Area of $\square ABCD = 5 \times 4$
Area of $\square ABCD = 20$

The area is **20**.

3. Since the x- and y-axes are perpendicular to one another, $\triangle OAB$ is a right triangle with legs $\overline{OA}$ and $\overline{OB}$.

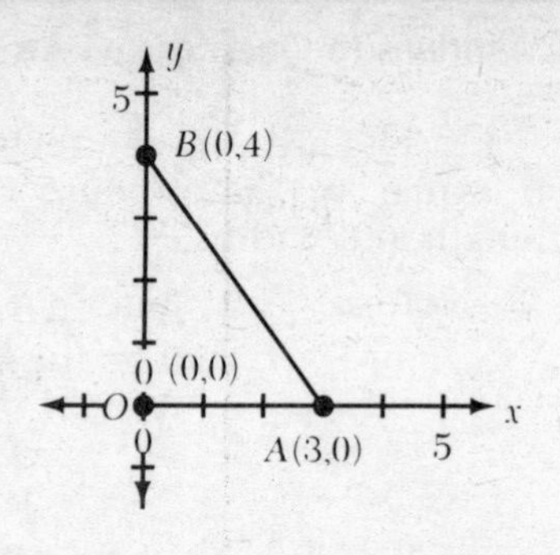

The area of a right triangle is equal to one-half the product of the lengths of its legs:

$$\text{Area of } \triangle\ OAB = \frac{1}{2}(OA)(OB)$$

$OA = 3$ and $OB = 4$:

$$\text{Area of } \triangle\ OAB = \frac{1}{2}(3)(4)$$

$$\text{Area of } \triangle\ OAB = 6$$

The area of the triangle is **6**.

4. Drop perpendiculars $\overline{AD}$, $\overline{CE}$ and $\overline{BF}$, to the x-axis.

The area of $\triangle ABC$ = the area of trapezoid $DACE$ + the area of trapezoid $ECBF$ − the area of trapezoid $DABF$.

The area, A, of a trapezoid whose altitude is h and the length of whose bases are b_1 and b_2 is given by the formula, $A = \frac{1}{2}h(b_1 + b_2)$.

Trapezoid $DACE$: $h = DE = 2$, $b_1 = DA = 1$, $b_2 = EC = 10$:

$$A = \frac{1}{2}(2)(1 + 10)$$
$$A = 1(11)$$
$$A = 11$$

Trapezoid $ECBF$: $h = EF = 6$, $b_1 = FB = 7$, $b_2 = EC = 10$:

$$A = \frac{1}{2}(6)(7 + 10)$$
$$A = 3(17)$$
$$A = 51$$

Trapezoid $DABF$: $h = DF = 8$, $b_1 = DA = 1$, $b_2 = FB = 7$:

$$A = \frac{1}{2}(8)(1 + 7)$$
$$A = 4(8)$$
$$A = 32$$

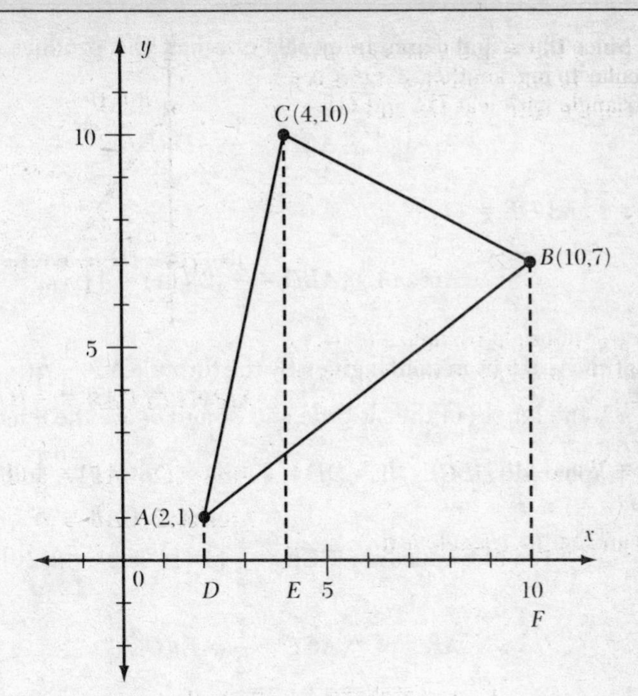

$$\begin{aligned}\text{The area of } \triangle ABC &= 11 + 51 - 32 \\ &= 62 - 32 \\ &= 30\end{aligned}$$

The area of $\triangle ABC$ *is* **30.**

5. Draw a line through A parallel to the x axis and drop $\overline{BD}$ and $\overline{CE}$ perpendicular to this line.

The area of $\triangle ABC$ = the area of right $\triangle ADB$ + the area of trapezoid $DBCE$ − the area of right $\triangle ACE$.

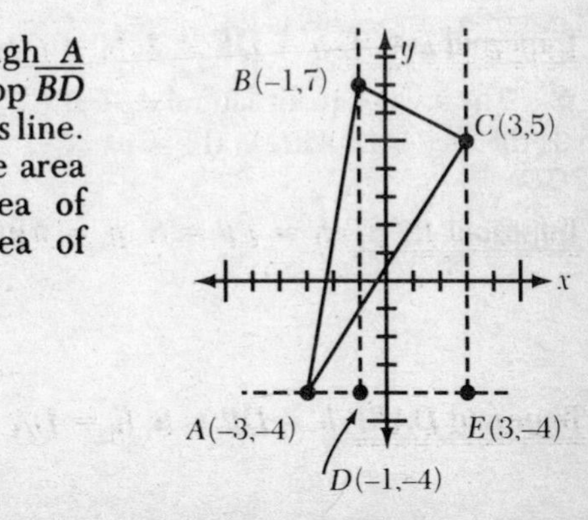

The coordinates of D are $(-1, -4)$ and the coordinates of E are $(3, -4)$.

The area of a right triangle equals one-half the product of the lengths of its legs:

$$\text{Area of } \triangle ABD = \frac{1}{2}(AD)(DB)$$

$AD = 2$ and $DB = 11$:

$$\text{Area of } \triangle ABD = \frac{1}{2}(2)(11) = 11$$

The area, A, of a trapezoid is given by the formula $A = \frac{1}{2}h(b_1 + b_2)$ where h is the length of the altitude and b_1 and b_2 are the lengths of the bases.

For trapezoid $DBCE$, $h = DE = 4$, $b_1 = DB = 11$, and $b_2 = CE = 9$:

$$\text{Area of trapezoid } DBCE = \frac{1}{2}(4)(11 + 9) = 2(20) = 40$$

$$\text{Area of } \triangle ACE = \frac{1}{2}(AE)(CE)$$

AE $= 6$ and $CE = 9$:

$$\text{Area of } \triangle ACE = \frac{1}{2}(6)(9) = 3(9) = 27$$

Area of $\triangle ABC = 11 + 40 - 27 = 51 - 27 = 24$
The area of $\triangle ABC$ is **24.**

6. The area of quadrilateral $ABCD$ equals the area of $\triangle ABD$ + the area of $\triangle BCD$.

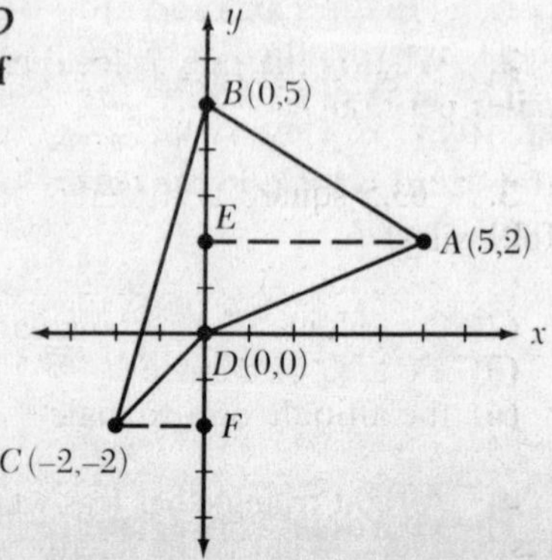

To compute these areas, use $\overline{BD}$ as the base for both triangles. Draw altitude $\overline{AE} \perp \overline{BD}$ and altitude $\overline{CF} \perp \overline{BD}$ extended.

From the graph, $BD = 5$, $AE = 5$, and $CF = 2$.

The area, A, of a triangle, the length of whose base is b and the length of whose altitude is h, is given by: $A = \frac{1}{2}bh$

For $\triangle ABD$, $b = BD$, $h = AE$: Area $\triangle ABD = \frac{1}{2}(BD)(AE)$

$BD = 5$, $AE = 5$: Area $\triangle ABD = \frac{1}{2}(5)(5)$

Area $\triangle ABD = \frac{25}{2} = 12\frac{1}{2}$

For $\triangle BCD$, $b = BD$, $h = CF$: Area $\triangle BCD = \frac{1}{2}(BD)(CF)$

$BD = 5$, $CF = 2$: Area $\triangle BCD = \frac{1}{2}(5)(2)$

Area $\triangle BCD = 5$

$$\text{Area of quadrilateral } ABCD = \text{Area } \triangle ABD + \text{Area } \triangle BCD$$
$$= 12\tfrac{1}{2} + 5 = 17\tfrac{1}{2}$$

The area of quadrilateral $ABCD$ is $\mathbf{17\frac{1}{2}}$.

30. QUESTIONS ON DIMENSIONAL ANALYSIS

1. If the circumference of a circle is doubled, the diameter of the circle

(1) increases by 2
(2) is doubled
(3) is multiplied by 4
(4) remains the same

2. What is the rate in feet per second of a car that is moving at 60 miles per hour?

3. "63.5 square centimeters" could possibly represent which of the following?

(1) the circumference of a circle
(2) the volume of a rectangular box
(3) the area of a circle
(4) the altitude of a triangle

4. A right triangle has legs whose lengths are one foot and 16 inches respectively. What is the length of the hypotenuse in feet?

5. The area of a parallelogram is 432 square inches. To convert this area to the equivalent area in square feet, 432 must be
(1) multiplied by 12
(2) divided by 12
(3) divided by 24
(4) divided by 144

6. The pressure of compressed air in a tank is measured at 10 ounces per square inch. Express this pressure in pounds per square foot.

Solutions to Questions on Dimensional Analysis

1. The formula for the circumference, C, of a circle whose diameter is of length d is:

$$C = \pi d$$

Assume that a particular circle has a circumference $= C_1$ and a diameter of length d_1. Then:

$$C_1 = \pi d_1$$

Suppose now that the length of the diameter is doubled to d_2 such that $d_2 = 2d_1$ and the circumference takes on a new value, C_2:

$$C_2 = \pi d_2$$

But we can substitute $2d_1$ for d_2: $C_2 = \pi(2d_1)$ or $C_2 = 2\pi d_1$.
Now replace πd_1 by C_1: $C_2 = 2C_1$.

Thus, by doubling the length of the diameter, we get a new circumference which is double the old one.

The correct choice is **(2)**.

ALTERNATE SOLUTION: Choose some arbitrary value for d, say $d = 3$. Since $C = \pi d$, then, for $d = 3$, $C = 3\pi$, that is, the circumference is 3π. If the diameter is now doubled, it becomes 6. Now $C = 6\pi$, which is double the old value of 3π.

2. Since 1 mile = 5,280 feet, $\frac{5{,}280 \text{ ft.}}{1 \text{ mi.}} = 1$
Since 1 hour = 60 minutes = 60 × 60 seconds = 3,600 seconds,

$$\frac{1 \text{ hr.}}{3{,}600 \text{ sec.}} = 1$$

Multiplying any quantity by the multiplicative identity, 1, does not change the value of the quantity:

$$\frac{60 \text{ mi.}}{1 \text{ hr.}} \times 1 \times 1 = \frac{60 \text{ mi.}}{1 \text{ hr.}} \times \frac{5{,}280 \text{ ft.}}{1 \text{ mi.}} \times \frac{1 \text{ hr.}}{3{,}600 \text{ sec.}}$$

Cancel like "factors" in numerator and denominator:

$$\frac{60 \text{ mi.}}{1 \text{ hr.}} = \frac{\overset{1}{\cancel{60}}\ \cancel{\text{mi.}}}{\cancel{1 \text{ hr.}}} \times \frac{\overset{88}{\cancel{5{,}280}} \text{ ft.}}{\cancel{1 \text{ mi.}}} \times \frac{\cancel{1 \text{ hr.}}}{\underset{\underset{1}{\cancel{60}}}{\cancel{3{,}600}} \text{ sec.}}$$

$$\frac{60 \text{ mi.}}{1 \text{ hr.}} = \frac{88 \text{ ft.}}{1 \text{ sec.}}$$

60 miles per hour is equal to **88 feet per second.**

3. The given measure is in square units. Only areas are measured in square units, and the only choice involving an area is (3), the area of a circle. Choice (1), the circumference of a circle, and choice (4), the altitude of a triangle, are both lines, and hence their measures must be in linear units. Choice (2), the volume of a rectangular box, must be measured in cubic units, as are all volumes.

The correct choice is (3).

4. Convert 1 foot to 12 inches.

Since 12 = 4(3) and 16 = 4(4), the triangle is a 3-4-5 right triangle with each dimension multiplied by 4. Therefore, the hypotenuse is 4(5) inches or 20 inches.

$$20 \text{ in.} = \frac{20}{12} \text{ ft.} = 1\frac{8}{12} \text{ ft.} = 1\frac{2}{3} \text{ ft.}$$

The length of the hypotenuse is $\mathbf{1\frac{2}{3}}$ feet.

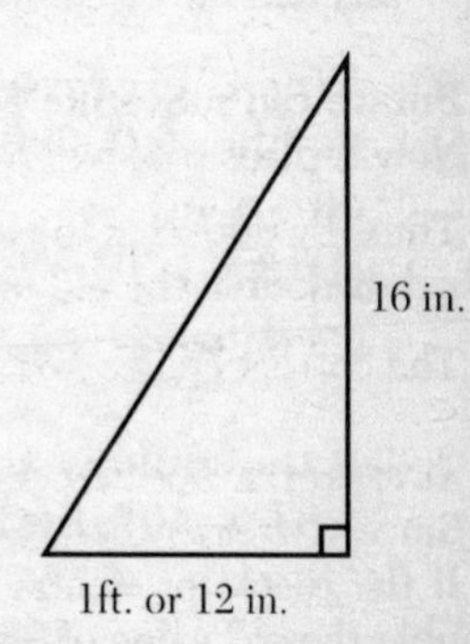

5. The parallelogram area, 432 square inches, would be the product of a base length expressed in inches by an altitude length expressed in inches. To convert the base length to feet, its measure in inches must be divided by 12. To convert the altitude length to feet, its measure in inches must also be divided by 12. The product of the two measures would thus be divided by 12 × 12, or by 144.

The correct choice is **(4).**

6. Since 16 ounces = 1 pound, $\frac{1 \text{ lb.}}{16 \text{ oz.}} = 1$

Since 144 square inches = 1 square foot, $\frac{144 \text{ sq. in.}}{1 \text{ sq. ft.}} = 1$

Multiplying any quantity by the multiplicative identity, 1, does not change the value of the quantity:

$$\frac{10 \text{ oz.}}{1 \text{ sq. in.}} \times 1 \times 1 = \frac{10 \text{ oz.}}{1 \text{ sq. in.}} \times \frac{1 \text{ lb.}}{16 \text{ oz.}} \times \frac{144 \text{ sq. in.}}{1 \text{ sq. ft.}}$$

Cancel like "factors" in numerator and denominator:

$$\frac{10 \text{ oz.}}{1 \text{ sq. in.}} = \frac{10 \cancel{\text{oz.}}}{\cancel{1 \text{ sq. in.}}} \times \frac{1 \text{ lb.}}{\underset{1}{\cancel{16}} \cancel{\text{oz.}}} \times \frac{\overset{9}{\cancel{144}} \cancel{\text{sq. in.}}}{1 \text{ sq. ft.}}$$

$$\frac{10 \text{ oz.}}{1 \text{ sq. in.}} = \frac{90 \text{ lb.}}{1 \text{ sq. ft.}}$$

The pressure is **90** pounds per square foot.

31. QUESTIONS ON SCIENTIFIC NOTATION, NEGATIVE AND ZERO EXPONENTS

1. If 0.00037 is expressed as 3.7×10^n, what is the value of n?

2. When expressed in scientific notation, the number $0.0000000364 = 3.64 \times 10^n$. The value of n is

(1) 8 (3) −10
(2) 10 (4) −8

3. If the number 0.00000467 is written in the form 4.67×10^n, find n.

4. If $f(x) = (3x)^{-2}$, find $f(2)$.

5. The value of $\frac{1}{2^{-3}}$ is

(1) $\frac{1}{8}$ (3) 6

(2) $\frac{1}{6}$ (4) 8

6. If $f(x) = x + x^{-1}$, find the value of $f(4)$.

Solutions to Questions on Scientific Notation, Negative and Zero Exponents

1. To change 3.7 to 0.00037, the decimal point must be moved 4 places to the left. Each move of one place to the left is equivalent to dividing by 10. Therefore, 3.7 must be divided by 10^4 to equal 0.00037. Division by 10^4 is equivalent to multiplying by $\frac{1}{10^4}$ or by 10^{-4}. Therefore, $0.00037 = 3.7 \times 10^{-4}$, or $n = -4$.

$n = \mathbf{-4}$.

2. The given equation is: $0.0000000364 = 3.64 \times 10^n$

In order for 3.64×10^n to become 0.0000000364, the decimal point in 3.64 must be moved eight places to the left. Each move of one place to the left is equivalent to dividing 3.64 by 10, or equivalent to multiplying it by 10^{-1}. Therefore, to move the decimal point eight places to the left, 3.64 must be multiplied by 10^{-8}, that is, $n = -8$.

The correct choice is **(4)**.

3. $0.00000467 = 4.67 \times 10^n$

Dividing 4.67 by 10 moves the decimal point one place to the left. Therefore, 4.67 must be divided by 10 six times to make it equal 0.00000467. Dividing by 10 six times is the same as multiplying by 10^{-6}: $0.00000467 = 4.67 \times 10^{-6}$

$n = \mathbf{-6}$.

4. The value of $f(2)$ is found by substituting 2 for x in the expression for $f(x)$:

$$f(x) = (3x)^{-2}$$
$$f(2) = (3 \cdot 2)^{-2}$$
$$f(2) = (6)^{-2}$$

Apply the definition of a negative exponent: $x^{-n} = \frac{1}{x^n}$:

$$f(2) = \frac{1}{(6)^2}$$
$$f(2) = \frac{1}{36}$$

$f(2) = \mathbf{\frac{1}{36}}$.

5. $\frac{1}{2^{-3}}$

By the definition of a negative exponent, $x^{-n} = \frac{1}{x^n}$; hence the expression becomes: $\frac{1}{\frac{1}{2^3}}$

$2^3 = 8$: $\frac{1}{\frac{1}{8}}$

Multiply the complex fraction by 1 in the form $\frac{8}{8}$: $\frac{8(1)}{8\left(\frac{1}{8}\right)}$

$\frac{8}{1}$

8

The correct choice is (4).

6. $f(x) = x + x^{-1}$

$x^{-1} = \frac{1}{x}$: $f(x) = x + \frac{1}{x}$

The value of $f(4)$ is obtained by substituting 4 for x in the expression for $f(x)$: $f(4) = 4 + \frac{1}{4}$

$f(4) = 4\frac{1}{4}$ or $\frac{17}{4}$

$f(4) = 4\frac{1}{4}$ or $\frac{17}{4}$.

Examination June 1988

Three-Year Sequence for High School Mathematics — Course I

PART ONE

DIRECTIONS: *Answer 30 questions from this part. Each correct answer will receive 2 credits. No partial credit will be allowed. Write your answers in the spaces provided. Where applicable, answers may be left in terms of π or in radical form.*

1 In the accompanying diagram, $\overleftrightarrow{AB}$ and $\overleftrightarrow{CD}$ intersect at E. If $m\angle BED = 25$ and $m\angle AEC = 3x + 10$, find the value of x.

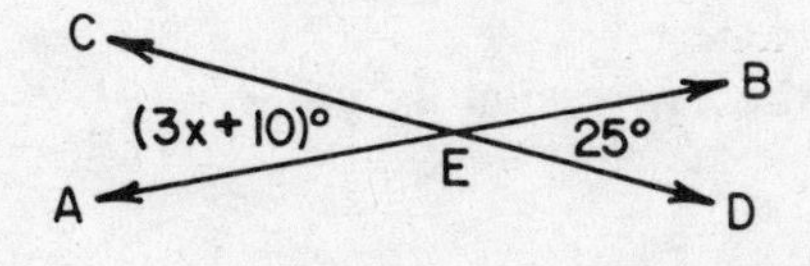

1_____

2 Solve for x: $0.02x + 4.1 = 6.3$ 2_____

3 In the accompanying diagram, parallel lines $\overleftrightarrow{HE}$ and $\overleftrightarrow{AD}$ are cut by transversal $\overleftrightarrow{BF}$ at points G and C, respectively. If $m\angle HGF = 5n$ and $m\angle BCD = 2n + 66$, find n.

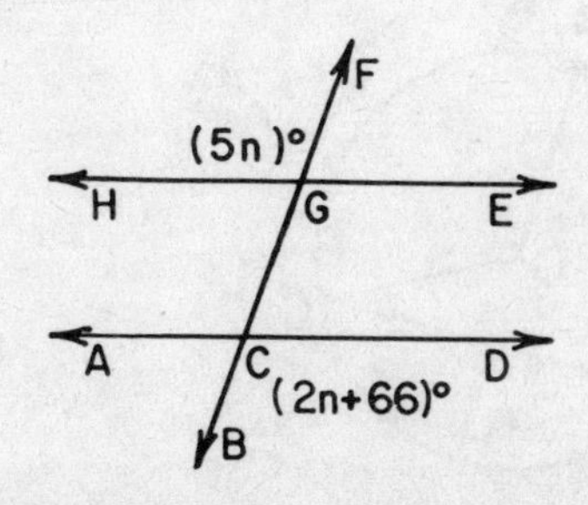

3_____

4 The ratio of similarity between two polygons is 2:3. If the perimeter of the smaller polygon is 14, find the perimeter of the larger polygon. 4_____

5 Solve the following system of equations for x:

$$y = 2x - 5$$
$$x + y = 4$$

5_____

6 The perimeter of an isosceles triangle is 14. If the length of the base of the triangle is one less than the length of each leg, find the length of a leg. 6_____

7 Find the value of the expression $2x^2y$ if $x = -1$ and $y = 3$. 7_____

8 The product of two factors is $x^2 - x - 20$. If one of the factors is $x - 5$, what is the other factor? 8_____

9 In the accompanying diagram, the measure of central angle AOC is 120. Find the measure of inscribed angle ABC.

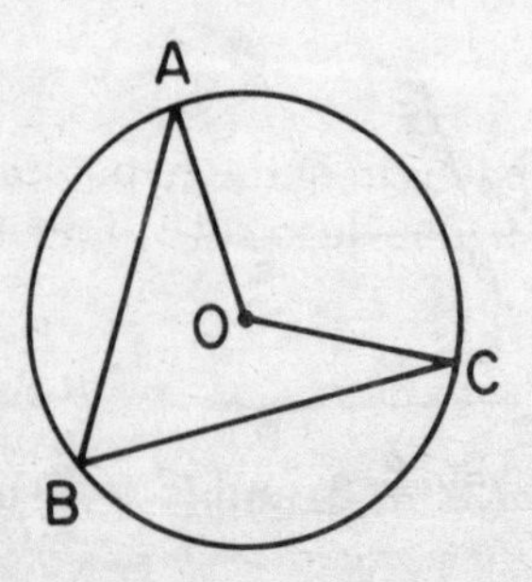

9_____

10 Solve for x: $ax + b = c$ 10_____

11 Write, in symbolic form, the contrapositive of $p \rightarrow \sim q$. 11_____

12 Solve for x: $6(x + 1) = 4x$ 12_____

13 The measures of the angles of a triangle are represented by $(3x - 20)$, $(7x + 30)$, and $(2x + 50)$. Find x. 13_____

14 Solve for x: $\frac{2}{5}x - 7 = -3$ 14_____

15 A box contains 3 green marbles, 2 red marbles, and 1 blue marble. If one marble is selected at random, what is the probability that a red marble was *not* selected? 15_____

16 Lois is four times as old as her son Dan. The sum of their ages is 40. How old is Dan? 16_____

17 Express $\frac{a}{6} + \frac{a}{4}$ as a single fraction in simplest form. 17_____

18 In a right triangle, if the length of the hypotenuse is 15 and the length of one leg is 12, find the length of the other leg. 18_____

19 Express the product of $(2x + 3)$ and $(x - 5)$ as a trinomial. 19_____

20 From $6x^2 - 3x + 9$ subtract $2x^2 - 5x + 8$. 20_____

Directions (21–35): For *each* question chosen, write in the *numeral* preceding the word or expression that best completes the statement or answers the question.

21 What is the median for the following set of data?

2, 7, 5, 4, 7, 2, 7

(1) 7
(2) 5
(3) $4\frac{6}{7}$
(4) 4

21_____

22 The quotient of $\frac{28x^4y^2}{14xy}$ is

(1) $2x^3y$
(2) $2x^4y^2$
(3) $14x^3y$
(4) $14x^4y^2$

22_____

23 Which sentence is an example of the distributive property?

(1) $ab = ba$
(2) $a(bc) = (ab)c$
(3) $a(b + c) = ab + ac$
(4) $a \cdot 1 = a$

23_____

24 The value of 5! is

(1) 120
(2) 25
(3) 15
(4) –5

24_____

25 Two supplementary angles are in the ratio 1:5. What is the measure of the smaller angle?

(1) 15°
(2) 30°
(3) 75°
(4) 150°

25_____

26 In the accompanying truth table, which statement should be the heading for column 3?

Column 1	Column 2	Column 3
p	q	?
T	T	T
T	F	T
F	T	T
F	F	F

(1) $p \rightarrow q$
(2) $p \leftrightarrow q$
(3) $p \wedge q$
(4) $p \vee q$

26_____

27 Which inequality is represented by the graph below?

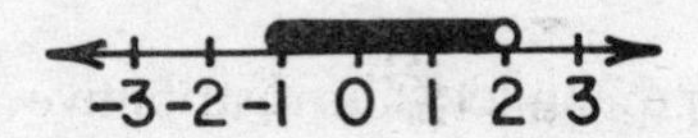

(1) $-1 < x < 2$
(2) $-1 \leq x < 2$
(3) $-1 < x \leq 2$
(4) $-1 \leq x \leq 2$

27_____

28 The area of a circle is 16π. What is the circumference of the circle?

(1) 8π
(2) 2π
(3) 16π
(4) 4π

28_____

29 Which expression represents the number of cents in d dimes and n nickels?

(1) $d + n$
(2) $15(d + n)$
(3) $10d + 5n$
(4) $\frac{d}{10} + \frac{n}{5}$

29_____

30 Which ordered pair is in the solution set of $y \geq 2x + 3$?

(1) (1,4) (3) (0,5)
(2) (3,2) (4) (0,0) **30**_____

31 For which value of x will the fraction $\frac{5}{2x - 8}$ be undefined?

(1) −4 (3) 8
(2) 0 (4) 4 **31**_____

32 What is the y-intercept of the line whose equation is $y - 2x = 4$?

(1) −2 (3) −4
(2) 2 (4) 4 **32**_____

33 Which is an irrational number?

(1) 5.7 (3) $\frac{8}{11}$
(2) $\sqrt{3}$ (4) $\sqrt{400}$ **33**_____

34 The expression $\sqrt{200}$ is equivalent to

(1) $25\sqrt{8}$ (3) $2\sqrt{10}$
(2) $100\sqrt{2}$ (4) $10\sqrt{2}$ **34**_____

35 What is the solution set of the equation $x^2 - 3x - 10 = 0$?

(1) {5,−2} (3) {5,2}
(2) {−5,−2} (4) {−5,2} **35**_____

PART TWO

DIRECTIONS: *Answer four questions from this part. Show all work unless otherwise directed.*

36 Solve the following system of equations graphically and check:

$$2x + y = -1$$
$$x + 2y = 4$$ [8,2]

37 In a math class, there are four students in the first row: three boys, Arthur, David, and Carlos, and one girl, Kim. The teacher will call one of these students to the board to solve a problem. When the problem is solved, the teacher will then call upon one of the remaining students in the first row to do a second problem at the board.

a Draw a tree diagram or list the sample space for calling two students to the board. [3]

b Find the probability that

(1) Kim will be one of the two students called to the board [2]

(2) at least one boy will be called to the board [2]

(3) Kim and Arthur will be the two students called to the board [2]

(4) two girls will be called to the board [1]

38 The length of the hypotenuse of a right triangle is 25 units. One of the legs is 5 units longer than the other leg.

a Find the length of each leg. [*Only an algebraic solution will be accepted.*] [8]

b Find in square units the area of the triangle. [2]

39 In the accompanying diagram, trapezoid $ABCD$ is inscribed in semicircle O. Triangles AOD, DOC, and COB are equilateral, $\overline{OE} \perp \overline{CD}$, $\overline{CD} \parallel \overline{AB}$, $OA = 6$, and $OE = 3\sqrt{3}$.

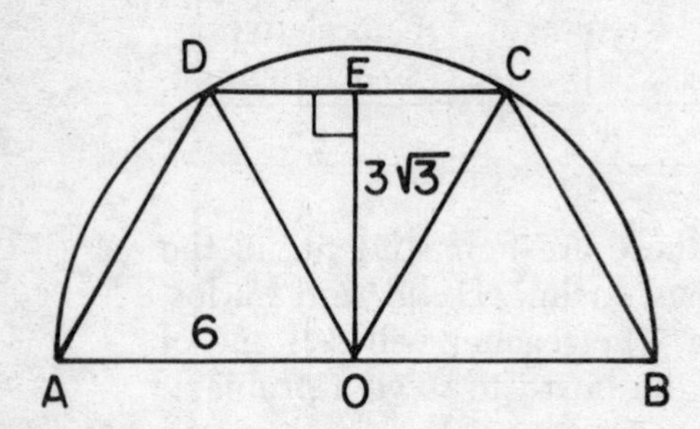

a Find AD. [1]

b Find the perimeter of trapezoid $ABCD$. [2]

c Find the area of $\triangle COD$. [Answer may be left in radical form.] [2]

d Find the area of trapezoid $ABCD$. [Answer may be left in radical form.] [2]

e Find the area of the semicircle. [Answer may be left in terms of π.] [3]

40 Write an equation or a system of equations that can be used to solve *each* of the following problems. In *each* case, state what the variable or variables represent. [*Solution of the equations is not required.*]

a One number is 4 more than three times a smaller number. If twice the larger number is decreased by three times the smaller number, the result is 32. Find the numbers. [5]

b Find three consecutive positive odd integers such that twice the sum of the second and the third is 2 less than six times the first. [5]

41 The table below shows the distribution of the total runs scored by a high school baseball team in each of 28 games.

Runs Scored	Frequency	Cumulative Frequency
0–1	4	4
2–3	11	
4–5	6	
6–7	5	
8–9	2	

a *On your answer paper*, copy the table and complete the cumulative frequency column. [2]

b Using the data in the cumulative frequency column, draw a cumulative frequency histogram. [4]

c Which interval contains the median number of runs scored in a game? [2]

d If one of these games is chosen at random, what is the probability that more than 5 runs were scored in the game chosen? [2]

42 Each part below consists of three sentences. *On your answer paper*, write the letters *a* through *e* and next to each letter, write the truth value for the third sentence in each part based on the truth values given for the first two sentences.

a	John will buy a car. John will get a raise. John will buy a car if and only if John will get a raise.	TRUE FALSE ?	[2]
b	Physics is a science. Jan plays piano. Physics is a science and Jan does *not* play piano.	TRUE FALSE ?	[2]
c	If a polygon is a square, then the polygon is a rectangle. Polygon *ABCD* is a square. Polygon *ABCD* is a rectangle.	TRUE TRUE ?	[2]
d	I like apples or it is Tuesday. I like apples. It is Tuesday.	FALSE FALSE ?	[2]
e	If it is snowing, the roads are slippery. The roads are not slippery. It is not snowing.	TRUE TRUE ?	[2]

Answers June 1988

Three-Year Sequence for High School Mathematics—Course I

ANSWER KEY

PART ONE

1.	5	**13.**	10	**24.**	(1)
2.	110	**14.**	10	**25.**	(2)
3.	22	**15.**	$\frac{4}{6}$	**26.**	(4)
4.	21	**16.**	8	**27.**	(2)
5.	3	**17.**	$\frac{5a}{12}$	**28.**	(1)
6.	5	**18.**	9	**29.**	(3)
7.	6	**19.**	$2x^2 - 7x - 15$	**30.**	(3)
8.	$x + 4$	**20.**	$4x^2 + 2x + 1$	**31.**	(4)
9.	60	**21.**	(2)	**32.**	(4)
10.	$\frac{c - b}{a}$	**22.**	(1)	**33.**	(2)
11.	$q \to \sim p$	**23.**	(3)	**34.**	(4)
12.	-3			**35.**	(1)

Part Two—*See* Answers Explained.

ANSWERS EXPLAINED

PART ONE

1. Vertical angles are congruent; hence, they are equal in measure: $m\angle AEC = m\angle BED$

$m\angle AEC = 3x + 10$ and $m\angle BED = 25$: $3x + 10 = 25$

Add -10 (the additive inverse of 10) to both sides of the equation:

$$\begin{array}{r} -10 = -10 \\ \hline 3x \quad = 15 \end{array}$$

Divide both sides of the equation by 3:

$$\frac{3x}{3} = \frac{15}{3}$$
$$x = 5$$

The value of x is **5**.

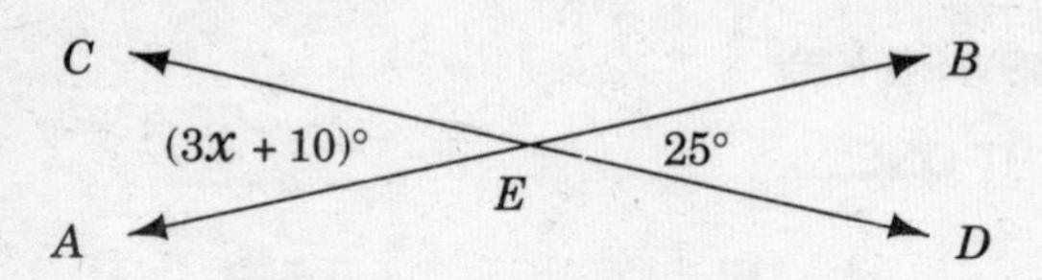

2. The given equation contains decimals:

$$0.02x + 4.1 = 6.3$$

Remove the decimals by multiplying each term on both sides of the equation by 100:

$$100(0.02x) + 100(4.1) = 100(6.3)$$
$$2x + 410 = 630$$

Add -410 (the additive inverse of 410) to both sides of the equation:

$$-410 = -410$$
$$2x \quad = 220$$

Divide both sides of the equation by 2:

$$\frac{2x}{2} = \frac{220}{2}$$
$$x = 110$$

$x =$ **110.**

3. Vertical angles are congruent; hence, they are equal in measure:

$$m\angle ACG = m\angle BCD$$

$m\angle BCD = 2n + 66$:

$$m\angle ACG = 2n + 66$$

If two lines are parallel, a transversal to them makes a pair of corresponding angles congruent:

$$m\angle HGF = m\angle ACG$$

$m\angle HGF = 5n$ and $m\angle ACG = 2n + 66$:

$$5n = 2n + 66$$

Add $-2n$ (the additive inverse of $2n$) to both sides of the equation:

$$-2n = -2n$$
$$3n = \quad 66$$

Divide both sides of the equation by 3:

$$\frac{3n}{3} = \frac{66}{3}$$
$$n = 22$$

$n =$ **22.**

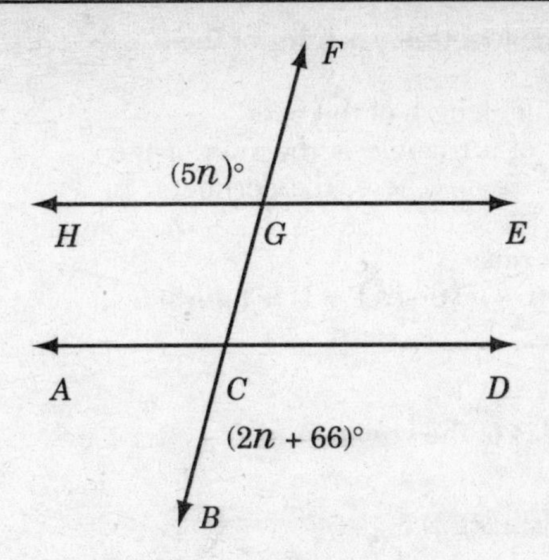

4. Let P = the perimeter of the larger polygon.

The ratio of the perimeters of two similar polygons is the same as the ratio of any two corresponding sides; this ratio (called the ratio of similarity of the two polygons) is given as 2:3 : $\frac{14}{P} = \frac{2}{3}$

In a proportion, the product of the means equals the product of the extremes (cross-multiply): $2P = 3(14)$

$2P = 42$

Divide both sides of the equation by 2: $\frac{2P}{2} = \frac{42}{2}$

$P = 21$

The perimeter of the larger polygon is **21**.

5. The given system of equations is: $y = 2x - 5$

$x + y = 4$

Substitute in the second equation the expression for y given by the first equation: $x + 2x - 5 = 4$

Combine like terms: $3x - 5 = 4$

Add 5 (the additive inverse of -5) to both sides of the equation: $5 = 5$

$3x = 9$

Divide both sides of the equation by 3: $\frac{3x}{3} = \frac{9}{3}$

$x = 3$

$x = $ **3**.

6. Let x = the length of each leg of the isosceles triangle.

Then $x - 1$ = the length of the base.

The perimeter of a triangle is the sum of the lengths of the three sides; here, the perimeter is 14: $x + x + x - 1 = 14$

Combine like terms: $3x - 1 = 14$

Add 1 (the additive inverse of -1) to both sides of the equation:

$$\begin{array}{r} 3x - 1 = 14 \\ 1 = 1 \\ \hline 3x \quad = 15 \end{array}$$

Divide both sides of the equation by 3: $\frac{3x}{3} = \frac{15}{3}$

$$x = 5$$

The length of each leg is **5**.

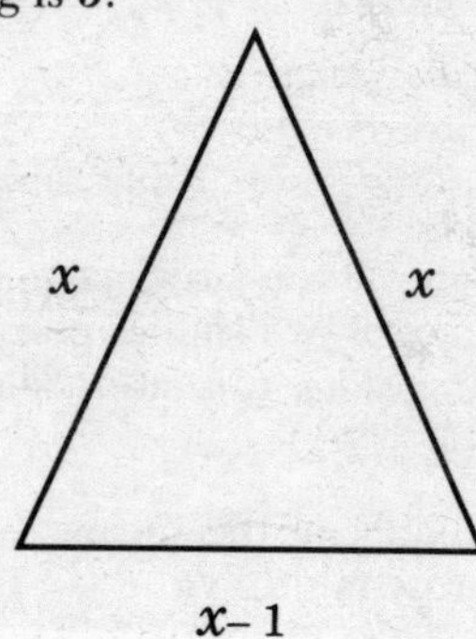

7. The given expression is: $2x^2y$

To evaluate, substitute -1 for x and 3 for y: $2(-1)^2(3)$

Square -1: $2(1)(3)$

Multiply the three factors together: 6

The value of the expression is **6**.

8. The given expression is a *quadratic trinomial:* $x^2 - x - 20$

A quadratic trinomial can be factored into the product of two binomials, one of which is given to be $(x - 5)$: $(x - 5)(\quad)$

The first two terms of the binomials are the factors of the first term, x^2, of the trinomial, so the first term of the second binomial must be x: $(x - 5)(x \quad)$

The second terms of the binomials are the factors of the last term, -20, of the trinomial.

Since we are given -5 as one of these factors, the other one must be $+4$:

$$(x - 5)(x + 4)$$

This result should be checked to see whether the sum of the inner cross product and the outer cross product equals the middle term, $-x$, of the original quadratic trinomial:

$$-5x = \text{inner product}$$
$$(x - 5)(x + 4)$$

Since $(-5x) + (+4x) = -x$, these are the correct factors:

$$+4x = \text{outer product}$$
$$(x - 5)(x + 4)$$

The other factor is $\mathbf{x + 4}$.

9. The measure of a *central angle* is equal to the measure of its intercepted arc:

$$m\angle AOC = m\,AC$$

$m\angle AOC = 120$:

$$120 = m\,AC$$

The measure of an *inscribed angle* is equal to one-half the measure of its intercepted arc:

$$m\angle ABC = \frac{1}{2} m\,AC$$
$$m\angle ABC = \frac{1}{2}(120)$$
$$m\angle ABC = 60$$

The measure of angle ABC is **60**.

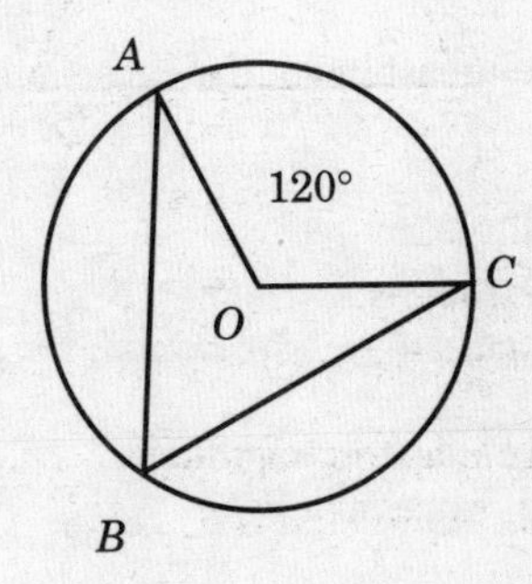

10. The given equation contains *literal coefficients:*

$$ax + b = c$$

Add $-b$ (the additive inverse of b) to both sides of the equation:

$$\underline{\quad -b = -b}$$
$$ax = c - b$$

Divide both sides of the equation by a:

$$\frac{ax}{a} = \frac{c - b}{a}$$
$$x = \frac{c - b}{a}$$

The solution for x is $\mathbf{\frac{c - b}{a}}$.

11. The given *conditional* or *implication* is: $p \rightarrow \sim q$

The *hypothesis* (*antecedent* or "*if* clause") of $p \rightarrow \sim q$ is p, and the *conclusion* (*consequent* or "*then* clause") is $\sim q$.

Form the *contrapositive of* $p \rightarrow \sim q$ by negating the original hypothesis and conclusion and then interchanging them. The negation of p is $\sim p$, and the negation of $\sim q$ is q. Therefore the contrapositive of $p \rightarrow \sim q$ is: $q \rightarrow \sim p$

The contrapositive is $\boldsymbol{q \rightarrow \sim p}$.

12. The given equation contains parentheses: $6(x + 1) = 4x$

Remove the parentheses by applying the distributive law, that is, by multiplying each term within the parentheses by the factor outside, 6: $6x + 6 = 4x$

Add -6 (the additive inverse of 6), and also add $-4x$ (the additive inverse of $4x$), to both sides of the equation:

$$\begin{array}{r} 6x + 6 = 4x \\ -4x - 6 = -4x - 6 \\ \hline 2x \quad = -6 \end{array}$$

Divide both sides of the equation by 2:

$$\frac{2x}{2} = \frac{-6}{2}$$

$$x = -3$$

$x = -3$.

13. The sum of the measures of the three angles of a triangle is 180: $3x - 20 + 7x + 30 + 2x + 50 = 180$

Combine like terms: $12x + 60 = 180$

Add -60 (the additive inverse of 60) to both sides of the equation:

$$\begin{array}{r} -60 = -60 \\ \hline 12x = 120 \end{array}$$

Divide both sides of the equation by 12:

$$\frac{12x}{12} = \frac{120}{12}$$

$$x = 10$$

$x =$ **10**.

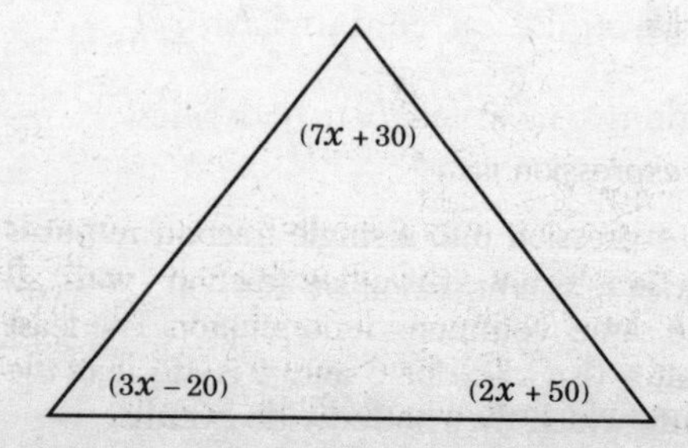

14. The given equation contains a fraction: $\frac{2}{5}x - 7 = -3$

Remove the fraction by multiplying each term on both sides of the equation by 5: $5\left(\frac{2}{5}x\right) - 5(7) = 5(-3)$

$$2x - 35 = -15$$

Add 35 (the additive inverse of -35) to both sides of the equation:

$$\begin{array}{r} 35 = 35 \\ \hline 2x \quad = 20 \end{array}$$

Divide both sides of the equation by 2: $\frac{2x}{2} = \frac{20}{2}$

$$x = 10$$

$x =$ **10**.

15. Probability of an event occurring $=$

$$\frac{\text{number of favorable outcomes}}{\text{total possible number of outcomes}}.$$

For selecting 1 marble at random from among 3 green marbles, 2 red marbles, and 1 blue marble, the total possible number of outcomes is $3 + 2 + 1$, or 6. For the probability that a red marble is *not* selected, the favorable outcomes result from the selection of a green or a blue marble; thus, there are $3 + 1$, or 4, favorable outcomes.

The probability of *not* selecting a red marble is $\frac{4}{6}$ or $\frac{2}{3}$.

The probability is $\mathbf{\frac{4}{6}}$ **or** $\mathbf{\frac{2}{3}}$.

16. Let $x =$ the age of the son, Dan.
Then $4x =$ the age of the mother, Lois.
The sum of their ages is 40: $x + 4x = 40$
Combine like terms: $5x = 40$
Divide both sides of the equation by 5: $\frac{5x}{5} = \frac{40}{5}$

$$x = 8$$

Dan is 8 years old.

17. The given expression is: $\frac{a}{6} + \frac{a}{4}$

Combining this expression into a single fraction requires changing each fraction to an equivalent fraction, with all fractions having the same (common) denominator. The least common denominator (L.C.D.) for 6 and 4 is 12; it is the smallest number into which they both divide evenly.

Multiply the first fraction by 1 in the form $\frac{2}{2}$ and the second fraction by 1 in the form $\frac{3}{3}$ to obtain equivalent fractions having the L.C.D.:

$$\frac{2a}{2(6)} + \frac{3a}{3(4)}$$

$$\frac{2a}{12} + \frac{3a}{12}$$

Fractions with a common denominator are combined by combining their numerators over the common denominator:

$$\frac{2a + 3a}{12}$$

Combine like terms in the numerator:

$$\frac{5a}{12}$$

The single fraction in simplest form is $\mathbf{\frac{5a}{12}}$.

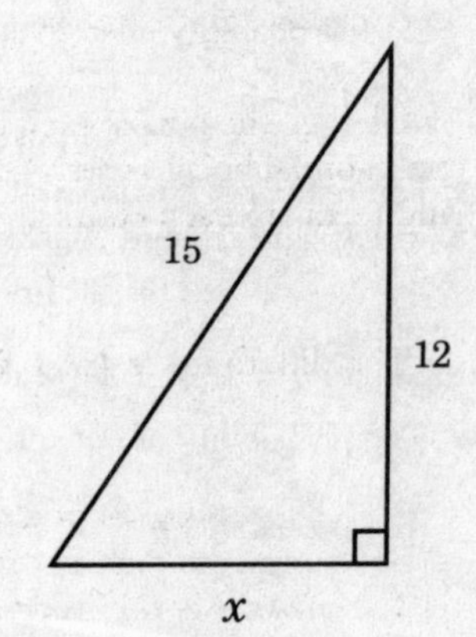

18. Let x = the length of the other leg.

By the Pythagorean Theorem, the sum of the squares of the lengths of the two legs of a right triangle is equal to the square of the length of the hypotenuse:

$$x^2 + 12^2 = 15^2$$

Square 12 and square 15:

$$x^2 + 144 = 225$$

Add -144 (the additive inverse of 144) to both sides of the equation:

$$\begin{aligned} -144 &= -144 \\ \hline x^2 \quad &= \quad 81 \end{aligned}$$

Take the square root of each side of the equation:

$$x = \pm\sqrt{81}$$

Reject the negative root as meaningless for a length:

$$x = 9$$

The length of the other leg is **9**.

ALTERNATIVE SOLUTION: The given triangle is a 3-4-5 right triangle with each term of the 3-4-5 ratio multiplied by 3: 12 = 3(4) and 15 = 3(5); therefore, the remaining leg must be 3(3), or **9**.

19. The given indicated product is: $(2x+3)(x-5)$

Multiply each term of $(2x + 3)$ by each term of $(x - 5)$, and combine the products:

$$\begin{array}{r} 2x+3 \\ x-5 \\ \hline 2x^2+3x \qquad \\ -10x-15 \\ \hline 2x^2-7x-15 \end{array}$$

The product as a trinomial is $\mathbf{2x^2 - 7x - 15}$.

20. Write the polynomial to be subtracted beneath the polynomial from which it is to be subtracted, keeping like terms in the same column:

$$\begin{array}{r} 6x^2-3x+9 \\ 2x^2-5x+8 \\ \hline \end{array}$$

Change the sign of each term in the subtrahend (the polynomial being subtracted), and combine the resulting terms in each column:

$$\begin{array}{r} 6x^2-3x+9 \\ -2x^2+5x-8 \\ \hline 4x^2+2x+1 \end{array}$$

The difference is $\mathbf{4x^2 + 2x + 1}$.

21. The given set of data is: 2, 7, 5, 4, 7, 2, 7

Rearrange the data in order of increasing size: 2, 2, 4, 5, 7, 7, 7

The *median* is the middle item when the data are arranged in order of size: The median is 5.

The correct choice is **(2)**.

22. The given expression is: $\frac{28x^4y^2}{14xy}$

To find the quotient of two monomials, first divide their numerical coefficients to obtain the numerical coefficient of the quotient: $28 \div 14 = 2$

Next, divide the literal factors to find the literal factors of the quotient. Powers of the same base are divided by subtracting their exponents. Remember that xy stands for x^1y^1: $x^4y^2 \div x^1y^1 = x^3y$

Combine the results of the two steps above: $\frac{28x^4y^2}{14xy} = 2x^3y$

The correct choice is **(1)**.

23. The distributive property states that, if a factor is to multiply the sum of two or more terms, it must be distributed over them, that is, it multiplies each of them.

Consider each choice in turn:

(1) $ab = ba$ represents the *commutative* property for multiplication.

(2) $a(bc) = (ab)c$ represents the *associative* property for multiplication.

(3) $a(b + c) = ab + ac$ represents the *distributive* property; this is the correct choice.

(4) $a \cdot 1 = a$ represents the *identity* property for multiplication.

The correct choice is **(3)**.

24. The symbol $n!$ stands for *factorial n*; $n! = n(n - 1)(n - 2) \ldots (3)(2)(1)$.

Therefore, $5! = 5(4)(3)(2)(1) = 20(6) = 120$.

The correct choice is **(1)**.

25. Let $x =$ the measure of the smaller angle.

Then $5x =$ the measure of the larger angle.

The sum of the measures of the two supplementary angles is 180: $x + 5x = 180$

Combine like terms: $6x = 180$

Divide both sides of the equation by 6: $\frac{6x}{6} = \frac{180}{6}$

$x = 30$

The measure of the smaller angle is 30°.

The correct choice is **(2)**.

26.

Column 1 p	Column 2 q	Column 3 ?
T	T	T
T	F	T
F	T	T
F	F	F

Choice (1), the *conditional* $p \rightarrow q$, can be ruled out because the second line of the table shows Column 3 to have the value T when p is T and q is F; $p \rightarrow q$ would have the value *F* in such a case.

Choice (2), the *biconditional* $p \leftrightarrow q$, can be ruled out because Column 3 has the value T in both lines 2 and 3 even though p and q have different truth values; the biconditional has the value T only if p and q have the same truth value.

Choice (3), the *conjunction* $p \wedge q$, can be ruled out from lines 2 and 3 because Column 3 has the value T even though *both* p and q do not have the value T; the conjunction has the value T only when both p and q are T.

Column 3 has the truth value F only when both p and q have the truth value F. Thus, if either p or q or both have the truth value T, Column 3 has the truth value T. This is the truth value pattern for the *disjunction* of p and q, $p \vee q$, which is choice (4).

The correct choice is (**4**).

27. The shaded line runs from -1 to 2, so the inequality it represents must include the numbers between -1 and 2.

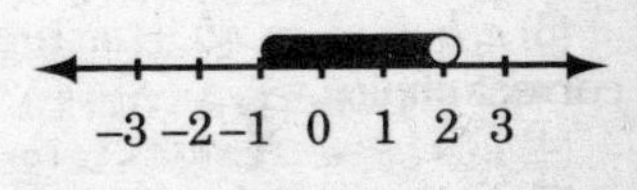

The shaded dot at -1 indicates that -1 is included as a member of the set, whereas the open, unshaded circle at 2 indicates that 2 is not a member of the set. The inequality represented by the graph is therefore $-1 \leq x < 2$.

The correct choice is (**2**).

28. If r represents the length of the radius of a circle, its area, A, is given by this formula: $A = \pi r^2$

The area of the circle is given as 16π: $16\pi = \pi r^2$

Divide both sides of the equation by π: $\frac{16\pi}{\pi} = \frac{\pi r^2}{\pi}$

$16 = r^2$

Take the square root of each side of the equation: $\pm\sqrt{16} = r$

Reject the negative root as meaningless for a length: $4 = r$

The circumference, C, of a circle is given by this formula: $C = 2\pi r$

Since $r = 4$: $C = 2\pi(4)$

Simplify: $C = 8\pi$

The correct choice is (**1**).

29. Each dime contains 10 cents and each nickel contains 5 cents. Therefore, the number of cents in d dimes and n nickels is $10d + 5n$.

The correct choice is (**3**).

30. The given inequality is: $y \geq 2x + 3$

To find which choice is in the solution set, substitute the values of x and

y given in each choice to see whether they satisfy the inequality:

(1) (1,4): $4 \overset{?}{\geq} 2(1) + 3$
$4 \overset{?}{\geq} 2 + 3$
$4 \ngeq 5$ (1,4) is not in.

(2) (3,2): $2 \overset{?}{\geq} 2(3) + 3$
$2 \overset{?}{\geq} 6 + 3$
$2 \ngeq 9$ (3,2) is not in.

(3) (0,5): $5 \overset{?}{\geq} 2(0) + 3$
$5 \overset{?}{\geq} 0 + 3$
$5 \geq 3 \checkmark$ (0,5) is in the set.

(4) (0,0): $0 \overset{?}{\geq} 2(0) + 3$
$0 \overset{?}{\geq} 0 + 3$
$0 \ngeq 3$ (0,0) is not in.

The correct choice is **(3)**.

31. The given fraction is: $\frac{5}{2x - 8}$

Since division by 0 is undefined, the fraction will be undefined if its denominator equals 0. Thus, we seek the value of x that makes the denominator equal to 0: $2x - 8 = 0$

Add 8 (the additive inverse of -8) to both sides of the equation:

$$\begin{array}{r} 8 = 8 \\ \hline 2x \quad = 8 \end{array}$$

Divide both sides of the equation by 2:

$$\frac{2x}{2} = \frac{8}{2}$$
$$x = 4$$

The fraction will be undefined if $x = 4$.
The correct choice is **(4)**.

32. The given equation is: $y - 2x = 4$

The y-intercept is the value of y where the graph of the equation crosses the y-axis, that is, where $x = 0$. Substitute 0 for x in the equation:

$$y - 2(0) = 4$$
$$y - 0 = 4$$
$$y = 4$$

The y-intercept is 4.
The correct choice is **(4)**.

ALTERNATIVE SOLUTION: If the equation of a straight line is in the form $y = mx + b$, b represents the y-intercept. To put $y - 2x = 4$ in the $y = mx + b$ form, add $2x$ to both sides of the equation:

$$\begin{array}{r} y - 2x = 4 \\ 2x = 2x \\ \hline y \quad = 2x + 4 \end{array}$$

The equation $y = 2x + 4$ is in the $y = mx + b$ form with $b = 4$; the y-intercept is **4**.

33. A rational number is a number that can be represented in the form $\frac{a}{b}$, where both a and b are integers and $b \neq 0$. An irrational number cannot be represented in this manner. An irrational number can be approximated by a decimal that is nonterminating and in which the pattern of successive digits does not repeat. Note that, if the pattern of digits repeats, a nonterminating decimal is a rational number (for example, $1.3333... = 1\frac{1}{3} = \frac{4}{3}$).

Consider each choice in turn:

(1) 5.7 can be represented in the form $\frac{57}{10}$ and hence is a rational number.

(2) $\sqrt{3}$ is approximately the nonterminating, nonrepeating decimal 1.732. Hence $\sqrt{3}$ is an irrational number.

(3) $\frac{8}{11}$ is in the form $\frac{a}{b}$, where a and b are integers and $b \neq 0$. Hence, $\frac{8}{11}$ is a rational number.

(4) $\sqrt{400} = 20$, which can be expressed as $\frac{20}{1}$ and hence is rational.

The correct choice is **(2)**.

34. The given radical expression is: $\sqrt{200}$

Factor out any perfect square factors in the radicand (the expression under the radical sign): $\sqrt{100(2)}$

Remove the perfect square factor from under the radical sign by taking its square root and writing it as a factor in front of the radical sign: $10\sqrt{2}$

The correct choice is **(4)**.

35. The given equation is a *quadratic equation*: $x^2 - 3x - 10 = 0$

The left side is a *quadratic trinomial* that can be factored into the product of two binomials. The factors of the first term, x^2, are x and x, and they become the first terms of the binomials: $(x \quad)(x \quad) = 0$

The factors of the last term, -10, become the second terms of the binomials, but they must be chosen in such a way that the product of the inner terms and the product of the outer terms add up to the middle term,

$-3x$, of the original quadratic trinomial. Try -5 and $+2$ as the factors of -10:

$$\begin{array}{r} -5x = \text{inner product} \\ (x-5)(x+2) = 0 \\ +2x = \text{outer product} \end{array}$$

Since $(-5x) + (+2x) = -3x$, these are the correct factors: $(x-5)(x+2) = 0$

If the product of two factors is 0, either factor may equal 0: $x - 5 = 0 \lor x + 2 = 0$

Add the appropriate additive inverse to both sides of each equation, 5 for the left equation, and -2 for the right equation:

$$\begin{array}{rr} 5 = 5 & -2 = -2 \\ \hline x \quad = 5 & x \quad = -2 \end{array}$$

The solution set is $\{5, -2\}$.

The correct choice is **(1)**.

PART TWO

36. The given system of equations is:

$$\begin{aligned} 2x + y &= -1 \\ x + 2y &= 4 \end{aligned}$$

STEP 1: Draw the graph of $2x + y = -1$. It is convenient to rearrange the equation so that it is solved for y; add $-2x$ to both sides of the equation:

$$\begin{array}{rcr} 2x + y & = & -1 \\ -2x \quad & = & -2x \\ \hline y & = & -1 - 2x \end{array}$$

Choose any three convenient values for x. Substitute them in the equation and calculate the corresponding values of y:

x	$-1-2x$	$= y$
-3	$-1-2(-3) = -1+6$	$= 5$
0	$-1-2(0) = -1-0$	$= -1$
3	$-1-2(3) = -1-6$	$= -7$

Plot the points $(-3,5)$, $(0,-1)$, and $(3,-7)$, and draw a straight line through them. This line is the graph of $2x + y = -1$.

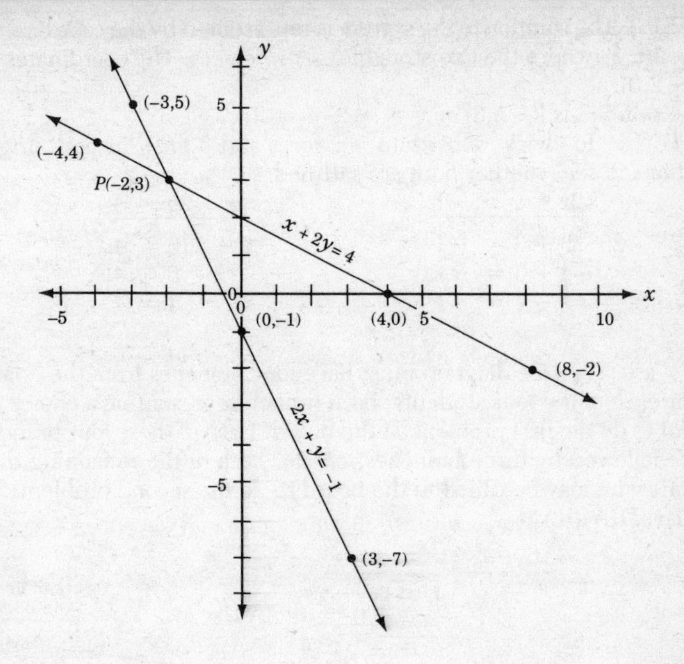

STEP 2: Draw the graph of $x + 2y = 4$. In this case, it is convenient to adopt the somewhat unusual procedure of rearranging the equation so that it is solved for x. And $-2y$ to both sides of the equation:

$$\begin{aligned} x + 2y &= 4 \\ -2y &= \quad -2y \\ \hline x \quad &= 4 - 2y \end{aligned}$$

Choose any three convenient values for y. Substitute them in the equation and calculate the corresponding values of x:

y	$4 - 2y$	$= x$
-2	$4 - 2(-2) = 4 + 4$	$= \quad 8$
0	$4 - 2(0) \quad = 4 - 0$	$= \quad 4$
4	$4 - 2(4) \quad = 4 - 8$	$= -4$

Plot the points $(-4,4)$, $4(0)$, and $(8, -2)$, and draw a straight line through them. This line is the graph of $x + 2y = 4$.

STEP 3: The solution to the system is represented by the coordinates of the point, P, where the two straight lines intersect. The coordinates of P are $(-2,3)$.

The solution is $\{(-2,3\}$ or $x = -2, y = 3$.

CHECK: To check, substitute -2 for x and 3 for y in both original equations to see whether both are satisfied:

$$\begin{array}{rcl} \underline{2x + y} & \underline{=} & \underline{-1} \\ 2(-2) + 3 & \stackrel{?}{=} & -1 \\ -4 + 3 & \stackrel{?}{=} & -1 \\ -1 & = & -1 \ \checkmark \end{array} \qquad \begin{array}{rcl} \underline{x + 2y} & \underline{=} & \underline{4} \\ -2 + 2(3) & \stackrel{?}{=} & 4 \\ -2 + 6 & \stackrel{?}{=} & 4 \\ 4 & = & 4 \ \checkmark \end{array}$$

37. a. The tree diagram must have four branches from the "start," one for each of the four students, each branch representing a choice of a student to do the first problem on the board. Each of these four branches will be followed by three branches, one for each of the remaining three students who may be called to the board to do the second problem.

TREE DIAGRAM:

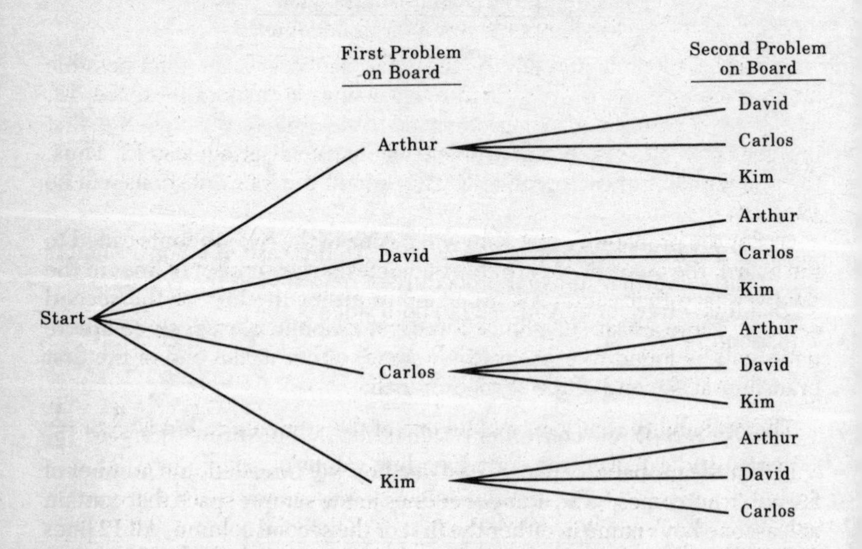

The sample space will consist of two columns, one indicating the names of students called to the board to do the first problem, and the other indicating the names of students chosen to do the second problem when the first problem has been solved. Since each student chosen to do the first problem may be followed by any one of three students for the second problem, each name in the first column must be repeated *three times* to provide lines for all the combinations of students that are possible.

SAMPLE SPACE:

First Problem on Board	Second Problem on Board
Arthur	David
Arthur	Carlos
Arthur	Kim
David	Arthur
David	Carlos
David	Kim
Carlos	Arthur
Carlos	David
Carlos	Kim
Kim	Arthur
Kim	David
Kim	Carlos

b. Probability of an event occurring =

$$\frac{\text{number of favorable outcomes}}{\text{total possible number of outcomes}}.$$

In each of the four probability questions that follow, the total possible number of outcomes is the total number of lines in the sample space, 12, or the total number of complete paths from "start" through the first problem and on to the second problem on the tree diagram, also 12. Thus, the denominator of each probability fraction in the four questions will be 12.

(1) For the probability that Kim will be one of the two students called to the board, the number of favorable outcomes is the number of lines in the sample space that contain Kim's name in either the first or the second column. There are six such lines. If the tree diagram is used, six complete paths will be found to contain Kim's name, either at the end of the first branch or at the end of the second branch.

The probability that Kim will be one of the students called is $\frac{6}{12}$ or $\frac{1}{2}$.

(2) For the probability that at least one boy will be called, the number of favorable outcomes is the number of lines in the sample space that contain at least one boy's name in either the first or the second column. All 12 lines will be found to contain at least one boy's name. Similarly, if the tree diagram is used, all 12 possible paths from "start" through the second problem will be found to contain at least one boy's name. (Note that, since there is only one girl, Kim, and she cannot be called on twice, it is impossible to have no boys when solving two problems.)

The probability that at least one boy will be called on is $\frac{12}{12}$ or **1**.

(3) For the probability that Kim and Arthur will be the two students called to the board, the number of favorable outcomes is the number of lines in the sample space that contain both Kim's and Arthur's names. There are two such lines: Arthur—Kim and Kim—Arthur. Similarly, if the tree diagram is used, two complete paths will be found to pass through the names of both Kim and Arthur.

The probability that Kim and Arthur will be the two students is $\mathbf{\frac{2}{12}}$ or $\mathbf{\frac{1}{6}}$.

(4) For the probability that two girls will be called to the board, we know without looking that we cannot find any lines in the sample space containing two girls' names or any branches in the tree diagram containing two girls' names since there is only one girl, Kim, in the first row.

The probability that two girls will be called to the board is $\mathbf{\frac{0}{12}}$ or **0**.

38.

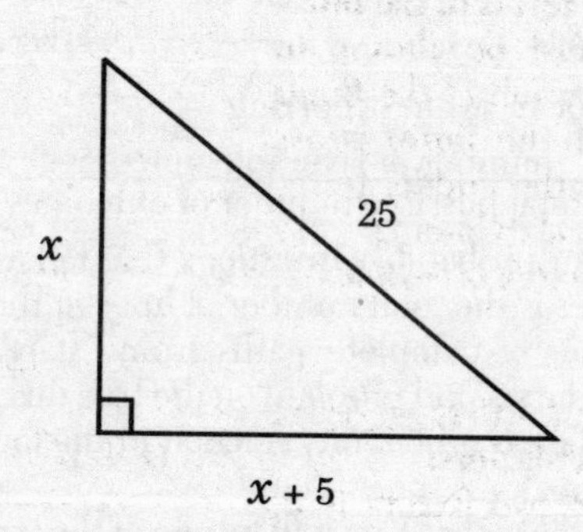

38. a. Let x = the length of one leg.

Then $x + 5$ = the length of the other leg.

By the Pythagorean Theorem, the sum of the squares of the lengths of the legs of a right triangle is equal to the square of the length of the hypotenuse:

$$x^2 + (x + 5)^2 = 25^2$$

Square $(x + 5)$ and square 25:

$$\begin{array}{r} x + 5 \\ x + 5 \\ \hline x^2 + 5x \qquad \\ 5x + 25 \\ \hline x^2 + 10x + 25 \end{array} \qquad \begin{array}{r} 25 \\ \times 25 \\ \hline 125 \\ 50\ \\ \hline 625 \end{array}$$

$$x^2 + x^2 + 10x + 25 = 625$$

Combine like terms:

$$2x^2 + 10x + 25 = 625$$

This is a *quadratic equation*. Rearrange it so that all terms are on one side equal to 0 by adding -625 to both sides:

$$\frac{\quad -625 = -625}{2x^2 + 10x - 600 = 0}$$

To simplify, divide each term of the equation by 2:

$$x^2 + 5x - 300 = 0$$

The left side of the equation is a *quadratic trinomial* that can be factored into the product of two binomials. The factors of the first term, x^2, are x and x, and they become the first terms of the binomials:

$$(x \quad)(x \quad) = 0$$

The factors of the last term, -300, become the second terms of the binomials, but they must be chosen in such a way that the sum of the inner cross product and the outer cross product is equal to the middle term, $+5x$, of the original trinomial. Try $+20$ and -15 as factors of -300:

$$+20x = \text{inner product}$$
$$(x + 20)(x - 15) = 0$$
$$-15x = \text{outer product}$$

Since $(+20x) + (-15x) = +5x$, these are the correct factors:

$$(x + 20)(x - 15) = 0$$

If the product of two factors is 0, either factor may equal 0:

$$x + 20 = 0 \quad \lor \quad x - 15 = 0$$

Add the appropriate additive inverse to both sides of each equation, -20 for the left equation, and 15 for the right equation:

$$\frac{\quad -20 = -20}{x \quad = -20} \qquad \frac{\quad 15 = 15}{x \quad = 15}$$

Reject the negative value as meaningless for a length:

$$x = 15$$
$$x + 5 = 20$$

The lengths of the legs are **15** and **20**.

b. The area of a right triangle is one-half the product of the lengths of its legs:

$$\text{Area} = \frac{1}{2}(20)(15)$$
$$\text{Area} = 10(15)$$
$$\text{Area} = 150$$

The area of the right triangle is **150** square units.

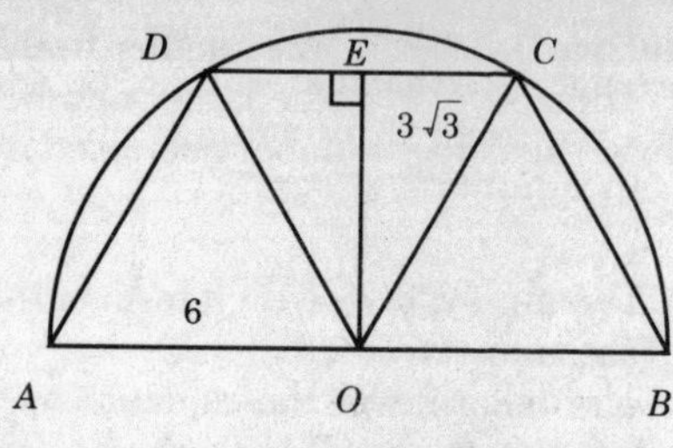

39. a. Since $\triangle AOD$ is a equilateral, $AD = AO = 6$.

$AD =$ **6**.

b. Since $\triangle AOD$, $\triangle DOC$, and $\triangle COB$ are all equilateral,

$AO = AD = DO = DC = CO = CB = OB = 6$.

The perimeter of trapezoid $ABCD$ is the sum of the lengths of its sides:

$$\text{Perimeter} = AB + BC + CD + DA$$
$$\text{Perimeter} = (6 + 6) + 6 + 6 + 6$$
$$\text{Perimeter} = 12 + 6 + 6 + 6$$
$$\text{Perimeter} = 30$$

The perimeter of trapezoid $ABCD$ is **30**.

c. The area of a triangle equals one-half the product of its base and the altitude to that base:

$$\text{Area of } \triangle COD = \frac{1}{2}(CD)(OE)$$
$$\text{Area of } \triangle COD = \frac{1}{2}(6)(3\sqrt{3})$$
$$\text{Area of } \triangle COD = 3(3\sqrt{3})$$
$$\text{Area of } \triangle COD = 9\sqrt{3})$$

The area of $\triangle COD$ is $\mathbf{9\sqrt{3}}$.

d. The area, A, of a trapezoid with altitude h and bases b_1 and b_2 is given by this formula:

$$A = \frac{1}{2}h(b_1 + b_2)$$

For trapezoid $ABCD$, $h = OE = 3\sqrt{3}$, $b_1 = AB = 12$, and $b_2 = CD = 6$:

$$A = \frac{1}{2}(3\sqrt{3})(12 + 6)$$
$$A = \frac{1}{2}(3\sqrt{3})(18)$$
$$A = 3\sqrt{3}(9)$$
$$A = 27\sqrt{3}$$

The area of trapezoid $ABCD$ is $\mathbf{27\sqrt{3}}$.

e. The area, A, of a circle the length of whose radius is r is given by this formula:

$$A = \pi r^2$$

For the circle of which the semicircle is half, $r = 6$:

$$A = \pi(6)^2$$
$$A = 36\pi$$

The area of the semicircle is one-half that of the circle:

$$\text{Semicircle's area} = \frac{1}{2}(36\pi) = 18\pi$$

The area of the semicircle is $\mathbf{18\pi}$.

40. a. Let n = the smaller number.

Then $3n + 4$ = the larger number.

Twice the larger number decreased by 3 times the smaller number is 32.

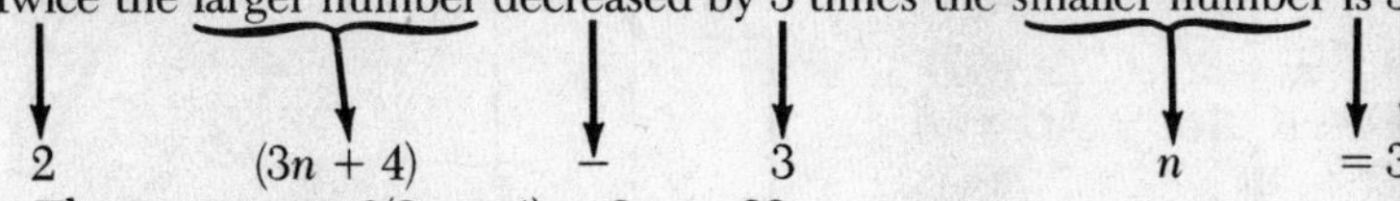

The equation is $\mathbf{2(3n + 4) - 3n = 32}$.

ALTERNATIVE SOLUTION:

Let x = the smaller number.

Let y = the larger number.

The larger number is 4 more than 3 times the smaller number: $y = 3x + 4$

Twice the larger number decreased by 3 times the smaller is 32: $2y - 3x = 32$

The use of two variables in this solution requires a system of two equations. The system is:

$$\begin{cases} y = 3x + 4 \\ 2y - 3x = 32 \end{cases}$$

b. Let x = the first (smallest) positive odd integer.

Then $x + 2$ = the second consecutive positive odd integer.

And $x + 4$ = the third consecutive positive odd integer

Twice the sum of the second and third integers is 6 times the first less 2.

2 $(x + 2 + x + 4)$ = 6 x − 2

The equation is $\mathbf{2(2x + 6) = 6x - 2}$.

41. a. The entries in the cumulative frequency column are obtained by adding the frequency shown on each line to the cumulative frequency of the preceding line. Note that the first frequency, 4, has no preceding line, so nothing is added to 4 to obtain the first cumulative frequency entry.

Runs Scored	Frequency	Cumulative Frequency
0–1	4	4
2–3	11	15
4–5	6	21
6–7	5	26
8–9	2	28

b.

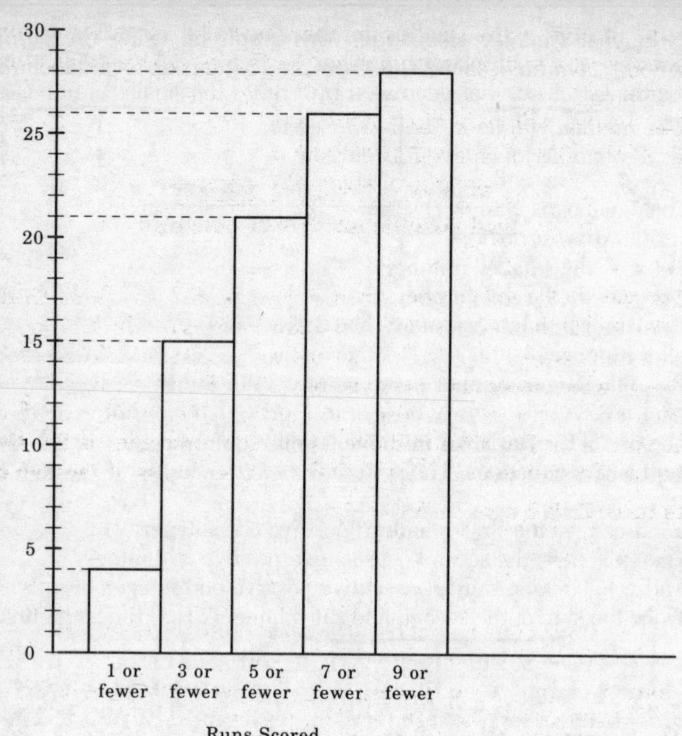

c. The *median* is the middle item if all the items in a set of data are arranged in order of increasing size. One half of all the items will be less than the median, and one half will be more than the median. Here, 28 games were played in all. If the number of runs in each of the 28 games were arranged in order of size, 14 (that is, half of 28), would be below the median and 14 would be above the median. Thus, the median must be between the 14th and 15th scores in the arrangement.

To find the position between the 14th and 15th scores, use the frequency column in the table. Four items are in the 0–1 interval. Ten more must be counted off to reach the 14th item, and 11 more must be counted off to reach the 15th item. Since the 2–3 interval contains 11 items, both the 14th and 15th items of the total distribution will lie in the 2–3 interval. The median, halfway between the 14th and the 15th item, will also lie in the 2–3 interval.

Note that we may also obtain this result by using the cumulative frequency column; it shows that there are 15 items up to and including the 2–3 interval.

The median will lie in the **2–3** interval.

d. Probability of an event occurring =

$$\frac{\text{number of favorable outcomes}}{\text{total possible number of outcomes}}.$$

The games involving more than 5 runs scored are those in the 6–7 interval and in the 8–9 interval. There are 5 games in which 6–7 runs were scored and 2 games in which 8–9 runs were scored, making a total of 7 games in which more than 5 runs were scored. For the probability of more then 5 runs in a game chosen at random, the number of favorable outcomes is 7. The total possible number of outcomes is 28, the total number of games played. The probability of choosing at random a game with more than 5 runs scored is $\frac{7}{28}$.

The probability is $\mathbf{\frac{7}{28}}$.

42. **a.** Let C represent the statement "John will buy a car." TRUE

Let R represent the statement "John will get a raise." FALSE

"John will buy a car if and only if John will get a raise" is the *biconditional*, or *equivalence relation*, represented by $C \leftrightarrow R$. The biconditional is TRUE only when both statements, C and R, have the *same* truth value, either both TRUE or both FALSE; otherwise, the biconditional is FALSE. Since C is TRUE and R is FALSE, $C \leftrightarrow R$ is FALSE. Therefore, "John will buy a car if and only if John will get a raise" is **FALSE**.

b. Let S represent the statement "Physics is a science." TRUE

Let P represent the statement "Jan plays piano." FALSE

"Physics is a science and Jan does *not* play piano" is the *conjunction* represented by $S \wedge \sim P$. Since P is FALSE, its *negation*, $\sim P$, is TRUE. The conjunction $S \wedge \sim P$ is TRUE if both of its *conjuncts*, S and $\sim P$, are TRUE. Since S is TRUE and $\sim P$ is TRUE, $S \wedge \sim P$ is TRUE. Thus, "Physics is a science and Jan does *not* play piano" is **TRUE**

c. Let $S \rightarrow R$ represent the *conditional* or *implication* "If a polygon is a square, then the polygon is a rectangle." TRUE.

S represents the statement "Polygon $ABCD$ is a square." TRUE

R represents the statement "Polygon $ABCD$ is a rectangle." Since the conditional, $S \rightarrow R$, is TRUE, and since its *hypothesis*, S, is TRUE, its *conclusion*, R, must be TRUE by the *Law of Detachment* (also called *modus ponens*). Thus, "Polygon $ABCD$ is a rectangle" is **TRUE**.

d. Let $A \vee T$ represent the *disjunction* "I like apples or it is Tuesday." FALSE

A represents the statement "I like apples." FALSE

T represents the statement "It is Tuesday." Since $A \vee T$ is FALSE, both A and T must be FALSE, since a disjunction is TRUE if either of its *disjuncts*, A and T, is TRUE or if both are TRUE. Thus, "It is Tuesday" must be **FALSE**.

e. Let $S \rightarrow R$ represent the *conditional* or *implication* "If it is snowing, the roads are slippery." TRUE

R represents the *conclusion* "The roads are slippery." $\sim R$ represents the *negation* of R, that is, "The roads are not slippery." TRUE

$\sim R \rightarrow \sim S$ is the *contrapositive* of $S \rightarrow R$, which is formed by interchanging the hypothesis S and the conclusion R of $S \rightarrow R$, and then negating them. Since $S \rightarrow R$ is TRUE, $\sim R \rightarrow \sim S$ must be TRUE by the *Law of the Contrapositive*. Since $\sim R$ is TRUE and $\sim R \rightarrow \sim S$ is TRUE, $\sim S$ must be TRUE by the *Law of Detachment* (also called *modus ponens*). $\sim S$ represents "It is not snowing," which we have proved to be **TRUE**.

The question requires that the answers be summarized in this form:

a. **False**
b. **True**
c. **True**
d. **False**
e. **True**

Topic	Question Numbers	Number of Points	Your Points	Your Percentage
1. Numbers (rat'l, irrat'l); Percent	33	2		
2. Properties of No. Systems	23, 31	2 + 2 = 4		
3. Operations on Rat'l Nos. and Monomials	17, 22	2 + 2 = 4		
4. Operations on Multinomials	19, 20	2 + 2 = 4		
5. Square root; Operations involving Radicals	34	2		
6. Evaluating Formulas and Expressions	7	2		
7. Linear Equations (simple cases incl. parentheses)	12	2		
8. Linear Equations containing Decimals or Fractions	2, 14	2 + 2 = 4		
9. Graphs of Linear Functions (slope)	32	2		
10. Inequalities	27, 30	2 + 2 = 4		
11. Systems of Eqs. & Inequal. (alg. & graphic solutions)	5, 36	2 + 10 = 12		
12. Factoring	8	2		
13. Quadratic Equations	35, 38a	2 + 8 = 10		
14. Verbal Problems	16, 40	2 + 5 + 5 = 12		
15. Variation	—	0		
16. Literal Eqs.; Expressing Relations Algebraically	10, 29	2 + 2 = 4		
17. Factorial *n*	24	2		
18. Areas, Perims., Circums., Vols. of Common Figures	6, 28, 38b, 39	2 + 2 + 2 + 10 = 16		
19. Geometry (≅, ∠ meas., ∥ lines, compls., suppls., const.)	1, 3, 9, 13, 25	2 + 2 + 2 + 2 + 2 = 10		
20. Ratio & Proportion (incl. similar triangles)	4	2		
21. Pythagorean Theorem	18	2		
22. Logic (symbolic rep., logical forms, truth tables)	11, 26, 42	2 + 2 + 10 = 14		
23. Probability (incl. tree diagrams & sample spaces)	15, 37, 41d	2 + 10 + 2 = 14		
24. Combinatorics (arrangements, permutations)	—	0		
25. Statistics (central tend., freq. dist., histograms)	21, 41a, b, c	2 + 2 + 4 + 2 = 10		

Examination January 1989

Three-Year Sequence for High School Mathematics—Course I

PART ONE

DIRECTIONS: *Answer 30 questions from this part. Each correct answer will receive 2 credits. No partial credit will be allowed. Write your answers in the spaces provided. Where applicable, answers may be left in terms of π or in radical form.*

1 On a restaurant menu, there are six sandwich choices and three beverage choices. How many different lunches may a person order consisting of one sandwich and one beverage? 1_____

2 A base angle of an isosceles triangle measures 50. Find the number of degrees in the measure of the vertex angle. 2_____

3 A 50-milliliter salt solution contains 4 milliliters of salt. How many milliliters of salt will a 75-milliliter solution of the same strength contain? 3_____

4 If 15 is 25% of a number, find the number. 4_____

5 In the accompanying figure, arc AB of circle O measures 70. If the measure of $\angle AOB$ is represented by x, find the value of x.

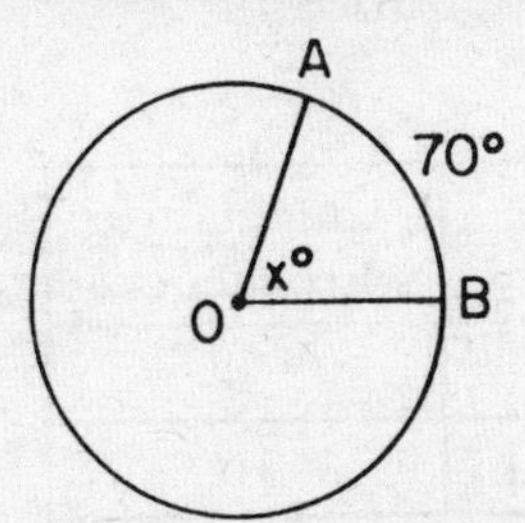

5_____

6 Two numbers are in the ratio 3:2. If the small number is 24, find the larger number. 6_____

7 Solve for x: $0.3x + 1 = 2.2$ 7_____

8 Solve for x: $\frac{x - 1}{4} = \frac{1}{2}$ 8_____

9 Twice the sum of a number and 4 is equal to 22. What is the number? 9_____

10 Solve for y in terms of x: $3y + 2 = x$ 10_____

11 In the accompanying diagram, parallel lines $\overleftrightarrow{AB}$ and $\overleftrightarrow{CD}$ are intersected by transversal $\overleftrightarrow{GH}$ at points E and F, respectively. If m$\angle AEG$ is $(3x + 7)$ and m$\angle CFE$ is $(4x - 2)$, find x.

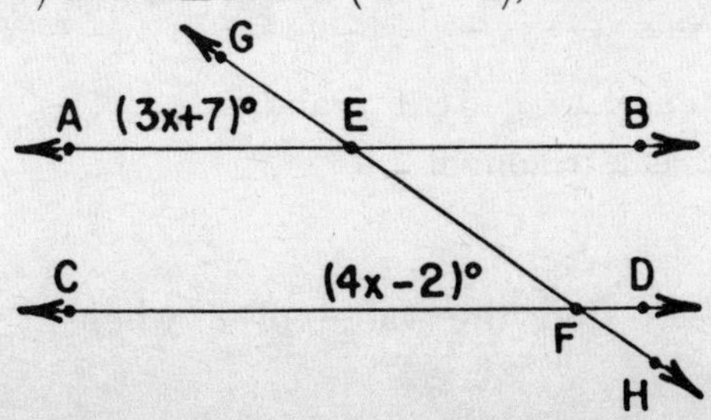

11_____

12 The table below shows the distribution of scores on a math test. Using the data in the table, determine the total number of students who took the test.

Interval	Tally	Cumulative Frequency
61–70	||||	4
71–80	~~||||~~ |	10
81–90	||	12
91–100	||||	16

12_____

13 In the accompanying diagram, the adjacent angles formed by intersecting lines $\overleftrightarrow{AB}$ and $\overleftrightarrow{CD}$ have measures of $3x + 50$ and $x + 10$. Find x.

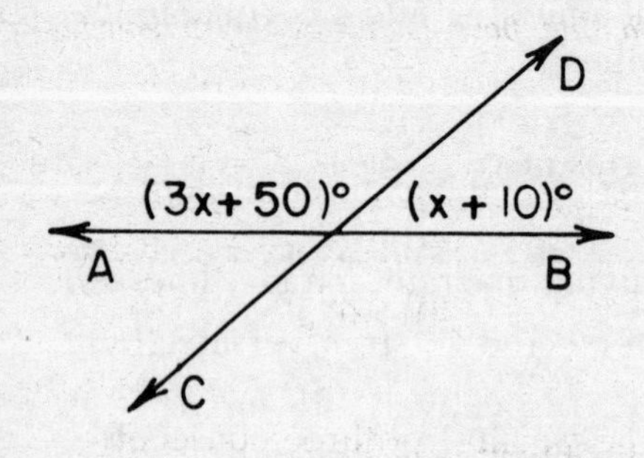

13_____

14 Express the sum of $x^2 - 3x + 5$ and $3x^2 - 2x - 2$ as a trinomial.

14_____

15 If $a = \dfrac{b^2 - c}{2}$, find the value of a when $b = 2$ and $c = -4$.

15_____

16 Solve the following system of equations for x:

$$3x + y = 9$$
$$-2x + y = -1$$

16_____

17 If $a + b = 5$ and $a - b = 3$, find the value of $a^2 - b^2$.

17_____

18 Express $\frac{3a}{4} - \frac{a}{3}$ as a single fraction in simplest form.

18_____

19 Express $(x + 1)(2x - 3)$ as a trinomial.

19_____

DIRECTIONS (20–34): *For each question chosen, write the* numeral *preceding the word or expression that best completes the statement or answers the question.*

20 The product of $3x^2y^3$ and $4xy^2$ is equivalent to

(1) $7x^2y^6$
(2) $7x^3y^5$
(3) $12x^2y^6$
(4) $12x^3y^5$

20_____

21 If x represents the smallest of three consecutive even integers, then the largest would be represented by

(1) $x + 2$
(2) $x + 3$
(3) $x + 4$
(4) $x + 6$

21_____

22 The value of 4! is

(1) 24
(2) 16
(3) 12
(4) 4

22_____

23 For the group of data 3, 3, 6, 7, 16, which is true?

(1) mean > median
(2) mode = mean
(3) median < mode
(4) median = mean

23_____

24 The expression $5\sqrt{3} - \sqrt{27}$ is equivalent to

(1) $8\sqrt{3}$
(2) $-8\sqrt{3}$
(3) $-2\sqrt{3}$
(4) $2\sqrt{3}$

24_____

25 In the accompanying truth table, which statement should be the heading for column 3?

Column 1	Column 2	Column 3
p	q	?
T	T	T
T	F	F
F	T	F
F	F	F

(1) $p \rightarrow q$
(2) $p \leftrightarrow q$
(3) $p \wedge q$
(4) $p \vee q$

25_____

26 What is the inverse of the statement "If it is sunny, I will go swimming"?

(1) If I go swimming, then it is sunny.
(2) If it is not sunny, I will not go swimming.
(3) If I do not go swimming, then it is not sunny.
(4) I will go swimming if, and only if, it is sunny.

26_____

27 Which graph represents the solution set of $2x + 1 \geq 3$?

(1)
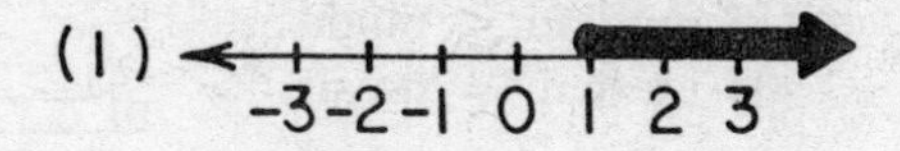

(2)

(3)
-3 -2 -1 0 1 2 3

(4)
-3 -2 -1 0 1 2 3

27_____

28 The sentence $3 + (5 + 2) = (5 + 2) + 3$ illustrates

(1) the commutative property of addition
(2) the associative property of addition
(3) the distributive property of multiplication over addition
(4) the additive identity element

28_____

29 What is the solution set of the equation $x^2 - 2x - 3 = 0$?

(1) $\{2,1\}$
(2) $\{2,-1\}$
(3) $\{-3,0\}$
(4) $\{3,-1\}$

29_____

30 Let p represent the statement "All sides are congruent," and let q represent the statement "All angles are congruent." The statement $p \land \sim q$ is true for a

(1) rectangle
(2) rhombus
(3) square
(4) trapezoid

30_____

31 The expression $\frac{1}{(x - 1)(x + 2)}$ is undefined if x is equal to

(1) -1 or 2 (3) 0
(2) 1 or -2 (4) -1

31_____

32 A single card is drawn from a standard deck of 52 cards. What is the probability the card is a five or a diamond?

(1) $\frac{17}{52}$ (3) $\frac{16}{52}$

(2) $\frac{15}{52}$ (4) $\frac{18}{52}$

32_____

33 What is the slope of the line whose equation is $y - 2x = 4$?

(1) -2 (3) -4
(2) 2 (4) 4

33_____

34 What is the area of a circle whose diameter is 10?

(1) 10π (3) 25π
(2) 20π (4) 100π

34_____

DIRECTIONS (35): *Leave all construction lines in the answer.*

35 Construct and label a segment $\overline{DE}$ on line ℓ congruent to segment $\overline{AB}$.

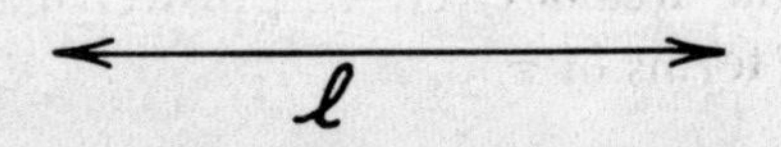

PART TWO

DIRECTIONS: *Answer* four *questions from this part. Show all work unless otherwise directed.*

36 *a* On the same set of coordinate axes, graph the following system of inequalities:

$$x < 5$$
$$2x + y \geq 6 \qquad [8]$$

b Write the coordinates of a point in the solution set of the inequalities graphed in part *a*. [2]

37 In the accompanying figure, $\triangle ABC$ is inscribed in circle O, $\overline{AB}$ is a diameter of circle O, $AB = 20$, $BC = 16$, and the measure of minor arc $AC = 74$.

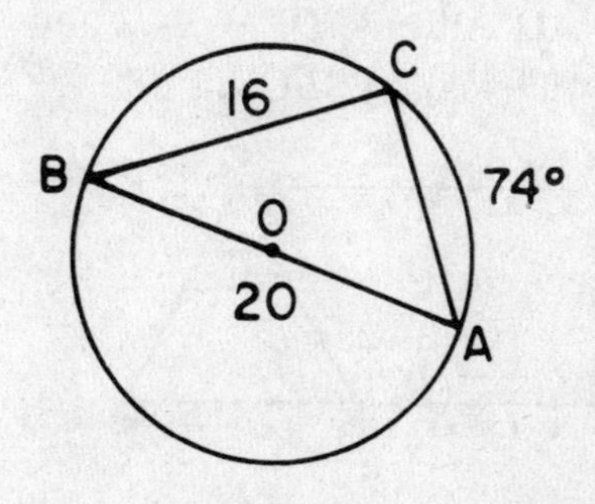

a Find the measure of $\angle ACB$. [2]

b Find the measure of $\angle BAC$. [2]

c Find the length of $\overline{AC}$. [2]

d Find the area of $\triangle ABC$. [2]

e Find the area of circle O. [Answer may be left in terms of π.] [2]

38 Mr. Walden has two square flower gardens. A side of the larger garden is 3 feet more than a side of the smaller garden. The sum of the areas of the two gardens is 269 square feet. Find the length of a side, in feet, of each garden. [*Only an algebraic solution will be accepted.*] [3,7]

39 Solve the following system of equations algebraically and check:

$$3x - 2y = 22$$
$$2x + 5y = 2$$ [8,2]

40 In the accompanying diagram, $ABCD$ is a parallelogram with $\overline{AB}$ extended through B to E. Segment EC is drawn forming equilateral triangle BEC. The length of $\overline{DC}$ is two units more than twice the length of $\overline{AD}$. Altitude $DF = 2\sqrt{3}$ and the perimeter of parallelogram $ABCD$ is 28.

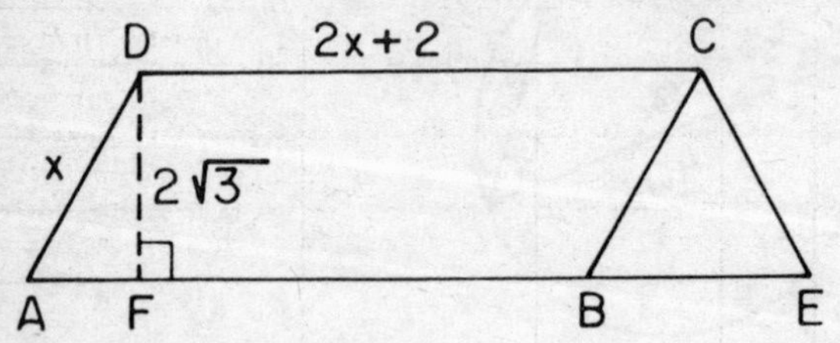

a If the length of $\overline{AD}$ is represented by x, find the measure of

(1) $\overline{AD}$ [4]

(2) $\overline{DC}$ [1]

(3) $\overline{AE}$ [1]

b Find the area of $\triangle BEC$. [Answer may be left in radical form.] [2]

c Find the area of trapezoid $AECD$. [Answer may be left in radical form.] [2]

41 Four chips, numbered 1, 2, 3, and 4, are in a hat. One chip is drawn and then, without replacement, a second chip is drawn.

a Draw a tree diagram or list the sample space showing all possible outcomes. [3]

b What is the probability that one of the two numbers drawn is odd and the other is even? [2]

c What is the probability of drawing two prime numbers? [2]

d What is the probability that the sum of the two numbers drawn is *not* less than 5? [2]

e What is the probability that the sum of the two numbers is greater than 7? [1]

42 *a* Complete the truth table for the statement $(p \rightarrow q) \leftrightarrow \sim(p \wedge \sim q)$. [8]

p	q	$p \rightarrow q$	$\sim q$	$p \wedge \sim q$	$\sim(p \wedge \sim q)$	$(p \rightarrow q) \leftrightarrow \sim(p \wedge \sim q)$
T	T					
T	F					
F	T					
F	F					

b Let p represent "Sue lives in Buffalo" and let q represent "Sue lives in New York State." Using the table completed in part *a*, which statement is equivalent to "If Sue lives in Buffalo, then Sue lives in New York State"? [2]

(1) Sue lives in Buffalo and Sue does not live in New York State.
(2) It is false that Sue lives in Buffalo and Sue does not live in New York State.
(3) Sue lives in Buffalo or Sue does not live in New York State.
(4) Sue does not live in Buffalo and Sue does not live in New York State.

Answers January 1989

Three-Year Sequence for High School Mathematics—Course I

ANSWER KEY

PART ONE

1. 18
2. 80
3. 6
4. 60
5. 70
6. 36
7. 4
8. 3
9. 7
10. $\frac{x-2}{3}$
11. 9
12. 16
13. 30
14. $4x^2 - 5x + 3$
15. 4
16. 2
17. 15
18. $\frac{5a}{12}$
19. $2x^2 - x - 3$
20. (4)
21. (3)
22. (1)
23. (1)
24. (4)
25. (3)
26. (2)
27. (1)
28. (1)
29. (4)
30. (2)
31. (2)
32. (3)
33. (2)
34. (3)
35. construction

Part Two—*See* Answers Explained.

ANSWERS EXPLAINED

PART ONE

1. Each of the six sandwich choices may be paired with each of the three beverage choices. Therefore there are 6×3 or 18 different lunches consisting of one sandwich and one beverage.

18 different lunches.

2. Let x = the number of degrees in the vertex angle.

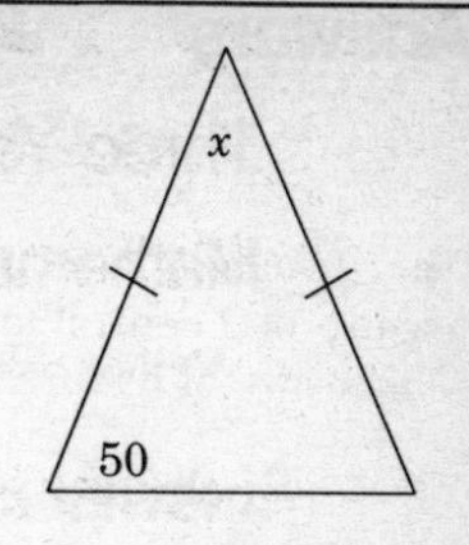

Since the base angles of an isosceles triangle are congruent, if one base angle measures 50°, so does the other one.

The sum of the measures of the three angles of a triangle is 180°: $x + 50 + 50 = 180$

Combine like terms: $x + 100 = 180$

Add -100 (the additive inverse of 100) to both sides of the equation: $-100 = -100$

$$x = 80$$

The number of degrees in the measure of the vertex angle is **80.**

3. Let x = the number of milliliters of salt in a 75-milliliter solution.

Since the 50-milliliter salt solution and the 75-milliliter salt solution are of the same strength, the amount of salt each contains is proportional to the amount of the solution: $\frac{x}{75} = \frac{4}{50}$

In a proportion, the product of the means equals the product of the extremes (cross-multiply): $50x = 4(75)$

$$50x = 300$$

Divide both sides of the equation by 50: $\frac{50x}{50} = \frac{300}{50}$

$$x = 6$$

The 75-milliliter solution has **6** milliliters of salt.

4. Let x = the number.

25% of the number equals 15.
↓ ↓ ↓ ↓ ↓
0.25 × x = 15

The equation to use is: $0.25x = 15$

Multiply both sides of the equation by 100 in order to clear decimals: $100(0.25x) = 100(15)$

$$25x = 1500$$

Divide both sides of the equation by 25: $\frac{25x}{25} = \frac{1500}{25}$

$$x = 60$$

The number is **60.**

5. $\angle AOB$ is a *central angle*. The measure of a central angle is equal to the measure of its intercepted arc:

$$m\angle AOB = x = 70$$

$x =$ **70.**

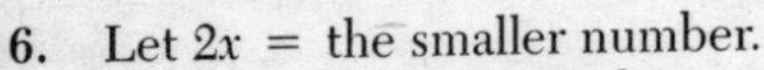

6. Let $2x$ = the smaller number.
Then $3x$ = the larger number since the two numbers are in the ratio 2:3.

The smaller number is 24: $2x = 24$

Divide both sides of the equation by 2: $\frac{2x}{2} = \frac{24}{2}$

$$x = 12$$
$$3x = 3(12) = 36$$

The larger number is **36.**

7. The given equation contains decimals: $0.3x + 1 = 2.2$

Multiply each term on both sides of the equation by 10 to clear the decimals:

$$10(0.3x) + 10(1) = 10(2.2)$$
$$3x + 10 = 22$$

Add -10 (the additive inverse of 10) to both sides of the equation:

$$\begin{aligned} -10 &= -10 \\ \hline 3x \quad &= 12 \end{aligned}$$

Divide both sides of the equation by 3:

$$\frac{3x}{3} = \frac{12}{3}$$
$$x = 4$$

$x =$ **4.**

8. The given equation contains fractions: $\frac{x-1}{4} = \frac{1}{2}$

Multiply both sides of the equation by the least common denominator (L.C.D.). The L.C.D. for 4 and 2 is 4:

$$4\left(\frac{x-1}{4}\right) = 4\left(\frac{1}{2}\right)$$
$$x - 1 = 2$$

Add 1 (the additive inverse of -1) to both sides of the equation:

$$\begin{aligned} 1 &= 1 \\ \hline x \quad &= 3 \end{aligned}$$

$x =$ **3.**

9. Let x = the number.

Twice the sum of the number and 4 is equal to 22.

$$2 \quad (\quad x \quad + \quad 4 \quad) \quad = \quad 22$$

The equation to use is: $2(x + 4) = 22$

Remove parentheses by applying the distributive law; multiply each term within the parentheses by 2: $2x + 8 = 22$

Add -8 (the additive inverse of 8) to both sides of the equation:

$$\begin{array}{rcr} -8 & = & -8 \\ \hline 2x & = & 14 \end{array}$$

Divide both sides of the equation by 2:

$$\frac{2x}{2} = \frac{14}{2}$$

$$x = 7$$

The number is **7**.

10. The given equation is: $3y + 2 = x$

Add -2 (the additive inverse of 2) to both sides of the equation:

$$\begin{array}{rcr} -2 & = & -2 \\ \hline 3y & = & x - 2 \end{array}$$

Divide both sides of the equation by 3:

$$\frac{3y}{3} = \frac{x-2}{3}$$

$$y = \frac{x-2}{3}$$

$$y = \frac{x-2}{3}.$$

11. Angles AEG and CFE are corresponding angles formed by transversal $\overleftrightarrow{GH}$ to parallel lines $\overleftrightarrow{AB}$ and $\overleftrightarrow{CD}$.

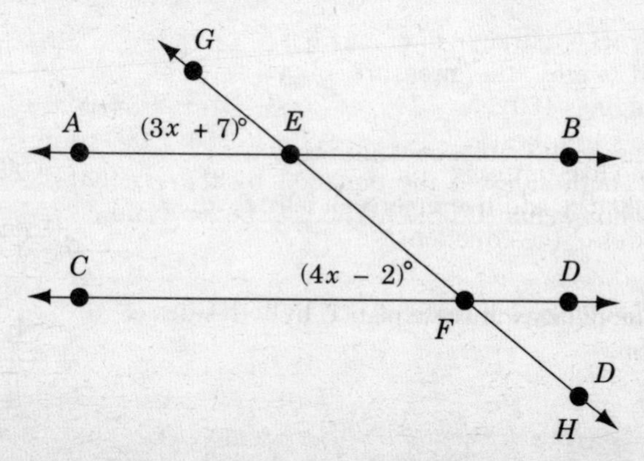

If two lines are parallel, a transversal makes a pair of corresponding angles equal in measure:

$$m\angle AEG = m\angle CFE$$
$$3x + 7 = 4x - 2$$

Add 2 (the additive inverse of -2) and also add $-3x$ (the additive inverse of $3x$) to both sides of the equation:

$$\underline{-3x + 2 = -3x + 2}$$
$$9 = x$$

$x = 9$.

12.

Interval	Tally	Cumulative Frequency
61–70	IIII	4
71–80	~~IIII~~ I	10
81–90	II	12
91–100	IIII	16

Since the cumulative frequency entries are obtained by adding the tally on each line to the cumulative frequency shown on the line above it, the last cumulative frequency entry, 16, must be the total number of students who took the test.

16 students took the test.

13.

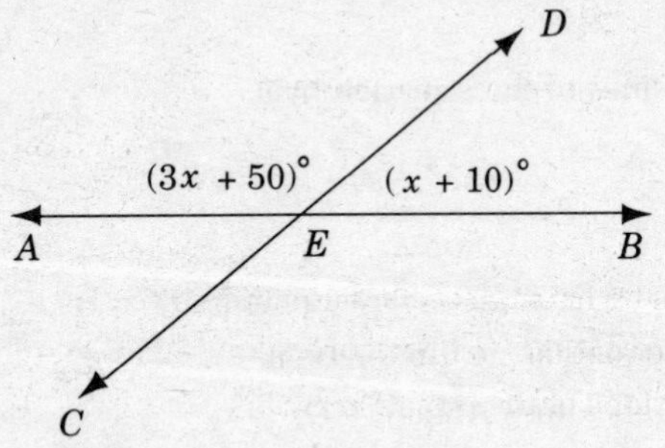

Since $\overleftrightarrow{AB}$ is a straight line, $\angle AEB$ is a straight angle; the measure of a straight angle is 180°:

$$(3x + 50) + (x + 10) = 180$$

Remove parentheses:

$$3x + 50 + x + 10 = 180$$

Combine like terms:

$$4x + 60 = 180$$

Add -60 (the additive inverse of 60) to both sides of the equation:

$$\underline{-60 = -60}$$
$$4x = 120$$

Divide both sides of the equation by 4:

$$\frac{4x}{4} = \frac{120}{4}$$
$$x = 30$$

$x = 30$.

14. To add the two given trinomials, write one under the other with like terms in the same column:

$$\begin{array}{r} x^2 - 3x + 5 \\ 3x^2 - 2x - 2 \\ \hline 4x^2 - 5x + 3 \end{array}$$

Combine the terms in each column algebraically:
The sum as a trinomial is $\mathbf{4x^2 - 5x + 3}$.

15. The given expression is: $a = \dfrac{b^2 - c}{2}$

Substitute 2 for b and -4 for c: $a = \dfrac{(2)^2 - (-4)}{2}$

Perform the operations indicated by the parentheses: $a = \dfrac{4 + 4}{2}$

Simplify by combining like terms: $a = \dfrac{8}{2}$

$a = 4$

$a = \mathbf{4}$.

16. The given system of equations is:

$$\begin{aligned} 3x + y &= 9 \\ -2x + y &= -1 \end{aligned}$$

Multiply each term of the second equation by -1:

$$\begin{array}{rcl} 3x + y &=& 9 \\ 2x - y &=& 1 \\ \hline 5x &=& 10 \end{array}$$

Add the two equations to eliminate y:
Divide both sides of the equation by 5: $\dfrac{5x}{5} = \dfrac{10}{5}$

$x = 2$

$x = \mathbf{2}$.

17. The expression $a^2 - b^2$ is the difference between two perfect squares. It can be factored by taking the square root of the two perfect squares; the factors are the sum and the difference of the square roots: $a^2 - b^2 = (a + b)(a - b)$

Substitute 5 for $(a + b)$ and 3 for $(a - b)$: $a^2 - b^2 = (5)(3)$

$a^2 - b^2 = 15$

The value of $a^2 - b^2$ is **15**.

18. The given fractions have different denominators: $\frac{3a}{4} - \frac{a}{3}$

In order to be combined, fractions must have a common denominator. The least common denominator (L.C.D.) is the smallest number into which each of the denominators will divide evenly; the L.C.D. for 4 and 3 is 12.

Convert each fraction to an equivalent fraction having the common denominator by multiplying the first by 1 in the form $\frac{3}{3}$ and the second by 1 in the form $\frac{4}{4}$: $\frac{3(3a)}{3(4)} - \frac{4a}{4(3)}$

$$\frac{9a}{12} - \frac{4a}{12}$$

Fractions having the same denominator may be combined by combining their numerators: $\frac{9a - 4a}{12}$

Combine like terms in the numerator: $\frac{5a}{12}$

The fraction in simplest form is $\mathbf{\frac{5a}{12}}$.

19. An easy way to express $(x + 1)(2x - 3)$ as a trinomial is to use the FOIL method for multiplying two binomials:

F is the product of the First terms of the binomials: $(x)(2x) = 2x^2$

O is the product of the Outer terms: $(x)(-3) = -3x$

I is the product of the Inner terms: $(+1)(2x) = 2x$

Combine O and I: $-x$

L is the product of the Last terms: $(+1)(-3) = -3$

Combine all the results; the product is: $2x^2 - x - 3$

The trinomial is $\mathbf{2x^2 - x - 3}$.

20. The product may be represented as: $(3x^2y^3)(4xy^2)$

To find the numerical coefficient of the product of two monomials, multiply their numerical coefficients together: $3 \times 4 = 12$

Multiply the literal factors together to find the literal factor of the product. Remember that, when powers of the same base are multiplied, the exponents are added to give the exponent of the product. Also remember that the exponent of x is understood to be 1: $x^2y^3 \times x^1y^2 = x^3y^5$

Combine the two results: $(3x^2y^3)(4xy^2) = 12x^3y^5$

The correct choice is **(4)**.

21. Consecutive even integers differ from each other by 2.

$x =$ the smallest even integer;
then $x + 2 =$ the next consecutive even integer,
and $x + 4 =$ the third consecutive even integer. This is the largest.
The correct choice is **(3)**.

22. *Factorial n*, symbolized as $n!$, is defined as the product

$$n(n-1)(n-2)(n-3)\ldots(3)(2)(1).$$

Therefore:

$$4! = 4(3)(2)(1)$$
$$4! = 24$$

The correct choice is **(1)**.

23. The given data are 3, 3, 6, 7, 16.

$$\text{Mean} = \frac{\text{sum of all items}}{\text{number of items}} = \frac{3+3+6+7+16}{5} = \frac{35}{5} = 7$$

Median = middle item when items are arranged in order of size = 6
Mode = item occurring most frequently = 3
Of the choices given, the only one that is true is "mean > median."
The correct choice is **(1)**.

24. The given expression is: $5\sqrt{3} - \sqrt{27}$

Simplify the *radicands* (numbers under the radical sign) by factoring out any perfect square factors: $5\sqrt{3} - \sqrt{9(3)}$

Remove any perfect square factor from under the radical sign by taking its square root and writing it as the numerical coefficient outside the radical sign: $5\sqrt{3} - 3\sqrt{3}$

The radicals are now *like radicals* since they have the same index (both are square roots) and the same radicand (in this case, 3). Like radicals may be combined by combining their numerical coefficients: $2\sqrt{3}$

The correct choice is **(4)**.

25. Prepare a column in the truth table for each choice:

Column 1	Column 2	Column 3	(1)	(2)	(3)	(4)
p	q	?	$p \rightarrow q$	$p \leftrightarrow q$	$p \wedge q$	$p \vee q$
T	T	T	T	T	T	T
T	F	F	F	F	F	T
F	T	F	T	F	F	T
F	F	F	T	T	F	F

(1) $p \to q$ is the *implication* or *conditional:* if p, then q. It has the truth value T whenever q is T and also when both p and q are F; it has the value F when p is T and q is F.

(2) $p \leftrightarrow q$ is the *biconditional* or *equivalence relation* between p and q. It has the truth value *T* only when p and q have the same truth value, that is, both T or both F; if p and q have different truth values, the truth value of the biconditional is F.

(3) $p \wedge q$ is the *conjunction* of p and q. It has the truth value T only when both p and q have the truth value T; in all other cases, the conjunction has the truth value F. The truth values in the column for $p \wedge q$ are the same as those in column 3. Therefore the heading for column 3 should be $p \wedge q$.

(4) $p \vee q$ is the *disjunction* of p and q. It has the truth value T when either p or q or both have the value T; if both p and q have the value F, then the disjunction has the value F.

The correct choice is **(3)**.

26. Let the given statement, "If it is sunny, I will go swimming," be represented symbolically by the conditional, $p \to q$. Then p represents the *hypothesis* or *antecedent*, "It is sunny," and q represents the *conclusion* or *consequent*, "I will go swimming."

The *inverse* of a conditional is formed by *negating* both its hypothesis and conclusion. Therefore the inverse of $p \to q$ is $\sim p \to \sim q$.

When the meanings of $\sim p$ and $\sim q$ are substituted, the inverse becomes "If it is not sunny, I will not go swimming."

The correct choice is **(2)**.

27.

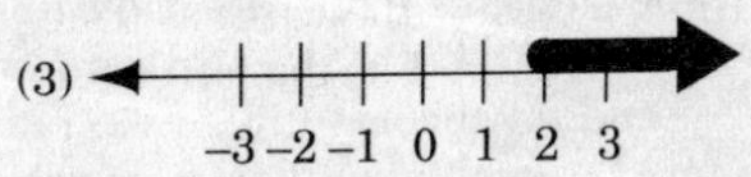

First solve the inequality: $2x + 1 \geq 3$

Add -1 (the additive inverse of 1) to both sides of the inequality: $\underline{\quad -1 = -1}$

$2x \geq 2$

Divide both sides of the inequality by 2: $\frac{2x}{2} \geq \frac{2}{2}$

$x \geq 1$

The value of x must be greater than or equal to 1. This is true in choice (1).
Choice (2) shows the set of numbers less than or equal to 1.
Choice (3) shows the set of numbers greater than or equal to 2.
Choice (4) shows the set of numbers less than or equal to 2.
The correct choice is **(1)**.

28. The given sentence, $3 + (5 + 2) = (5 + 2) + 3$, says that, when we add the quantity $(5 + 2)$ to 3, we get the same result as when we add 3 to the quantity $(5 + 2)$. This is the commutative property of addition.
The correct choice is **(1)**.

29. The given equation is a *quadratic equation*: $x^2 - 2x - 3 = 0$

The left side is a *quadratic trinomial* that can be factored into the product of two binomials. The factors of the first term, x^2, are x and x, and they become the first terms of the binomials: $(x \quad)(x \quad) = 0$

The factors of the last term, -3, become the second terms of the binomials, but they must be chosen in such a way that the sum of the inner and outer products is equal to the middle term, $-2x$, of the original trinomial. Try -3 and $+1$ as the factors of -3:

$-3x$ = inner product
$(x - 3)(x + 1) = 0$
$+x$ = outer product

Since $(-3x) + (+x) = -2x$, these are the correct factors: $(x - 3)(x + 1) = 0$

If the product of two factors equals 0, either factor may equal 0: $x - 3 = 0 \vee x + 1 = 0$

Add the appropriate additive inverse to both sides of the equations, 3 for the left equation, and -1 for the right:

$$\begin{array}{rr} 3 = 3 & -1 = -1 \\ \hline x \quad = 3 & x \quad = -1 \end{array}$$

The solution set is $\{3, -1\}$.
The correct choice is **(4)**.

30. p represents "All sides are congruent."

q represents "All angles are congruent."

$\sim q$ is the *negation* of q, that is, "All angles are *not* congruent." Note that some may be congruent.

$p \wedge \sim q$ is the *conjunction* of p and $\sim q$, that is, "All sides are congruent and all angles are not congruent." This is true for a rhombus but not true for any of the other choices:

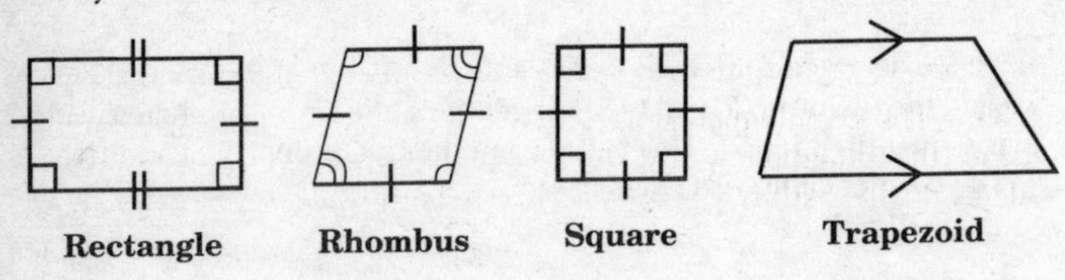

The correct choice is **(2)**.

31. Since division by 0 is undefined, the given expression:

$$\frac{1}{(x-1)(x+2)}$$

will be undefined if the denominator, $(x-1)(x+2)$, is equal to 0. The denominator will be equal to 0 if either $(x-1)$ or $(x+2)$ equals 0; $(x-1)$ will equal 0 if $x = 1$ and $(x+2)$ will equal 0 if $x = -2$.

The correct choice is **(2)**.

32. Let A be the set of all fives in the deck, and let B be the set of all diamonds. Then $P(A)$ is the probability of drawing a five, and $P(B)$ is the probability of drawing a diamond. The probability of drawing a five *or* a diamond is $P(A \cup B)$, where $A \cup B$ is the *union* of sets A and B.

$$P(A \cup B) = P(A) + P(B) - P(A \cap B)$$

where $(A \cap B)$ is the *intersection* of sets A and B.

The probability of an event occurring =

$$\frac{\text{number of successful outcomes}}{\text{total possible number of outcomes}}.$$

Since there are 4 fives in the deck, 1 in each suit, there are 4 successful outcomes for $P(A)$. The total possible number of outcomes is 52 since any one of the 52 cards in the deck may be drawn. Therefore, $P(A) = \frac{4}{52}$.

Since there are 13 diamonds in the deck, there are 13 successful outcomes for $P(B)$. Therefore, $P(B) = \frac{13}{52}$.

The intersection set, $A \cap B$, consists of cards that are both fives and diamonds. There is only 1 card in this set—the five of diamonds. The number of successful outcomes for $P(A \cap B)$ is 1. Therefore, $P(A \cap B) = \frac{1}{52}$

$$P(A \cup B) = P(A) + P(B) - P(A \cap B) = \frac{4}{52} + \frac{13}{52} - \frac{1}{52} = \frac{16}{52}$$

The correct choice is **(3)**.

33. The given equation is: $y - 2x = 4$

Put the equation in the $y = mx + b$ form by adding $2x$ (the additive inverse of $-2x$) to both sides:

$$\begin{aligned} y - 2x &= 4 \\ 2x &= 2x \\ \hline y \quad &= 2x + 4 \end{aligned}$$

If an equation of a line is in the form $y = mx + b$, then m represents its slope and b represents its y-intercept. The equation $y = 2x + 4$ is in the form $y = mx + b$, with $m = 2$ and $b = 4$. Therefore, the slope of the line is 2.

The correct choice is **(2)**.

34. The area, A, of a circle whose radius is r is given by this formula: $A = \pi r^2$

The radius of a circle is one-half its diameter. If the diameter is 10, then the radius, r, is 5:

$$A = \pi (5)^2$$
$$A = 25\pi$$

The correct choice is **(3)**.

35.

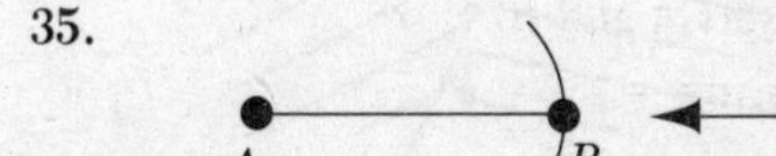
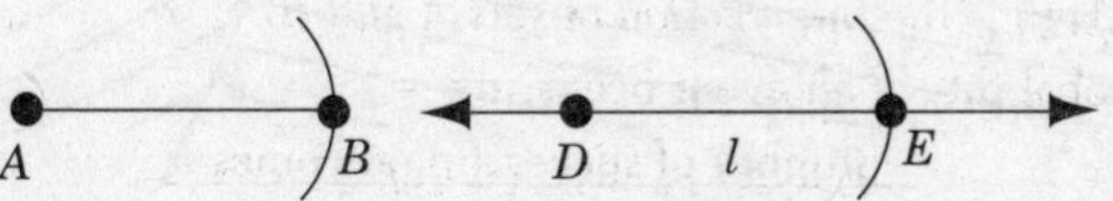

STEP 1: With the point of the compass on A, draw an arc passing through B.

STEP 2: Choose any point on line l and label it D. With the point of the compass on D and a radius equal to that used in Step 1, draw an arc that intersects line l in a point that should now be labeled E.

$\overline{DE}$ is the required segment on line l congruent to segment $\overline{AB}$.

PART TWO

36. a. STEP 1: To draw the graph of the inequality $x < 5$, first draw the graph of the equation $x = 5$. The graph of $x = 5$ is a line every point on which has an x-coordinate of 5. This is a line parallel to the y-axis and 5 units to the right of it. Draw it as a *broken* line to signify that the points on it are *not* part of the graph of $x < 5$.

The inequality $x < 5$ is satisfied by all points whose x-coordinates are less than 5. All such points lie in the region to the left of the line $x = 5$. Shade this region with cross-hatching extending to the left and down.

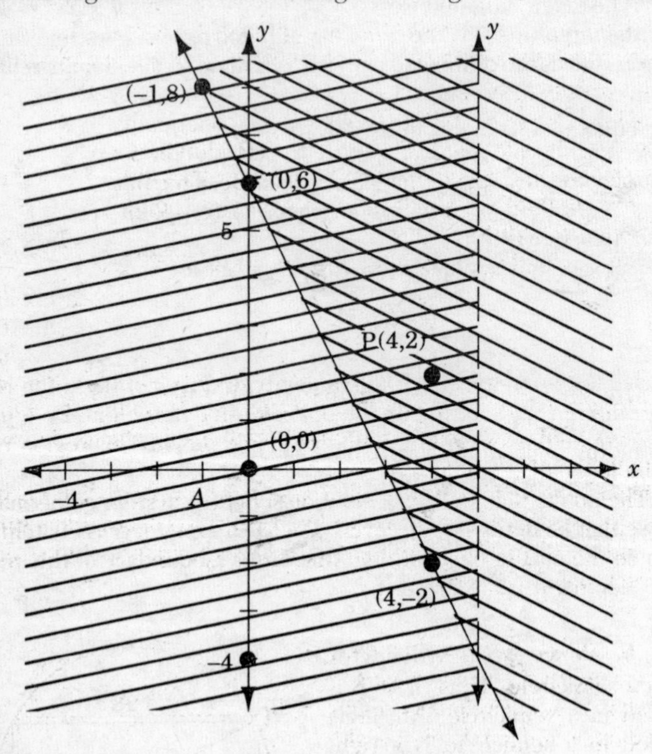

STEP 2: To draw the graph of the inequality $2x + y \geq 6$, first draw the graph of the line $2x + y = 6$. It is convenient to rearrange this equation so that it is solved for y by adding $-2x$ to both sides:

$$\begin{aligned} 2x + y &= 6 \\ -2x \quad &= -2x \\ \hline y &= 6 - 2x \end{aligned}$$

Prepare a table of values for x and y by selecting any three convenient values for x and substituting in the equation to compute the corresponding values of y:

x	$6 - 2x$	$= y$
-1	$6 - 2(-1) = 6 + 2$	$= 8$
0	$6 - 2(0) = 6 - 0$	$= 6$
4	$6 - 2(4) = 6 - 8$	$= -2$

Plot the points $(-1,8)$, $(0,6)$, and $(4,-2)$, and draw a *solid* line through them. This line is the graph of $2x + y = 6$. The solid line denotes that it is part of the graph of $2x + y \geq 6$.

The remaining part of the graph of the inequality $2x + y \geq 6$ is the inequality $2x + y > 6$. Its solution set consists of points in a region that lies on one side of the line $2x + y = 6$. To find out which side, choose a test point, say $(0,0)$, and substitute its coordinates in the inequality to see whether it is satisfied:

$$2x + y > 6$$
$$2(0) + 0 \overset{?}{>} 6$$
$$0 + 0 \overset{?}{>} 6$$
$$0 \not> 6$$

Since $(0,0)$ does not satisfy the inequality, it does not lie in the region representing $2x + y > 6$. Shade the *opposite* side of the line $2x + y = 6$ with cross-hatching extending to the right and down. This region represents the inequality $2x + y > 6$.

b. The points that are in the solution set of the system of inequalities are those that lie in the region covered by both types of cross-hatching or that lie on the part of the solid line that forms a boundary of this region.

One such point is P**(4,2)**.

37. a. Since $\overline{AB}$ is a diameter, $\overparen{ACB}$ is a semicircle. Thus, $\angle ACB$ is inscribed in a semicircle. An angle inscribed in a semicircle is a right angle.

$m\angle ACB =$ **90**.

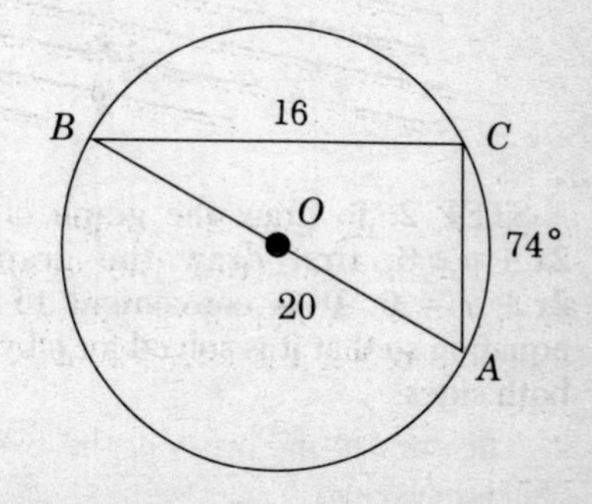

b. Since $\overparen{ACB}$ is a semicircle, $m\overparen{ACB} = 180$.

$m\overparen{BC} = m\overparen{ACB} - m\overparen{AC} = 180 - 74 = 106$.

Angle BAC is an inscribed angle; the measure of an inscribed angle is equal to one-half the measure of its intercepted arc:

$$m\angle BAC = \frac{1}{2}\, m\widehat{BC}$$
$$m\angle BAC = \frac{1}{2}(106)$$
$$m\angle BAC = 53$$

$m\angle BAC =$ **53**.

c. From part **a,** $\angle C$ is a right angle. Triangle ABC is a 3-4-5 right triangle with hypotenuse $AB = 4(5)$ and leg $BC = 4(4)$; the length of the other leg AC must be 4(3) or 12.

The length of $\overline{AC}$ is **12**.

d. The area, A, of a right triangle is one-half the product of the lengths of its legs. From part **c,** $AC = 12$ and $BC = 16$:

$$A = \frac{1}{2}(12)(16)$$
$$A = 6(16)$$
$$A = 96$$

The area of $\triangle ABC$ is **96**.

e. The area, A, of a circle whose radius is r is given by this formula:

$$A = \pi r^2$$

The radius of a circle is one-half its diameter.

Since diameter $AB = 20$, $r = 10$:

$$A = \pi(10)^2$$
$$A = 100\pi$$

The area of circle O is **100π**.

38. Let x = the length in feet of a side of the smaller garden.
Then $x + 3$ = the number of feet in a side of the larger garden.

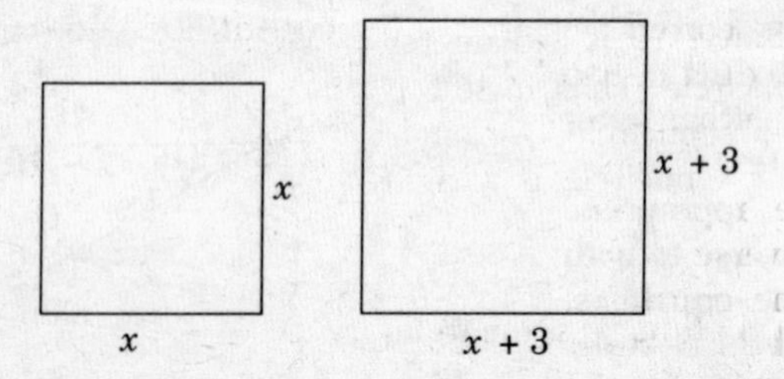

The sum of the areas of the two gardens is 269 square feet:

$$x^2 + (x + 3)^2 = 269$$

Multiply out $(x + 3)^2$:

$$\begin{array}{r} x + 3 \\ x + 3 \\ \hline x^2 + 3x \\ + 3x + 9 \\ \hline x^2 + 6x + 9 \end{array}$$

$$x^2 + x^2 + 6x + 9 = 269$$

Combine like terms:

$$2x^2 + 6x + 9 = 269$$

This is a *quadratic equation*. Rearrange it so that all terms are on one side equal to 0 by adding -269 (the additive inverse of 269) to both sides:

$$\begin{array}{r} -269 = -269 \\ \hline 2x^2 + 6x - 260 = 0 \end{array}$$

To simplify, divide all terms on both sides of the equation by 2:

$$x^2 + 3x - 130 = 0$$

The left side is a *quadratic trinomial* that can be factored into the product of two binomials. The factors of the first term, x^2, are x and x, and they become the first terms of the binomials:

$$(x \quad)(x \quad) = 0$$

The factors of the last term, -130, become the second terms of the binomials, but they must be chosen in such a way that the sum of the inner and outer products of the binomials equals the middle term, $+3x$, of the quadratic trinomial. Try $+13$ and -10 as the factors of -130:

$$+13x = \text{inner product}$$
$$(x + 13)(x - 10) = 0$$
$$-10x = \text{outer product}$$

Since $(+13x) + (-10x) = +3x$, these are the correct factors:

$$(x + 13)(x - 10) = 0$$

If the product of two factors is 0, either factor may equal 0:

$$x + 13 = 0 \quad \vee \quad x - 10 = 0$$

Add the appropriate additive inverse to both sides of the equations, -13 for the left equation and $+10$ for the right:

$$\begin{array}{r} -13 = -13 \\ \hline x \quad = -13 \end{array} \qquad \begin{array}{r} +10 = +10 \\ \hline x \quad = 10 \end{array}$$

Reject -13 as meaningless for a length:
$$x = 10$$
$$x + 3 = 13$$

A side of the smaller garden is **10** feet; a side of the larger garden is **13** feet.

39. a. The given system of equations is:
$$3x - 2y = 22$$
$$2x + 5y = 2$$

Multiply each term of the first equation by 5: $15x - 10y = 110$

Multiply each term of the second equation by 2: $4x + 10y = 4$

Add the equations to eliminate y: $19x = 114$

Divide both sides of the equation by 19: $\frac{19x}{19} = \frac{114}{19}$
$$x = 6$$

Substitute 6 for x in the original second equation:
$$2(6) + 5y = 2$$
$$12 + 5y = 2$$

Add -12 (the additive inverse of 12) to both sides of the equation:
$$-12 \quad = -12$$
$$5y = -10$$

Divide both sides of the equation by 5: $\frac{5y}{5} = \frac{-10}{5}$
$$y = -2$$

The solution is **(6, −2)** or $\boldsymbol{x = 6, y = -2}$.

CHECK: Substitute 6 for x and -2 for y in *both* of the *original* equations to see whether both are satisfied:

$$\underline{3x - 2y = 22} \qquad \underline{2x + 5y = 2}$$
$$3(6) - 2(-2) \stackrel{?}{=} 22 \qquad 2(6) + 5(-2) \stackrel{?}{=} 2$$
$$18 + 4 \stackrel{?}{=} 22 \qquad 12 - 10 \stackrel{?}{=} 2$$
$$22 = 22 \checkmark \qquad 2 = 2 \checkmark$$

40. a. (1) Since $ABCD$ is a parallelogram, its opposite sides are congruent. Therefore,

$$AB = DC = 2x + 2 \quad \text{and} \quad BC = AD = x.$$

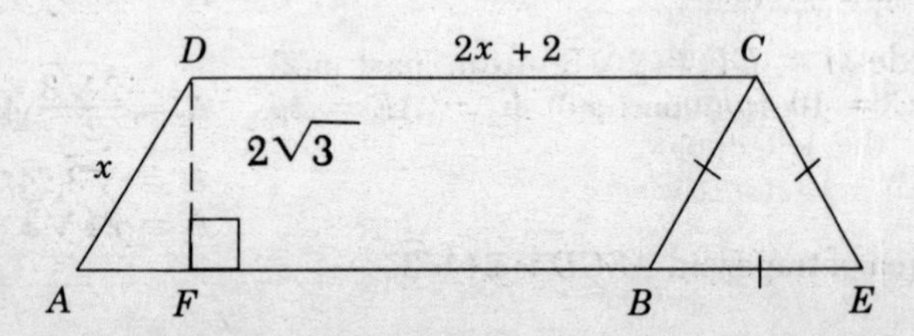

The perimeter of parallelogram $ABCD$ is 28: $x + 2x + 2 + x + 2x + 2 = 28$

Combine like terms: $6x + 4 = 28$

Add -4 (the additive inverse of 4) to both sides of the equation: $-4 = -4$

$6x = 24$

Divide both sides of the equation by 6: $\frac{6x}{6} = \frac{24}{6}$

$x = 4$

The measure of $\overline{AD}$ is **4**.

(2) $DC = 2x + 2 = 2(4) + 2 = 8 + 2 = 10$

The measure of DC is **10**.

(3) Since $\triangle BEC$ is equilateral, $BE = BC = x = 4$.

$AB = DC = 10$

$AE = AB + BE = 10 + 4 = 14$

The measure of AE is **14**.

b. The area, A, of an equilateral triangle whose side is s is given by this formula: $A = \frac{s^2\sqrt{3}}{4}$

Here, $s = BE = 4$: $A = \frac{(4)^2\sqrt{3}}{4}$

$A = \frac{16\sqrt{3}}{4}$

$A = 4\sqrt{3}$

The area of $\triangle BEC$ is $\mathbf{4\sqrt{3}}$.

ALTERNATIVE SOLUTION: The area of $\triangle BEC$ may also be calculated by using the formula for the area, A, of any triangle with base $= b$ and altitude $= h$: $A = \frac{1}{2}bh$

The altitude of $\triangle BEC$ from C is equal to DF or $h = 2\sqrt{3}$; the base $BE = b = 4$: $A = \frac{1}{2}(4)(2\sqrt{3})$

$A = 4\sqrt{3}$

c. The area, A, of a trapezoid whose altitude is h and the lengths of whose bases are b_1 and b_2 is given by this formula: $A = \frac{h}{2}(b_1 + b_2)$

Altitude $h = DF = 2\sqrt{3}$; from part **a**(2), $b_1 = DC = 10$; from part **a**(3), $b_2 = AE = 14$: $A = \frac{2\sqrt{3}}{2}(10 + 14)$

$A = \sqrt{3}(24)$

$A = 24\sqrt{3}$

The area of trapezoid $AECD$ is $\mathbf{24\sqrt{3}}$.

41. a. The tree diagram will have four primary branches radiating from START to represent the first selection from chips numbered 1, 2, 3, and 4. Each primary branch will have three secondary branches representing the second selection from among the three remaining chips.

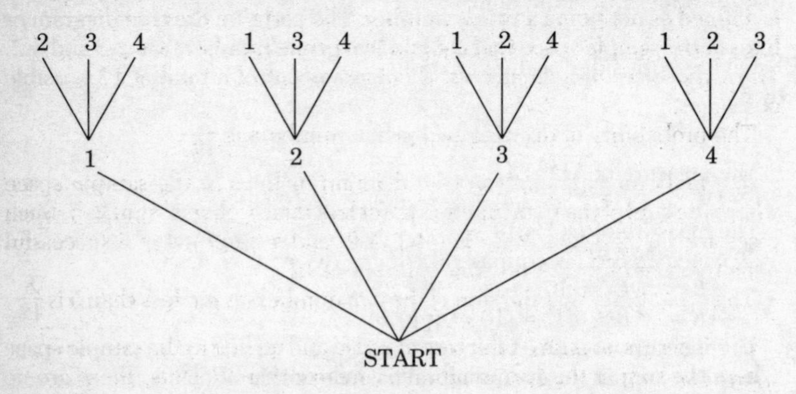

The sample space will have 12 lines since each of the four possible first choices may be followed by any one of the three remaining chips for the second choice:

First Choice	Second Choice
1	2
1	3
1	4
2	1
2	3
2	4
3	1
3	2
3	4
4	1
4	2
4	3

b. Probability of an event occurring =

$$\frac{\text{number of successful outcomes}}{\text{total possible number of outcomes}}.$$

Examine either paths on the tree diagram or lines in the sample space to discover possible combinations of an even and an odd number in the same path or line. These combinations are 1-2, 1-4, 2-1, 2-3, 3-2, 3-4, 4-1, and 4-3, making 8 successful outcomes. Since there are 12 possible paths through the two selections on the tree diagram, or 12 lines in the sample space, the total possible number of outcomes is 12.

The probability of one number odd and the other even $= \frac{8}{12}$.

c. The number 4 is not a prime number since it is divisible by 2, and 1 is defined as not being a prime number. The paths on the tree diagram or lines in the sample space that contain two prime numbers are 2-3 and 3-2. Thus, there are only 2 successful outcomes out of a total of 12 possible outcomes.

The probability of drawing two prime numbers is $\frac{2}{12}$.

d. Look for paths on the tree diagram or lines in the sample space where the sum of the two numbers is *not* less than 5, that is, sum ≥ 5. Such cases are 1-4, 2-3, 2-4, 3-2, 3-4, 4-1, 4-2, and 4-3, a total of 8 successful outcomes out of a total possible number of 12.

The probability that the sum of the two numbers is *not* less than 5 is $\frac{8}{12}$.

e. There is no path on the tree diagram and no line in the sample space where the sum of the two numbers is greater than 7. Thus, there are no successful outcomes.

The probability that the sum of the two numbers is greater than 7 is $\frac{0}{12}$ or **0**.

42. **a.**

p	q	$p \to q$	$\sim q$	$p \wedge \sim q$	$\sim(p \wedge \sim q)$	$(p \to q) \leftrightarrow \sim(p \wedge \sim q)$
T	T	T	F	F	T	T
T	F	F	T	T	F	T
F	T	T	F	F	T	T
F	F	T	T	F	T	T

Fill in each line according to the values for p and q shown on the corresponding line in the first two columns.

$p \to q$ is the *implication* or *conditional:* if p, then q. It has the truth value T whenever q has the value T and also when both p and q are F; it has the value F when p is T and q is F.

$\sim q$ is the *negation* of q. The truth value of $\sim q$ is always the opposite of that for q.

$p \wedge \sim q$ is the *conjunction* of p and $\sim q$. The conjunction has the truth value T when both p and $\sim q$ have the truth value T; if either p or $\sim q$ or both are F, the conjunction is F.

$\sim(p \wedge \sim q)$ is the *negation* of the conjunction just discussed above. The truth value of the negation is always the opposite of that of $(p \wedge \sim q)$.

$(p \rightarrow q) \leftrightarrow \sim(p \wedge \sim q)$ is the *biconditional* or *equivalence relation* between $(p \rightarrow q)$ and $\sim(p \wedge \sim q)$. When the truth values of $(p \rightarrow q)$ and $\sim(p \wedge \sim q)$ are the same, that is, when both are T or both are F, the truth value of the biconditional is T; otherwise it is F.

b. p represents "Sue lives in Buffalo."
q represents "Sue lives in New York State."

"If Sue lives in Buffalo, then Sue lives in New York State" is the conditional represented by $p \rightarrow q$.

Consider each choice in turn:

(1) "Sue lives in Buffalo and Sue does not live in New York State" is the conjunction $(p \wedge \sim q)$, where $\sim q$ is the negation of q.

(2) "It is false that Sue lives in Buffalo and Sue does not live in New York State" is the negation of the conjunction above, that is, $\sim(p \wedge \sim q)$.

(3) "Sue lives in Buffalo or Sue does not live in New York State" is the disjunction $(p \vee \sim q)$.

(4) "Sue does not live in Buffalo and Sue does not live in New York State" is the conjunction $(\sim p \wedge \sim q)$.

Since the equivalence relation $(p \rightarrow q) \leftrightarrow \sim(p \wedge \sim q)$ was shown in part **a** to have the truth value T for all possible combinations of values of p and q, $(p \rightarrow q)$ is equivalent to $\sim(p \wedge \sim q)$.

The correct choice is **(2)**.

Topic	Question Numbers	Number of Points	Your Points	Your Percentage
1. Numbers (rat'l, irrat'l); Percent	4	2		
2. Properties of No. Systems	28, 31	2 + 2 = 4		
3. Operations on Rat'l Nos. and Monomials	18, 20	2 + 2 = 4		
4. Operations on Multinomials	14, 19	2 + 2 = 4		
5. Square root; Operations involving Radicals	24	2		
6. Evaluating Formulas and Expressions	15	2		
7. Linear Equations (simple cases incl. parentheses)	—	0		
8. Linear Equations Containing Decimals or Fractions	7, 8	2 + 2 = 4		
9. Graphs of Linear Functions (slope)	33, 36a	2 + 8 = 10		
10. Inequalities	27	2		
11. Systems of Eqs. & Inequal. (alg. & graphic solutions)	16, 36b, 39	2 + 2 + 10 = 14		
12. Factoring	17	2		
13. Quadratic Equations	29	2		
14. Verbal Problems	6, 9, 38	2 + 2 + 10 = 14		
15. Variation	—	0		
16. Literal Eqs.; Expressing Relations Algebraically	10, 21	2 + 2 = 4		
17. Factorial n	22	2		
18. Areas, Perims., Circums., Vols. of Common Figures	34, 37d, e, 40a, b, c	2 + 2 + 2 + 6 + 2 + 2 = 16		
19. Geometry ($\cong$, $\angle$ meas., $\parallel$ lines, compls., suppls., const.)	2, 5, 11, 13, 35, 37a, b	2 + 2 + 2 + 2 + 2 + 2 + 2 = 14		
20. Ratio & Proportion (incl. similar triangles)	3	2		
21. Pythagorean Theorem	37c	2		
22. Logic (symbolic rep., logical forms, truth tables)	25, 26, 30, 42a, b	2 + 2 + 2 + 8 + 2 = 16		
23. Probability (incl. tree diagrams & sample spaces)	32, 41a, b, c, d, e	2 + 3 + 2 + 2 + 2 + 1 = 12		
24. Combinatorics (arrangements, permutations)	1	2		
25. Statistics (central tend., freq. dist., histograms)	12, 23	2 + 2 = 4		

Examination June 1989

Three-Year Sequence for High School Mathematics—Course I

PART ONE

DIRECTIONS: *Answer 30 questions from this part. Each correct answer will receive 2 credits. No partial credit will be allowed. Write your answers in the spaces provided. Where applicable, answers may be left in terms of π or in radical form.*

1 Find the measure of a base angle of an isosceles triangle if the measure of the vertex angle is 100. 1_____

2 If 0.000023 is expressed in the form 2.3×10^n, what is the value of n? 2_____

3 If a letter is selected at random from the word "PARALLEL," what is the probability that the letter selected will *not* be an L? 3_____

4 In the accompanying diagram, $\overleftrightarrow{AB}$ and $\overleftrightarrow{CD}$ are parallel and $\overleftrightarrow{EF}$ intersects $\overleftrightarrow{AB}$ at G and $\overleftrightarrow{CD}$ at H. If m∠AGH = 80, what is m∠CHG? 4_____

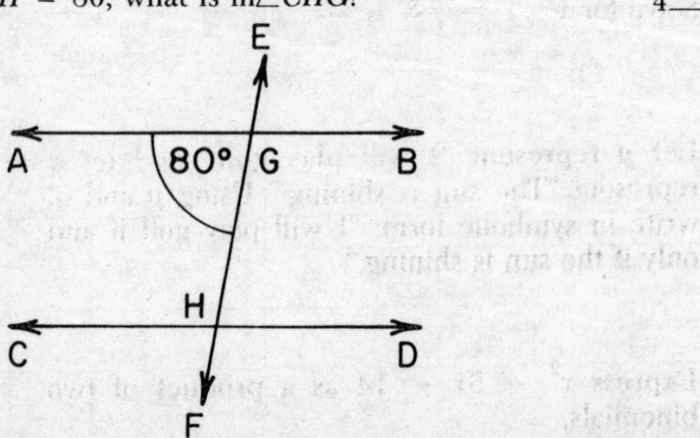

5 If y varies directly as x and $y = 32$ when $x = 4$, find the value of y when $x = 5$. 5_____

6 Find the value of $3(ab)^2$ if $a = 2$ and $b = -1$. 6_____

7 In rectangle $ABCD$, diagonal $AC = x + 10$ and diagonal $BD = 2x - 30$. Find the value of x. 7_____

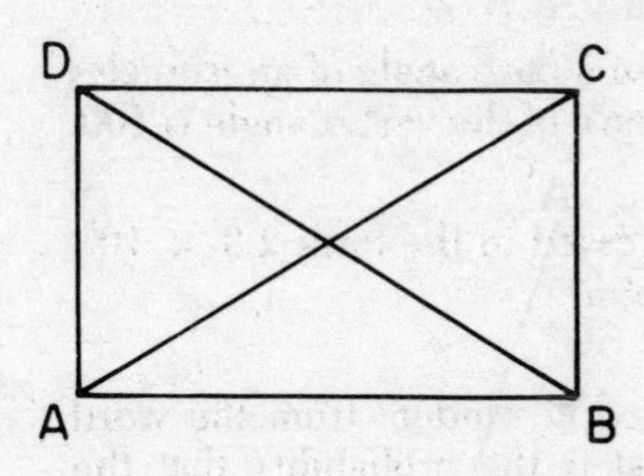

8 Solve for h: $2.3h - 1.9 = 5$ 8_____

9 Solve for x: $\frac{x - 5}{4} = \frac{1}{2}$ 9_____

10 Let p represent "I will play golf" and let q represent "The sun is shining." Using p and q, write in symbolic form: "I will play golf if and only if the sun is shining." 10_____

11 Express $x^2 - 5x - 14$ as a product of two binomials. 11_____

12 Two angles are complementary. The measure of one angle is twice the measure of the other angle. What is the measure of the *smaller* of the two angles? 12_____

13 In the accompanying diagram, $\angle ACD$ is an exterior angle of $\triangle ABC$. If $m\angle A = 60$ and $m\angle B = 50$, find $m\angle ACD$.

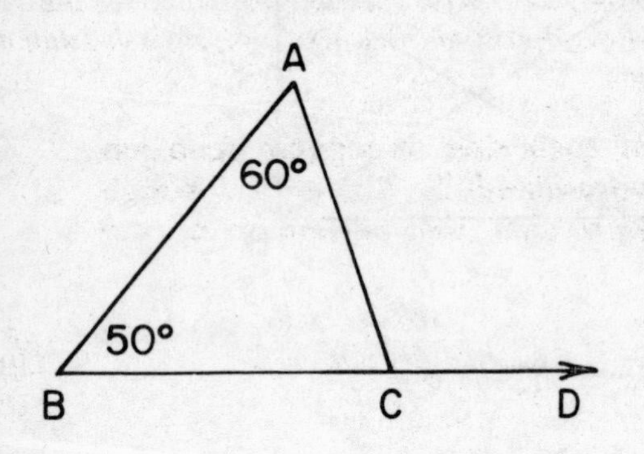

13_____

14 Solve for x in terms of b: $2x + b = 3$ 14_____

15 How many different two-digit numbers greater than 50 can be formed from the digits 3, 6, 7, and 9 if repetition of digits is allowed? 15_____

16 Find the area of the square whose vertices are (0,0), (3,0), (3,3), and (0,3). 16_____

17 The probability of a particular manufactured product being defective is $\frac{1}{100}$. How many defective products would be expected in a random sample of 1500 products? 17_____

18 Express the sum $\frac{1}{2x} + \frac{3}{8x}$, $x \neq 0$, as a single fraction in lowest terms. 18_____

DIRECTIONS: **(19–35)**: *For* each *question chosen, write the* numeral *preceding the word or expression that best completes the statement or answers the question.*

19 The length of each side of regular pentagon *ABCDE* is represented by $(3x + 1)$. Which expression represents the perimeter of the pentagon?

(1) $15x + 5$ (3) $3x + 5$
(2) $18x + 6$ (4) $15x + 1$ 19_____

20 The fraction $\frac{-12x^3y^5}{3xy^2}$, $x \neq 0$, $y \neq 0$, is equivalent to

(1) $-4x^4y^7$ (3) $4x^2y^3$
(2) $\frac{-4x^2}{y^3}$ (4) $-4x^2y^3$ 20_____

21 The product of $3a^2$ and $2a^4$ is

(1) $5a^8$ (3) $6a^6$
(2) $6a^8$ (4) $5a^6$ 21_____

22 Which is the *smallest* integer that makes the inequality $2x + 3 > 5$ true?

(1) 1
(2) 2
(3) 5
(4) −4

22_____

23 Which binomial is equivalent to $3(x - 1) - 2(x - 3)$?

(1) $x - 7$
(2) $5x - 7$
(3) $x + 5$
(4) $x + 3$

23_____

24 In which graph does the slope of line ℓ equal zero?

(1)

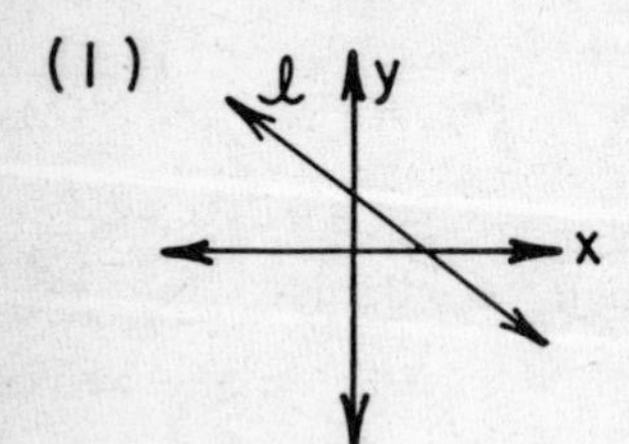

(3)

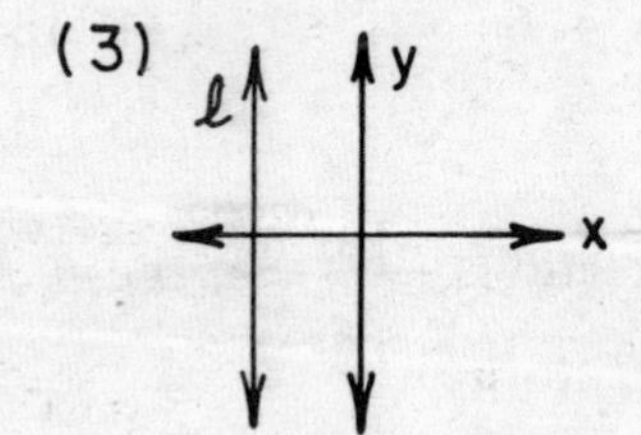

(2)

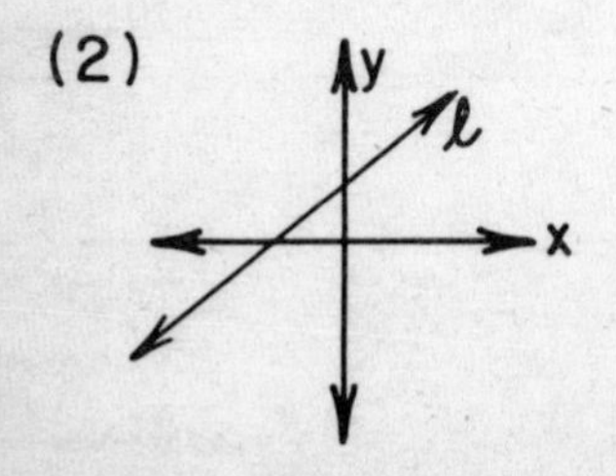

(4)

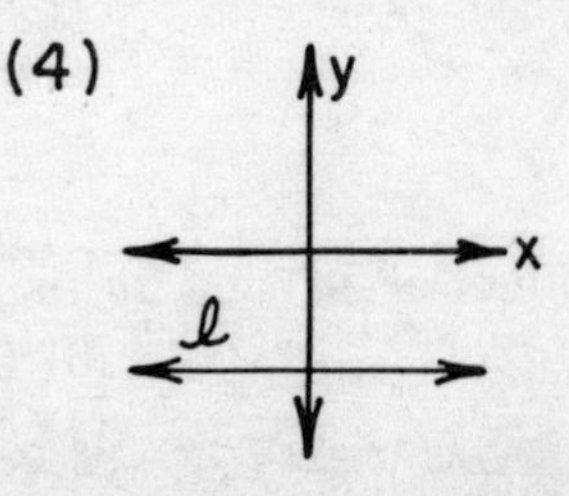

24_____

25 Triangle ABC is a right triangle with legs that measure 7 and 8. The length of the hypotenuse is

(1) $\sqrt{15}$ (3) 9
(2) $\sqrt{113}$ (4) 15

25______

26 What is the solution set of the equation $(x - a)(x + b) = 0$?

(1) $\{a,-b\}$ (3) $\{-a,-b\}$
(2) $\{-a,b\}$ (4) ϕ

26______

27 Which statement would be a correct heading for the last column in the table below?

p	q	?
T	T	F
T	F	F
F	T	T
F	F	F

(1) $p \rightarrow q$ (3) $\sim p \wedge q$
(2) $p \vee q$ (4) $p \leftrightarrow q$

27______

28 In the accompanying diagram, which point may be the image of point A after a line reflection in the x-axis?

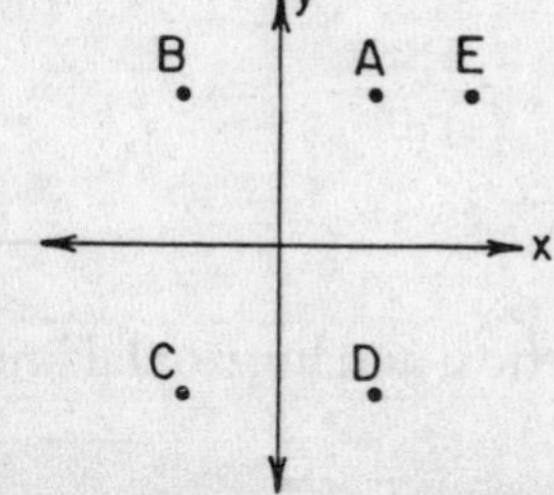

(1) B (3) D
(2) C (4) E

28______

29 Let p represent "The number is an even integer" and let q represent "Three times the number is 12." Which statement is true if the number is 6?

(1) $p \wedge q$ (3) q
(2) $p \vee q$ (4) $p \rightarrow q$ 29_____

30 Larry has 7 more dimes than nickels, for a total value of $1.45. If n represents the number of nickels, which equation could be used to find the number of nickels Larry has?

(1) $n + (n + 7) = 145$
(2) $5n + 5(n + 7) = 145$
(3) $5n + 10(n + 7) = 145$
(4) $15(n + n + 7) = 145$ 30_____

31 Which fraction would be undefined if $x = -3$?

(1) $\frac{3}{2x + 6}$ (3) $\frac{x + 3}{x - 3}$
(2) $\frac{2x + 6}{3}$ (4) $\frac{3 + x}{3 - x}$ 31_____

32 The sum of two consecutive integers is −1. The smaller integer is

(1) 1 (3) −1
(2) −2 (4) 0 32_____

33 Which letter has both vertical and horizontal line symmetry?

(1) E (3) T
(2) M (4) X 33_____

34 The scores on a test were 70, 75, 75, 85, and 90. Which statement about these scores is true?

(1) The mean, median, and mode have the same value.
(2) Only the mean and the median have the same value.
(3) Only the mode and the median have the same value.
(4) The mean, median, and mode have different values.

34_____

35 The expression $2\sqrt{5}$ is equivalent to

(1) $\sqrt{10}$ (3) $\sqrt{50}$
(2) $\sqrt{20}$ (4) $\sqrt{100}$

35_____

PART TWO

DIRECTIONS: *Answer* four *questions from this part. Show all work unless otherwise directed.*

36 *a* On the same set of coordinate axes, graph the lines of the following equations:

(1) $x + y = 8$ [3]
(2) $x = 1$ [2]
(3) $y = 3$ [2]

b Find the area of the triangle formed by the three lines graphed in part *a*. [3]

37 If five times the square of a certain positive number is decreased by twice the number, the result is 16. Find the number. [*Only an algebraic solution will be accepted.*] [5,5]

38 A bag contains only red, blue, and white marbles. The ratio of red marbles to white marbles is 3:1. There are 8 more blue marbles than white marbles.

a If x represents the number of white marbles, express, in terms of x:

(1) the number of red marbles [1]
(2) the number of blue marbles [1]

b Express, as a binomial in terms of x, the total number of marbles in the bag. [2]

c If there are 38 marbles in the bag,
(1) how many of them are red? [4]
(2) what is the probability that a marble, selected at random, is *not* blue? [2]

39 In $\triangle ABC$, the measure of $\angle B$ is 13 more than the measure of $\angle A$, and the measure of $\angle C$ is 9 less than twice the measure of $\angle A$. Find the measure of each angle in $\triangle ABC$. [*Only an algebraic solution will be accepted.*] [5,5]

40 Solve the following system of equations algebraically and check:

$$2x + 5y = -1$$
$$-3x + y = 10$$

[8,2]

41 The table below represents the distribution of the SAT scores of 60 students at State High School.

Scores	Frequency	Cumulative Frequency
710–800	4	
610–700	10	
510–600	15	
410–500	18	
310–400	11	
210–300	2	2

a *On your answer paper*, copy the table and complete the cumulative frequency column. [2]

b Using the table completed in part *a*, draw a cumulative frequency histogram. [4]

c Which interval contains the upper quartile? [2]

d If a student is selected at random, what is the probability of choosing a student who scored higher than 500? [2]

42 Answer *both a* and *b*.

a In each diagram below, quadrilateral $A'B'C'D'$ is the image of quadrilateral $ABCD$ under a transformation in the plane. *On your answer paper*, write the numerals 1 through 3, and after each numeral, identify the type of transformation as a dilation, a translation, a rotation or a line reflection. [3]

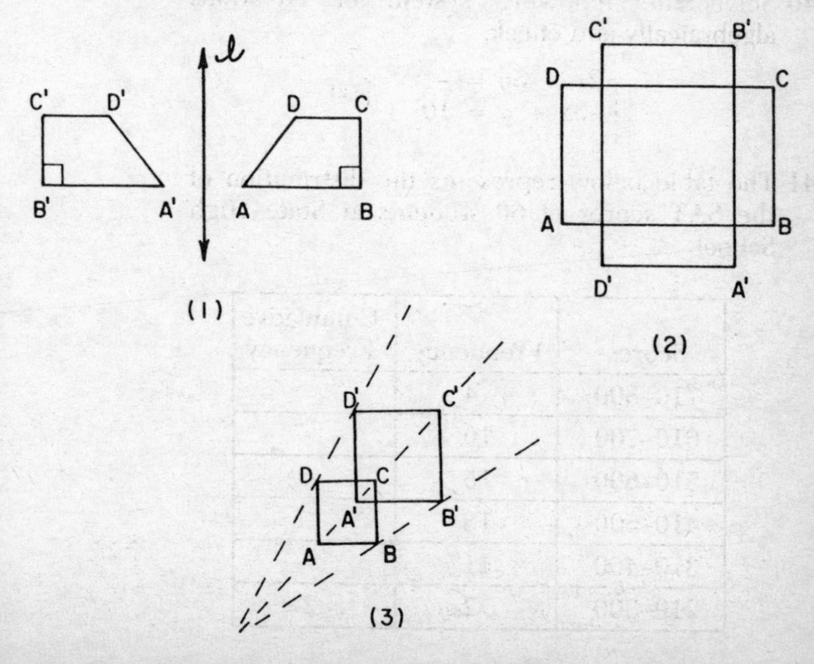

b (1) *On your answer paper,* copy and complete the truth table for the statement $[p \wedge (p \rightarrow q)] \rightarrow q$. [6]

p	q	$p \rightarrow q$	$p \wedge (p \rightarrow q)$	$[p \wedge (p \rightarrow q)] \rightarrow q$
T	T			
T	F			
F	T			
F	F			

(2) Is $[p \wedge (p \rightarrow q)] \rightarrow q$ a tautology? [1]

Answers June 1989

Three-Year Sequence for High School Mathematics—Course I

ANSWER KEY

PART ONE

1. 40
2. –5
3. $\frac{5}{8}$
4. 100
5. 40
6. 12
7. 40
8. 3
9. 7
10. $p \leftrightarrow q$
11. $(x - 7)(x + 2)$
12. 30
13. 110
14. $\frac{3 - b}{2}$
15. 12
16. 9
17. 15
18. $\frac{7}{8x}$
19. (1)
20. (4)
21. (3)
22. (2)
23. (4)
24. (4)
25. (2)
26. (1)
27. (3)
28. (3)
29. (2)
30. (3)
31. (1)
32. (3)
33. (4)
34. (3)
35. (2)

Part Two—*See* Answers Explained.

ANSWERS EXPLAINED

PART ONE

1. Let x = the measure of one of the base angles.

The base angles of an isosceles triangle are equal in measure; therefore, the measure of the other base angle is also x.

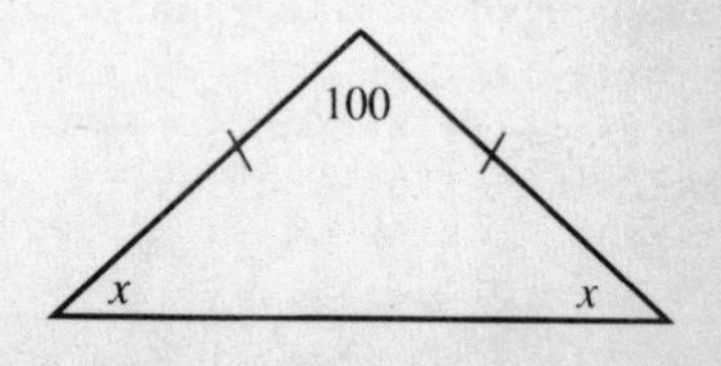

The sum of the measures of the three angles of a triangle is 180°:	$x + x + 100 = 180$
Combine like terms:	$2x + 100 = 180$
Add –100 (the additive inverse of 100) to both sides of the equation:	$-100 = -100$
	$2x = 80$
Divide both sides of the equation by 2:	$\frac{2x}{2} = \frac{80}{2}$
The measure of a base angle is **40:**	$x = 40$

2. Dividing a number by 10 moves the decimal point one place to the left. For 2.3×10^n to equal 0.000023, the decimal point in 2.3 must be moved five places to the left, that is, 2.3 must be divided by 10 five times, or multiplied by 10^{-5}.

$0.000023 = 2.3 \times 10^{-5}$ or $n = -5$.

$n =$ **–5.**

3. The probability of an event occurring

$$= \frac{\text{the possible number of successes}}{\text{the total possible number of outcomes}}.$$

For the selection from the word "PARALLEL" of a letter that will *not* be an L, there can be 5 successes: P, A, R, A, or E. The total possible number of outcomes is 8, the total number of letters in the word "PARALLEL."

The probability of selecting a letter that is *not* an L is $\frac{5}{8}$.

The probability is $\mathbf{\frac{5}{8}}$.

4. If a transversal cuts two parallel lines, two interior angles on the same side of the transversal will be supplementary: $\angle AGH$ is the supplement of $\angle CHG$.

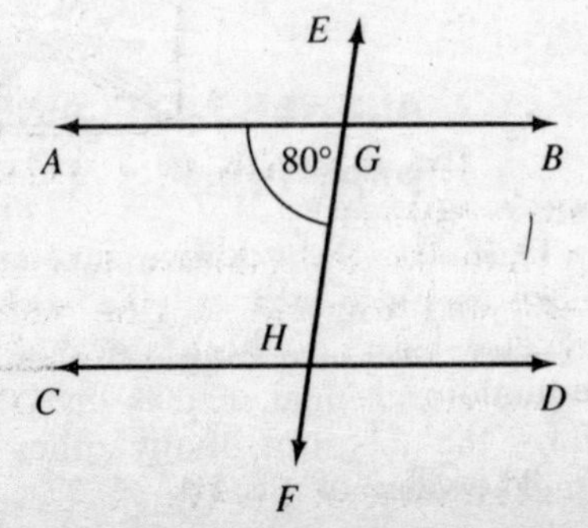

The sum of the measures of two supplementary angles is 180°: $m\angle AGH + m\angle CHG = 180$

Since m∠AGH = 80, we have: $80 + m\angle CHG = 180$

Add −80 (the additive inverse of 80) to both sides of the equation: $-80 \quad = -80$

$m\angle CHG = 100$

m∠*CHG* = **100.**

5. If y varies directly as x, then the relationship between y and x can be written as: $y = kx$

where k is a constant.

Given that $y = 32$ when $x = 4$: $32 = 4k$

Divide both sides of the equation by 4: $\frac{32}{4} = \frac{4k}{4}$

$8 = k$

Since k is a constant, the value of k remains 8 when 5 is substituted for x: $y = 8(5)$

y = **40** when $x = 5$. $y = 40$

6. The given expression is: $3(ab)^2$

Substitute 2 for a and −1 for b: $3([2][-1])^2$

Multiply the inner brackets first: $3(-2)^2$

Since $(-2)^2 = 4$, we have: $3(4)$

The value of $3(ab)^2$ is **12.** 12

7.

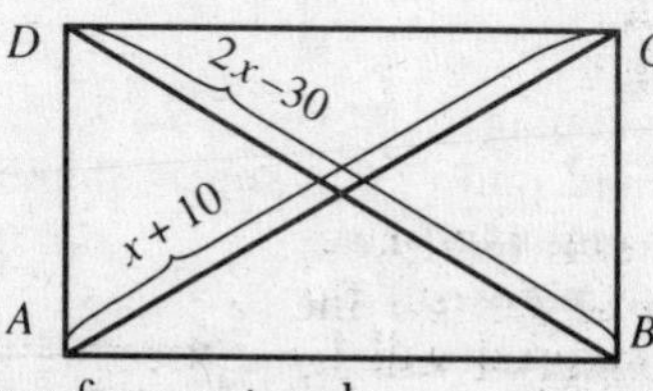

The diagonals of a rectangle are congruent: $2x - 30 = x + 10$

Add 30 (the additive inverse of −30) and also add $-x$ (the additive inverse of x) to both sides of the equation: $-x + 30 = -x + 30$

$x = 40$

The value of x is **40.**

8. The given equation contains decimals:

$$2.3h - 1.9 = 5$$

To clear decimals, multiply each term on both sides of the equation by 10:

$$10(2.3h) - 10(1.9) = 10(5)$$

$$23h - 19 = 50$$

Add 19 (the additive inverse of −19) to both sides of the equation:

$$\underline{\quad 19 = 19}$$

$$23h = 69$$

Divide both sides of the equation by 23:

$$\frac{23h}{23} = \frac{69}{23}$$

$$h = 3$$

h = **3.**

9. The given equation contains fractions:

$$\frac{x-5}{4} = \frac{1}{2}$$

Clear fractions by multiplying each term on both sides of the equation by the least common multiple (L.C.M.) of the denominators. The L.C.M is the smallest number into which all the denominators will divide evenly: the L.C.M. for 4 and 2 is 4:

$$4\left(\frac{x-5}{4}\right) = 4\left(\frac{1}{2}\right)$$

$$x - 5 = 2$$

Add 5 (the additive inverse of −5) to both sides of the equation:

$$\underline{5 = 5}$$

$$x = 7$$

x = **7.**

10. p = "I will play golf."
q = "The sun is shining."

"I will play golf if and only if the sun is shining" is the *biconditional* or *equivalence relation*, $p \leftrightarrow q$.

The symbolic form is $\mathbf{p \leftrightarrow q}$.

11. The given expression is a *quadratic trinomial:*

$$x^2 - 5x - 14$$

A quadratic trinomial can be factored into the product of two binomials. The factors of the first term, x^2, are x and x, and they become the first terms of the binomials:

$$(x \quad)(x \quad)$$

The factors of the last term, –14, become the second terms of the binomials, but they must be chosen in such a way that the sum of the inner and outer cross products equals the middle term, $-5x$, of the original trinomial. Try –7 and +2 as the factors of –14:

$$-7x = \text{inner product}$$
$$(x - 7)(x + 2)$$
$$+2x = \text{outer product}$$

Since $(-7x) + (+2x) = -5x$, these are the correct factors:

$$(x - 7)(x + 2)$$

The trinomial $x^2 - 5x - 14$ can be expressed as the product of two binomials $\mathbf{(x - 7)(x + 2)}$.

12. Let x = the measure of the smaller angle.

Then $2x$ = the measure of the larger angle.

The sum of the measures of two complementary angles is 90°:

$$x + 2x = 90$$

Combine like terms:

$$3x = 90$$

Divide both sides of the equation by 3:

$$\frac{3x}{3} = \frac{90}{3}$$

$$x = 30$$

The measure of the *smaller* angle is **30.**

13. Angle ACD is an exterior angle of ΔABC.

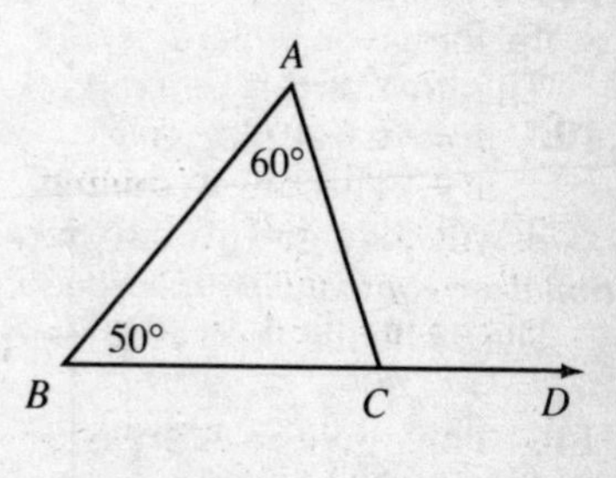

The measure of an exterior angle of a triangle is equal to the sum of the measures of the two remote interior angles:

$$m\angle ACD = m\angle A + m\angle B$$

Since $m\angle A = 60$, and $m\angle B = 50$ we have:

$$m\angle ACD = 60 + 50$$
$$m\angle ACD = 110$$

$m\angle ACD$ = **110.**

14. The given equation is a literal equation:

$$2x + b = 3$$

Add $-b$ (the additive inverse of b) to both sides of the equation:

$$\begin{array}{r} 2x + b = 3 \\ -b = -b \\ \hline 2x \quad = 3 - b \end{array}$$

Divide both sides of the equation by 2:

$$\frac{2x}{2} = \frac{3-b}{2}$$

$$x = \frac{3-b}{2}$$

In terms of b, $x = \mathbf{\frac{3-b}{2}}$.

15. The problem of forming a two-digit number greater than 50 by using the digits 3, 6, 7, and 9 is equivalent to making choices to fill each of 2 places: — — .

Since 6, 7, or 9 (but not 3) may be used to fill the first place, there are 3 possible ways to fill it. Any one of the numbers 3, 6, 7, or 9 may be used for the second place so there are 4 possible ways to fill it.

Any of the 3 possible ways for the first place may be paired with any of the 4 possible ways for the second place, so the total possible number of two-digit numbers greater than 50 is 3×4 or 12.

12 different two-digit numbers greater than 50 are possible.

16. The area, A, of a square whose side has a length s is given by this formula: $A = s^2$

In the given square, $s = 3$: $A = 3^2$

The area of the square is **9.** $A = 9$

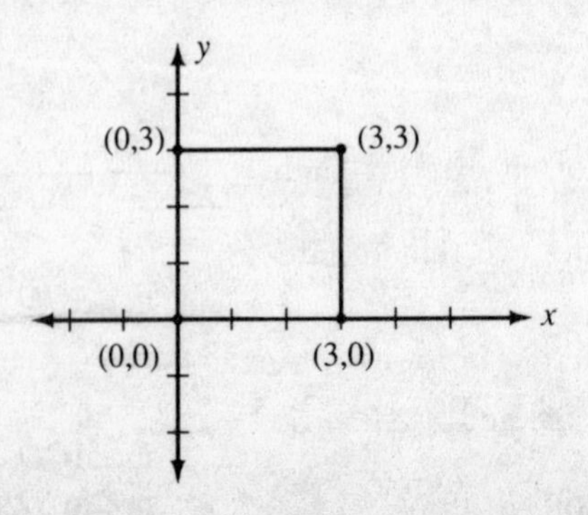

17. The probability of an event occurring

$$= \frac{\text{the possible number of successes}}{\text{the total possible number of outcomes}}.$$

If the probability that a product is defective is $\frac{1}{100}$, this means that there is one success (a defective product) out of every 100 outcomes (number of products manufactured). In 1,500 products it could be expected that

$$1{,}500 \div 100, \text{ or } 15,$$

would be defective.

15 products would be expected to be defective.

18. The given expression is: $\frac{1}{2x} + \frac{3}{8x} \quad (x \neq 0)$

To be combined, fractions must have a common denominator. The least common denominator (L.C.D.) is the smallest number into which all of the denominators will divide evenly. The L.C.D. for $2x$ and $8x$ is $8x$. Change the first fraction into an equivalent fraction having the L.C.D. by multiplying it by 1 in the form $\frac{4}{4}$:

$$\frac{4(1)}{4(2x)} + \frac{3}{8x}$$

$$\frac{4}{8x} + \frac{3}{8x}$$

Fractions having the same denominator may be combined by combining their numerators:

$$\frac{4+3}{8x}$$

The single fraction in lowest terms is $\mathbf{\frac{7}{8x}}$.

$$\frac{7}{8x}$$

19. A regular pentagon has five congruent sides.

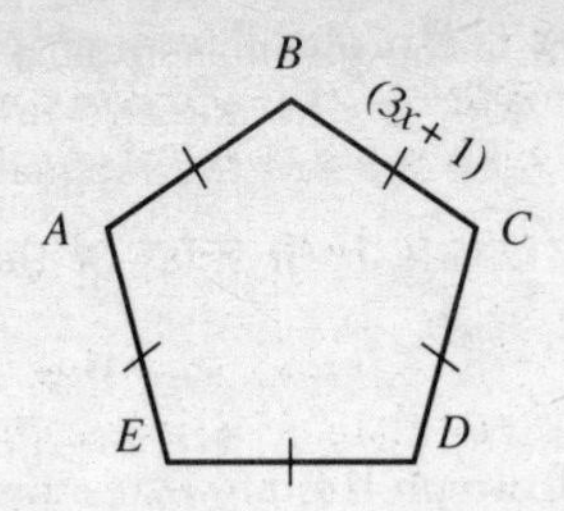

The perimeter equals the sum of the lengths of all the sides:

$$\text{Perimeter} = 5(3x + 1)$$

Remove parentheses by applying the distributive law (multiply each term within the parentheses by 5, the factor in front of the parentheses):

$$\text{Perimeter} = 15x + 5$$

The correct choice is **(1)**.

20. The given fraction is:

$$\frac{-12x^3y^5}{3xy^2} \quad (x \neq 0), (y \neq 0)$$

The fraction may be reduced by dividing both numerator and denominator by the same nonzero factor. In this fraction, numerator and denominator may both be divided by $3xy^2$. In dividing powers of the same base, remember to subtract the exponents:

$$\frac{\overset{-4x^2y^3}{\cancel{-12x^3y^5}}}{\underset{1}{\cancel{3xy^2}}}$$

$$\frac{-4x^2y^3}{1} = -4x^2y^3$$

The correct choice is **(4)**.

21. The product in symbolic form is:

$$(3a^2)(2a^4)$$

To multiply two monomials, first multiply their numerical coefficients to obtain the numerical coefficient of the product:

$$3 \times 2 = 6$$

Next multiply the literal factors to obtain the literal factor of the product; remember that powers of the same base are multiplied by adding their exponents:

$$a^2 \times a^4 = a^6$$

Combine the two results:

$$(3a^2)(2a^4) = 6a^6$$

The correct choice is **(3)**.

22. The given inequality is: $2x + 3 > 5$

Add –3 (the additive inverse of 3) to both sides of the inequality: $-3 = -3$

$2x > 2$

Divide both sides of the inequality by 2: $\frac{2x}{2} > \frac{2}{2}$

The solution says that x must be greater than 1. The smallest integer for which this is true is 2. $x > 1$

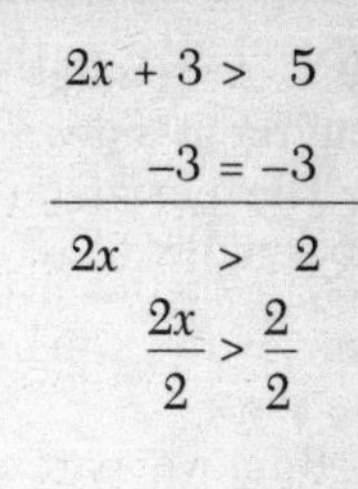

The correct choice is **(2)**.

23. The given expression is: $3(x - 1) - 2(x - 3)$

Remove the parentheses by applying the distributitive law (multiply each term within the parentheses by 3, the factor in front of the parentheses): $3x - 3 - 2x + 6$

Combine like terms: $x + 3$

The correct choice is **(4)**.

24. If the slope of a line is zero, that line must be horizontal, that is, parallel to the x-axis. Only choice (4) shows a line parallel to the x-axis.

The correct choice is **(4)**.

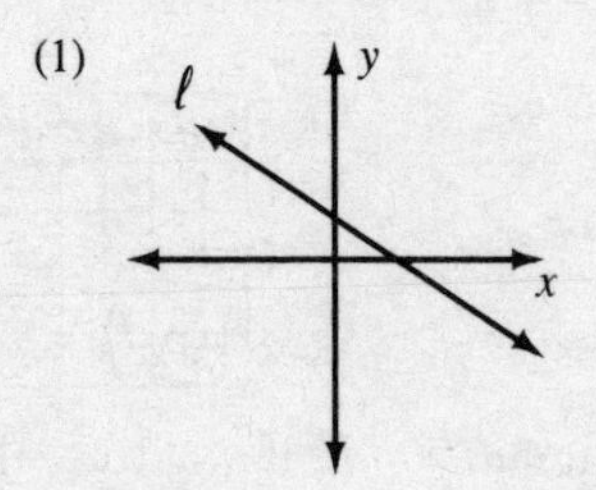

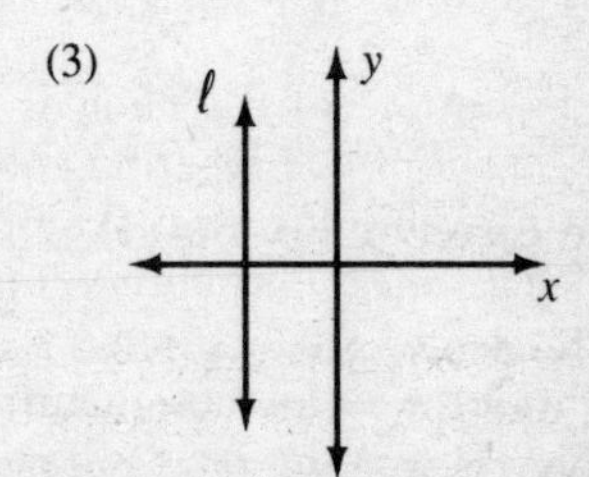

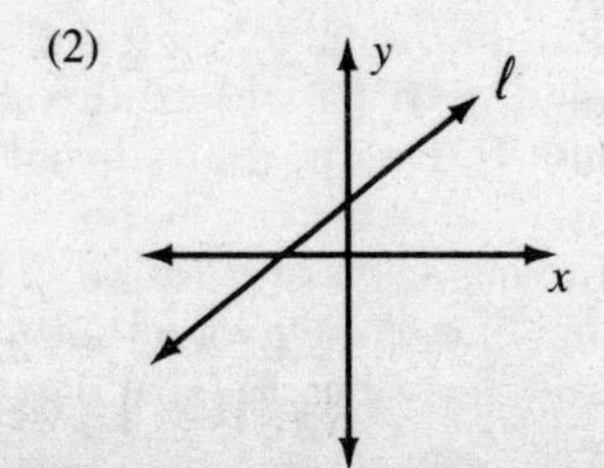

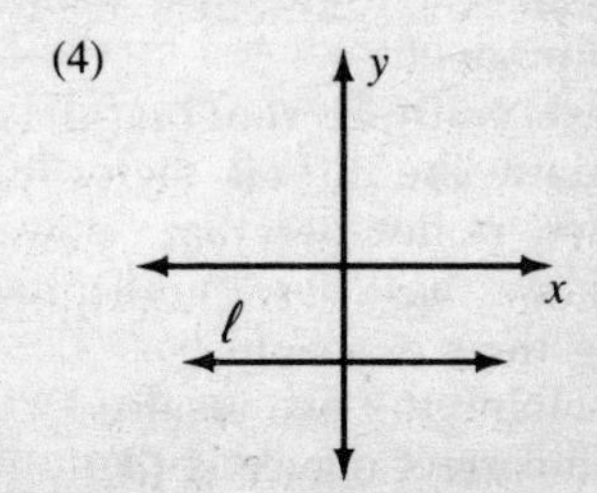

25. Let x = the length of the hypotenuse.

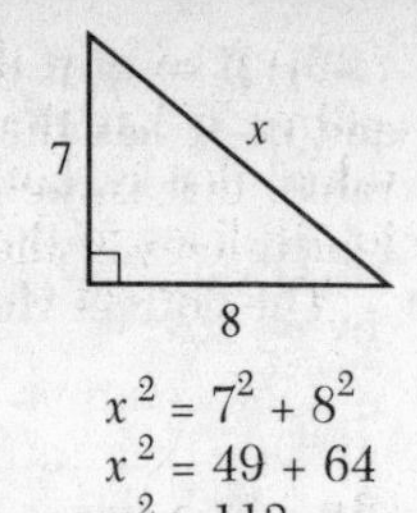

By Pythagorean Theorem, in a right triangle the square of the length of the hypotenuse equals the sum of the squares of the lengths of the legs:

$$x^2 = 7^2 + 8^2$$
$$x^2 = 49 + 64$$
$$x^2 = 113$$

Take the square root of each side of the equation:

$$x = \pm\sqrt{113}$$

Reject the negative value as meaningless for a length:

$$x = \sqrt{113}$$

The correct choice is **(2)**.

26. The given equation is:

$$(x - a)(x + b) = 0$$

If the product of two factors equals 0, then either factor may equal 0:

$$x - a = 0 \vee x + b = 0$$

Add the appropriate additive inverse to both sides of the equation, a in the case of the left-hand equation, and $-b$ in the case of the right-hand one:

$$\frac{\quad a = a \quad}{x \quad = a} \qquad \frac{\quad -b = -b \quad}{x \quad = -b}$$

The solution set is $\{a, -b\}$.

The correct choice is **(1)**.

27. Since one of the choices involves $\sim p$, the *negation* of p, prepare a column for $\sim p$; each truth value for $\sim p$ is the opposite of the truth value on the same line in the column for p.

~p	p	q	?
F	T	T	F
F	T	F	F
T	F	T	T
T	F	F	F

Consider each choice in turn:

(1). $p \rightarrow q$ is the *implication* or *conditional*, if p is true, then q is true. If p and q both have the truth value T, then $p \rightarrow q$ has the value T . This is contradicted by the first line in the table.

(2) $p \vee q$ is the *disjunction* of p and q. It has the truth value T if either p or q or both have the value T. This is contradicted by the first line in the table.

(3) $\sim p \wedge q$ is the *conjunction* of the *negation* of p with q. The conjunction, $\sim p \wedge q$, will have the truth value T only when both $\sim p$ and q are T; otherwise, $\sim p \wedge q$, will have the value F. This agrees with the truth values in the column headed "?."

(4) $p \leftrightarrow q$, is the *biconditional* or *equivalence relation* between p and q. It has the truth value T when p and q have the same truth value, that is, both T or both F. This is contradicted by the first and fourth lines in the table.

The correct choice is **(3)**.

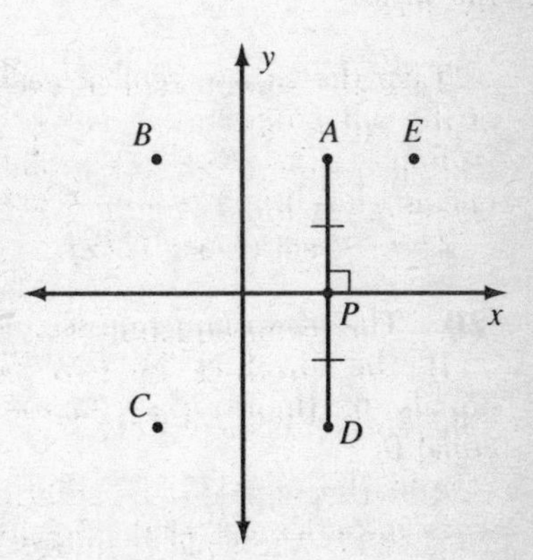

28. The image of a point A after a reflection in the x-axis is a point A' such that AA' is perpendicular to the x-axis and intersects it at a point P such that $AP = PA'$.

Here, AD is perpendicular to the x-axis and intersects it at point P such that $AP = PD$. Therefore the image of point A after a reflection in the x-axis would be point D.

The correct choice is **(3)**.

29. p = "The number is an even integer."
q = "Three times the number is 12."

If the number is 6, then p is true, for 6 is an even integer. However since $3(6) \neq 12$, q is false.

Consider each choice in turn:

(1) $p \wedge q$ is the *conjunction* of p and q. It is true only if both p and q are true. Since q is false, $p \wedge q$ is false.

(2) $p \vee q$ is the disjunction of p and q. It is true if either p or q or both are true. Since p is true and q is false, $p \vee q$ is true.

(3) The statement q is false when $p = 6$, since $3(6) \neq 12$.

(4) $p \rightarrow q$ is the *implication* or *conditional*, if p is true, then q is true. $p \rightarrow q$ is false if p is true and q is false, as is the case here.

The correct choice is **(2)**.

30. Here n = the number of nickels.
Then $n + 7$ = the number of dimes.
Also, $5n$ = the value in cents of n nickels,
and $10(n + 7)$ = the value in cents of $(n + 7)$ dimes.

Since the total value is \$1.45 (or 145 cents); we have the equation

$$5n + 10(n + 7) = 145$$

The correct choice is **(3)**.

31. A fraction is undefined if its denominator equals zero, since division by zero is undefined. Examine each choice if –3 is substituted for x:

(1) $\dfrac{3}{2x+6}$ becomes $\dfrac{3}{2(-3)+6}$ or $\dfrac{3}{-6+6}$ or $\dfrac{3}{0}$, which is undefined.

(2) $\dfrac{2x+6}{3}$ becomes $\dfrac{2(-3)+6}{3}$ or $\dfrac{-6+6}{3}$ or $\dfrac{0}{3}$, which equals 0.

(3) $\dfrac{x+3}{x-3}$ becomes $\dfrac{-3+3}{-3-3}$ or $\dfrac{0}{-6}$, which equals 0.

(4) $\dfrac{3+x}{3-x}$ becomes $\dfrac{3-3}{3-(-3)}$ or $\dfrac{0}{3+3}$ or $\dfrac{0}{-6}$, which equals 0.

The correct choice is **(1)**.

32. Let x = the smaller integer.
Then $x + 1$ = the next consecutive integer.
The sum of the two consecutive integers is –1: $x + x + 1 = -1$
Combine like terms: $2x + 1 = -1$
Add –1 (the additive inverse of 1) to both sides of the equation: $\underline{\quad -1 = -1}$
$2x = -2$
Divide both sides of the equation by 2: $\dfrac{2x}{2} = \dfrac{-2}{2}$
The smaller integer is –1. $x = -1$

The correct choice is **(3)**.

33. A figure has horizontal symmetry if there exists a horizontal line such that, when the figure is folded along that line, every point above the line will correspond to a point below the line, and every point below the line will correspond to a point above the line.

A figure has vertical symmetry if there exists a vertical line such that, when the figure is folded along that line, every point to the left of the line will correspond to a point to the right of the line and every point to the right of the line will correspond to a point to the left of the line.

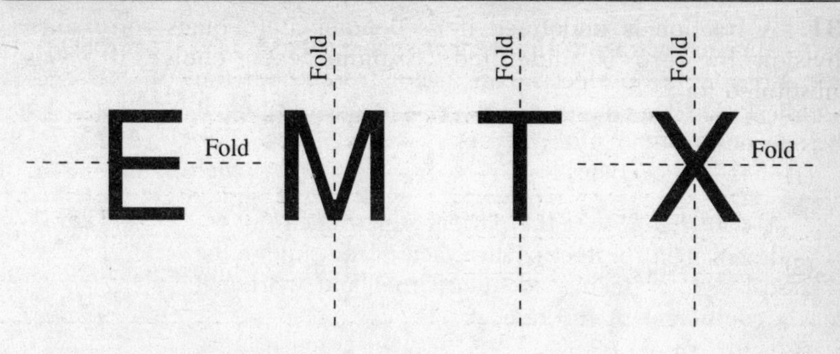

Consider each choice in turn:
(1) E has horizontal symmetry only.
(2) M has vertical symmetry only.
(3) T has vertical symmetry only.
(4) X has both horizontal and vertical symmetry.
The correct choice is **(4).**

34. The given scores are 70, 75, 75, 85, and 90.

$$\text{The mean} = \frac{\text{the sum of all scores}}{\text{the number of scores}}$$

$$= \frac{70 + 75 + 75 + 85 + 90}{5} = \frac{395}{5} = 79$$

The median is the middle score when the scores are arranged in order of size. The median here is 75.

The mode is the score that appears most frequently. The mode here is 75.

The median and the mode are both 75, but the mean is 79. Therefore, only the mode and the median have the same value.

The correct choice is **(3).**

35. The given expression is: $2\sqrt{5}$

A radical expression can be simplified by factoring out any perfect square factor in the radicand and writing the square root of that perfect square as a coefficient of the radical. Consider each choice in turn:

(1) $\sqrt{10}$: The radicand, 10, has no perfect square factor, so the radical cannot be simplified.

(2) $\sqrt{20}$: Factor out the perfect square factor: $\sqrt{4(5)}$
Remove the perfect square factor from under the radical sign by taking its square root and writing it as a coefficient of the radical: $2\sqrt{5}$

(3) $\sqrt{50}$: Factor out the perfect square factor: $\sqrt{25(2)}$
Remove the perfect square factor from under the radical sign by taking its square root and writing it as a coefficient of the radical: $5\sqrt{2}$

(4) $\sqrt{100}$: 100 is a perfect square: $\sqrt{100} = 10$

The correct choice is **(2)**.

PART TWO

36. a. (1) To graph $x + y = 8$, it is convenient to rearrange the equation so that it is solved for y by adding $-x$ (the additive inverse of x) to both sides:

$$\begin{array}{rcl} x + y &=& 8 \\ -x &=& -x \\ \hline y &=& 8 - x \end{array}$$

Prepare a table of values for x and y by choosing any three convenient values for x and substituting them in the equation to calculate the corresponding values of y:

x	$8 - x$	$= y$
–2	$8 - (-2) = 8 + 2$	$= 10$
0	$8 - 0$	$= 8$
3	$8 - 3$	$= 5$

Plot the points (–2, 10), (0, 8), and (3, 5), and draw straight line through them. This line is the graph of $x + y = 8$.

(2) $x = 1$ is a line parallel to the y-axis and 1 unit to the right of it.
(3) $y = 3$ is a line parallel to the x-axis and 3 units above it.

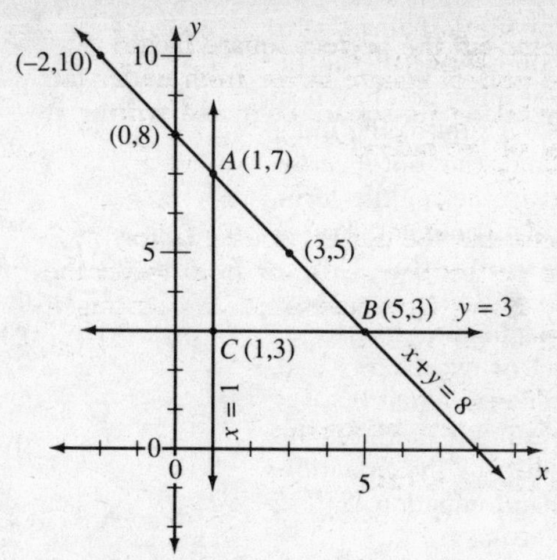

b. The three plotted lines form a triangle whose vertices are labeled A, B, and C. The coordinates of A are (1,7), the coordinates of B are (5,3), and the coordinates of C are (1,3).

ΔABC is a right triangle with base $CB = 4$ and altitude $AC = 4$.

The area, A, of a triangle with base b and altitude h is $A = \frac{1}{2}bh$.

The area of $\Delta ABC = \frac{1}{2}(CB)(AC) = \frac{1}{2}(4)(4) = \frac{1}{2}(16) = 8$.

The area is **8**.

37. Let x = the positive number.

If five times the square of the number is decreased by twice the number, the result is 16. In symbols, this is the equation:

$$5x^2 - 2x = 16$$

This is a *quadratic equation*. Rearrange it so that all terms are on one side equal to 0 by adding –16 (the additive inverse of 16) to both sides:

$$-16 = -16$$

$$5x^2 - 2x - 16 = 0$$

The left side is a *quadratic trinomial* that can be factored into the product of two binomials. The factors of the first term, $5x^2$, are $5x$ and x, and they become the first terms of the binomials:

$$(5x \quad)(x \quad) = 0$$

The factors of the last term, –16, become the second terms of the binomials, but they must be chosen in such a way that the sum of the inner cross product and the outer cross product is equal to the middle term, –2x, of the original trinomial. Try +8 and –2 as the factors of –16:

$+8x$ = inner product

$$(5x + 8)(x - 2) = 0$$

$-10x$ = outer product

Since $(+8x) + (-10x) = -2x$, these are the correct factors:

$$(5x + 8)(x - 2) = 0$$

If the product of two factors is 0, then either factor may equal 0:

$$5x + 8 = 0 \lor x - 2 = 0$$

Add the appropriate additive inverse to both sides of the equation, –8 for the left-hand equation and +2 for the right-hand one:

$$\begin{array}{rl} -8 = -8 & \quad +2 = +2 \\ \hline 5x = -8 & \quad x = 2 \end{array}$$

In the left-hand equation, divide both sides by 5:

$$\frac{5x}{5} = \frac{-8}{5}$$

Reject $\frac{-8}{5}$ since it is neither positive nor an integer:

$$x = \frac{-8}{5}$$

The number is **2**.

$$x = 2$$

38. a. Here x represents the number of white marbles.

(1) Since the ratio of red marbles to white marbles is 3:1, the number of red marbles is $\mathbf{3x}$.

(2) There are 8 more blue marbles than white marbles; therefore the number of blue marbles is $\mathbf{x + 8}$.

b. The total number of marbles is the sum of the numbers of red, white and blue marbles: $3x + x + x + 8 = \mathbf{5x + 8}$.

c. (1) If there are 38 marbles in the bag, then:

$$5x + 8 = 38$$

Add –8 (the additive inverse of 8) to both sides of the equation:

$$\begin{array}{r} -8 = -8 \\ \hline 5x = 30 \end{array}$$

Divide both sides of the equation by 5:

$$\frac{5x}{5} = \frac{30}{5}$$

$$x = 6$$

The number of red marbles is $3x$: $3x = 18$
There are **18** red marbles.

(2) Probability of an event occurring =

$$\frac{\text{the possible number of successes}}{\text{the total possible number of outcomes}}.$$

The successes for selecting a marble that is *not* blue are the selections of red or white marbles. There are 18 + 6, or 24, marbles that are red or white.

The total possible number of marbles is 38.

The probability of selecting a marble that is *not* blue is $\frac{24}{38}$.

The probability is $\mathbf{\frac{24}{38}}$.

39. Let x = the measure of $\angle A$.
Then $x + 13$ = the measure of $\angle B$,
and $2x - 9$ = the measure of $\angle C$.

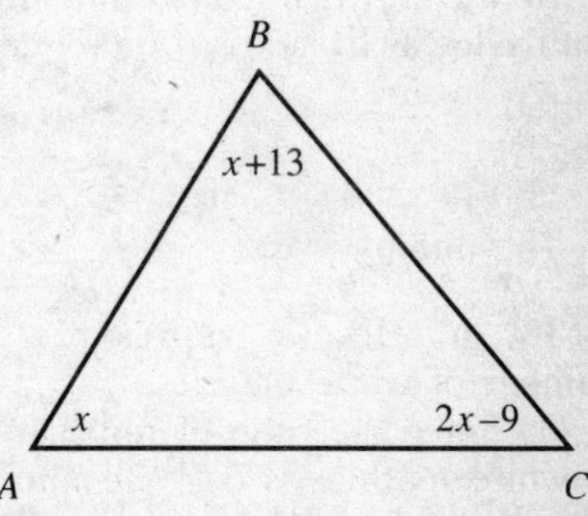

The sum of the measures of the three angles of a triangle is 180°: $x + x + 13 + 2x - 9 = 180$

Combine like terms: $4x + 4 = 180$

Add −4 (the additive inverse of +4) to both sides of the equation: $-4 = -4$

$4x = 176$

Divide both sides of the equation by 4: $\frac{4x}{4} = \frac{176}{4}$

$x = 44 \ (\angle A)$

$x + 13 = 57 \ (\angle B)$

$2x - 9 = 2(44) - 9 = 88 - 9 = 79 \ (\angle C)$

The measures of the angles are as follows: $\angle A$: **44**, $\angle B$: **57**, $\angle C$: **79**.

40. The given system of equations is:

$$2x + 5y = -1$$
$$-3x + y = 10$$

Eliminate y by multiplying each term of the second equation by 5 and subtracting it from the first equation:

$$\begin{array}{r} 2x + 5y = -1 \\ \underline{-15x + 5y = 50} \\ 17x \quad = -51 \end{array}$$

Divide both sides of the equation by 17:

$$\frac{17x}{17} = \frac{-51}{17}$$
$$x = -3$$

Substitute −3 for x in the first equation:

$$2(-3) + 5y = -1$$

Add 6 (the additive inverse of −6) to both sides of the equation:

$$\begin{array}{r} -6 + 5y = -1 \\ \underline{6 \qquad = \ \ 6} \\ 5y = \ \ 5 \end{array}$$

Divide both sides of the equation by 5:

$$\frac{5y}{5} = \frac{5}{5}$$
$$y = 1$$

The solution is **(−3, 1)** or $\boldsymbol{x = -3}$, $\boldsymbol{y = 1}$.

CHECK: The solution must satisfy *both original* equations. Substitute −3 for x and 1 for y in *both* of the *original* equations to see whether they are satisfied:

$$\underline{2x + 5y = -1} \qquad \underline{-3x + y = 10}$$

$$2(-3) + 5(1) \stackrel{?}{=} -1 \qquad -3(-3) + 1 \stackrel{?}{=} 10$$

$$6 + 5 \stackrel{?}{=} -1 \qquad 9 + 1 \stackrel{?}{=} 10$$

$$-1 = -1 \ \checkmark \qquad 10 = 10 \ \checkmark$$

41. a. To fill in the cumulative frequency column, begin by copying the frequency, 2, of the lowest scores, 210-300, into the cumulative frequency column. The entry for each line above is obtained by adding the frequency on that line to the last cumulative frequency entry immediately below it.

Scores	Frequency	Cumulative Frequency
710–800	4	60
610–700	10	56
510–600	15	46
410–500	18	31
310–400	11	13
210–300	2	2

b.

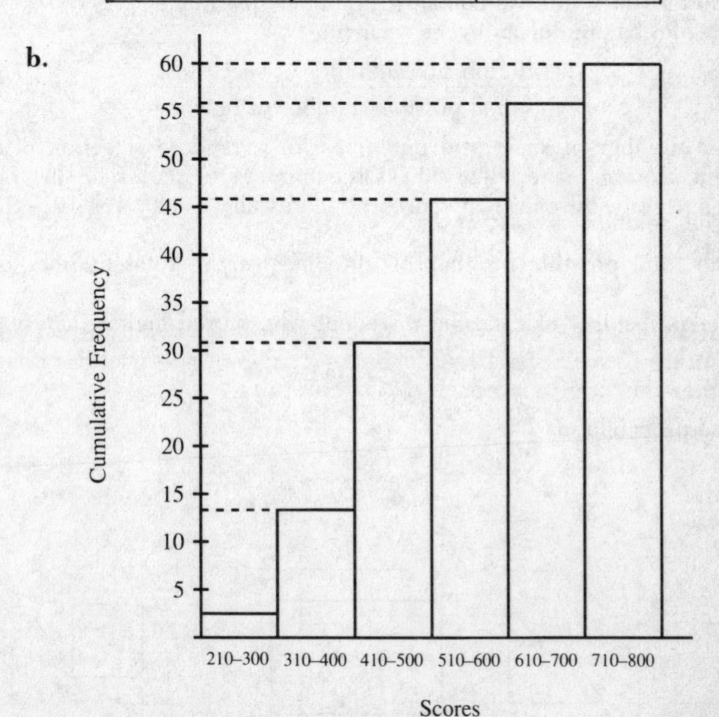

c. The upper quartile is the score that is exceeded by $\frac{1}{4}$ of the scores, and that itself exceeds $\frac{3}{4}$ of the scores. Since there are 60 scores in all, and $\frac{1}{4}(60) = 15$, the upper quartile will be midway between the 15th and the 16th scores from the top; 15 scores will exceed it, and it will exceed the 45 others.

Counting down from the top of the table, there are 4 scores in the highest interval and 10 in the second highest, making a total of 14 scores in the two top intervals. One more score is needed to reach the 15th; it will lie in the 510-600 interval. The 16th score will also lie in the same interval. Therefore the upper quartile, midway between the 15th and the 16th score, will lie in the 510-600 interval.

The **510-600** interval contains the upper quartile.

d. Probability of an event occurring

$$= \frac{\text{the possible number of successes}}{\text{the total possible number of outcomes}}.$$

The number of successful outcomes for a random selection of a student scoring higher than 500 is the number of scores in the top three intervals, all of which represent scores above 500: $4 + 10 + 15 = 29$.

The total possible number of outcomes is the total number of scores: 60.

The probability of choosing a student who scored higher than 500 is $\frac{29}{60}$.

The probability is $\mathbf{\frac{29}{60}}$.

42.a.

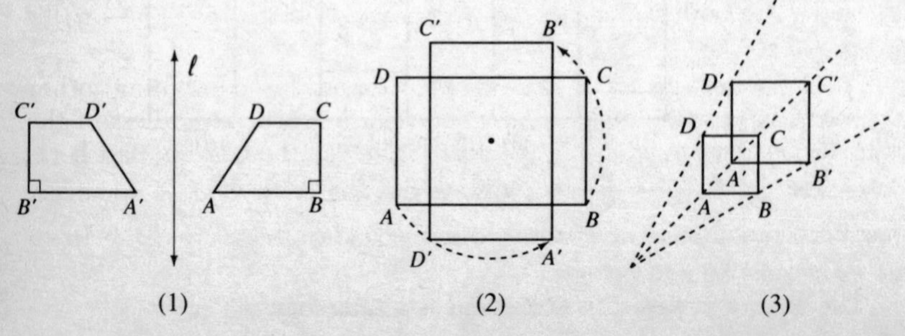

(1) **line reflection**: $A'B'C'D'$ is the image of $ABCD$ under a reflection in line l.

(2) **rotation**: If $ABCD$ is rotated counterclockwise 90° about its center its image will be $A'B'C'D'$.

(3) **dilation**: $A'B'C'D'$ is the image of $ABCD$ after a dilation involving rays coming from a common point O.

b.(1)

p	q	$p \rightarrow q$	$p \wedge (p \rightarrow q)$	$[p \wedge (p \rightarrow q)] \rightarrow q$
T	T	T	T	T
T	F	F	F	T
F	T	T	F	T
F	F	T	F	T

For each column in the table, the truth values, T and F, are determined in accordance with the truth values of p and q on the same line.

The statement $p \rightarrow q$ is the *implication* or *conditional*, if p, then q. The conditional has the truth value T whenever q is T or when p and q are both F, but has the truth value F if p is T and q is F.

The statement $p \wedge (p \rightarrow q)$ is the *conjunction* of p with the conditional, $(p \rightarrow q)$. The conjunction has the truth value T when p and $(p \rightarrow q)$ are both T, but has the truth value F if either p or $(p \rightarrow q)$ or both are F.

The statement $[p \wedge (p \rightarrow q)] \rightarrow q$ is the *implication* or *conditional*, if the conjunction of p with $(p \rightarrow q)$ is true, then q is true. The conditional has the truth value T whenever q is T or when $[p \wedge (p \rightarrow q)]$ and q are both F, but has the truth value F when $[p \wedge (p \rightarrow q)]$ is T but q is F.

(2) A *tautology* is a statement formed by combining other propositions in statements $(p, q, r, \ldots)$ which is true regardless of the truth or falsity of $p, q, r, \ldots$. The truth values in the table in part **b** (1) show that $[p \wedge (p \rightarrow q)] \rightarrow q$ always has the truth value T for every possible combination of truth values of p and q; therefore $[p \wedge (p \rightarrow q)] \rightarrow q$ must be a tautology.

The answer is **yes**; this statement is a tautology.

Topic	Question Numbers	Number of Points	Your Points	Your Percentage
1. Numbers (rat'l, irrat'l); Percent	—	0		
2. Properties of No. Systems	31	2		
3. Operations of Rat'l Nos. and Monomials	20, 21	2 + 2 = 4		
4. Operations on Multinomials	18, 23	2 + 2 = 4		
5. Square root; Operations involving Radicals	35	2		
6. Evaluating Formulas and Expressions	6	2		
7. Linear Equations (simple cases incl. parentheses)	—	0		
8. Linear Equations containing Decimals or Fractions	8, 9	2 + 2 = 4		
9. Graphs of Linear Functions (slope)	24, 36a	2 + 7 = 9		
10. Inequalities	22	2		
11. Systems of Eqs. & Inequalities (alg. & graphic solutions)	40	10		
12. Factoring	11	2		
13. Quadratic Equations	26	2		
14. Verbal Problems	12, 32, 37, 39	2 + 2 +10 + 10 = 24		
15. Variation	5	2		
16. Literal Eqs.; Expressing Relations Algebraically	14, 30, 38a, b	2 + 2 + 2 + 2 = 8		
17. Factorial *n*	—	0		
18. Areas, Perims., Circums., Vols. of Common Figures	19	2		
19. Geometry (≅, ∥ lines, compls., suppls.)	4	2		
20. Ratio & Proportion (incl. similar triangles)	—	0		
21. Pythagorean Theorem	25	2		
22. Logic (symbolic rep., logical forms, truth tables)	10, 27, 29, 42b	2 + 2 + 2 + 7 = 13		
23. Probability (incl. tree diagrams & sample spaces)	3, 17, 38c, 41d	2 + 2 + 6 + 2 = 12		
24. Combinatorics (arrangements, permutations)	15	2		
25. Statistics (central tend., freq. dist., histograms, quartiles, percentiles)	34, 41a, b, c	2 + 2 + 4 + 2 = 10		

Topic	Question Numbers	Number of Points	Your Points	Your Percentage
*26. Properties of Triangles and Quadrilaterals	1, 7, 13	2 + 2 + 2 = 6		
*27. Transformations (reflect., translat., rotat., dilat.)	28, 42a	2 + 3 = 5		
*28. Symmetry	33	2		
*29. Area from Coord. Geom.	16, 36b	2 + 3 = 5		
*30. Dimensional Analysis	—	0		
*31. Scientific Notation, Neg. Exponents	2	2		

* These topics were added to the Course I syllabus beginning with the June 1989 Regents examination. Questions on these topics did not appear on earlier examinations.

Examination January 1990

Three-Year Sequence for High School Mathematics—Course I

PART ONE

DIRECTIONS: *Answer 30 questions from this part. Each correct answer will receive 2 credits. No partial credit will be allowed. Write your answers in the spaces provided. Where applicable, answers may be left in terms of π or in radical form.*

1 A set of geometric figures consists of a square, a trapezoid, an obtuse triangle, an equilateral triangle, and a right triangle. If one of the figures is selected at random, what is the probability that all of its sides are congruent? 1____

2 Solve for x: $5(x + 2) - 3x = 12$ 2____

3 In the accompanying diagram, $\overleftrightarrow{AB} \parallel \overleftrightarrow{CD}$ and $\overleftrightarrow{EF}$ intersects $\overleftrightarrow{AB}$ at G and $\overleftrightarrow{CD}$ at H. If $m\angle AGH = 80$ and $m\angle DHG = 5x$, find the value of x.

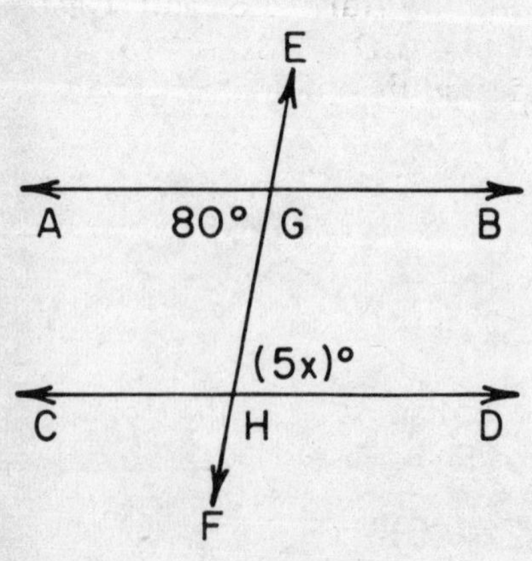

3____

4 If x varies directly as y and $x = 3$ when $y = 15$, find the value of x when $y = 25$. 4_____

5 Solve for x: $0.5x - 4 = 0.5$ 5_____

6 In the accompanying diagram, $\angle ACD$ is an exterior angle of $\triangle ABC$. If $m\angle A = 35$ and $m\angle B = 65$, find $m\angle ACD$.

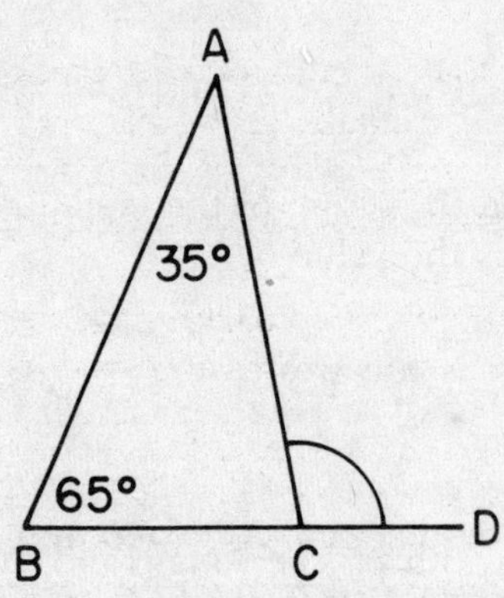

6_____

7 In the accompanying diagram, $\overleftrightarrow{AB}$ and $\overleftrightarrow{CD}$ intersect at E, and $m\angle AED = 3x + 15$. If $m\angle CEB = 2x + 45$, find the value of x.

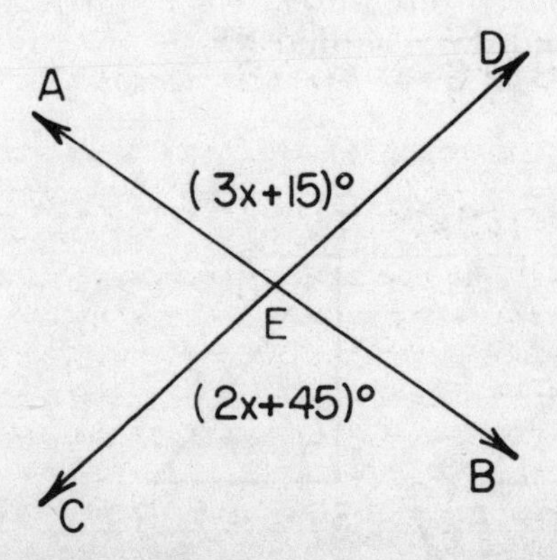

7_____

8 The measure of the vertex angle of an isosceles triangle is 50°. Find the measure of a base angle of the triangle. 8____

9 Solve for x: $\frac{x}{5} - 3 = 7$ 9____

10 Express $x^2 + 2x - 15$ as the product of two binomials. 10____

11 What is the value of $-2x^2y$ if $x = -1$ and $y = -3$? 11____

12 If the point $(3,k)$ is on the graph of the equation $x + 2y = 15$, what is the value of k? 12____

13 For which value of x is the expression $\frac{1}{x - 2}$ undefined? 13____

14 If $3x$ represents the width of a rectangle and $5x$ represents the length of the rectangle, express the perimeter of the rectangle as a monomial in terms of x. 14____

15 In the accompanying diagram, right triangle ABC is similar to right triangle RST with $\angle A \cong \angle R$. If $AB = 6$, $AC = 9$, and $RS = 4$, find RT.

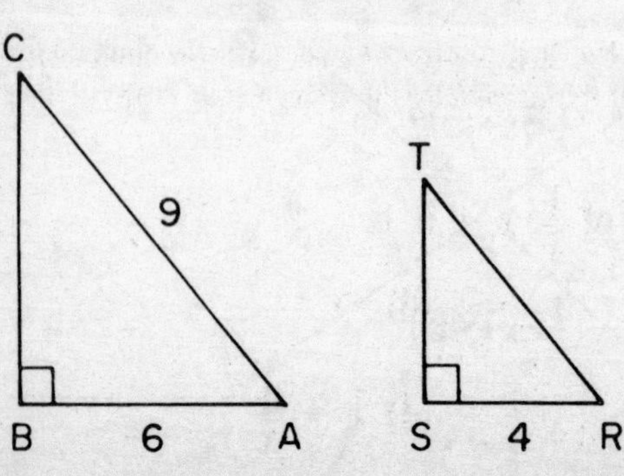

15____

16 The probability that the XYZ Light Bulb Factory will produce a defective bulb is 5%. How many defective light bulbs might be expected out of each 1,000 bulbs produced? 16_____

17 If 198,000,000 is written in the form 1.98×10^n, what is the value of n? 17_____

18 In the accompanying diagram, $MINT$ is a rectangle and $MPKT$ is a square. If $MT = 6$ and $TN = 14$, find PN.

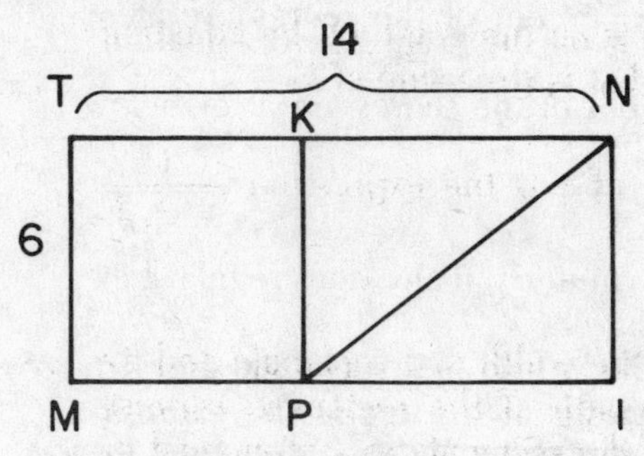

18_____

19 If the sum of $\sqrt{50}$ and $x\sqrt{2}$ is $8\sqrt{2}$, find the value of x. 19_____

DIRECTIONS (**20–35**): *For each question chosen, write the* numeral *preceding the word or expression that best completes the statement or answers the question.*

20 The reciprocal of $\frac{1}{x}$, $x \neq 0$, is

(1) $-\frac{1}{x}$

(2) x

(3) $-x$

(4) $1 - \frac{1}{x}$

20_____

21 The product of $\frac{1}{2}x^3$ and $\frac{1}{4}x^5$ is

(1) $\frac{1}{8}x^{15}$
(2) $\frac{1}{8}x^8$
(3) $\frac{1}{6}x^8$
(4) $\frac{3}{4}x^8$

21_____

22 For a set of scores, 80 is the score for the 75th percentile. Which statement is true?

(1) Eighty scores are at or below 75.
(2) Seventy-five scores are at or below 80.
(3) Seventy-five percent of the scores are at or below 80.
(4) Eighty percent of the scores are at or below 75.

22_____

23 Which type of symmetry, if any, does regular hexagon *ABCDEF* have?

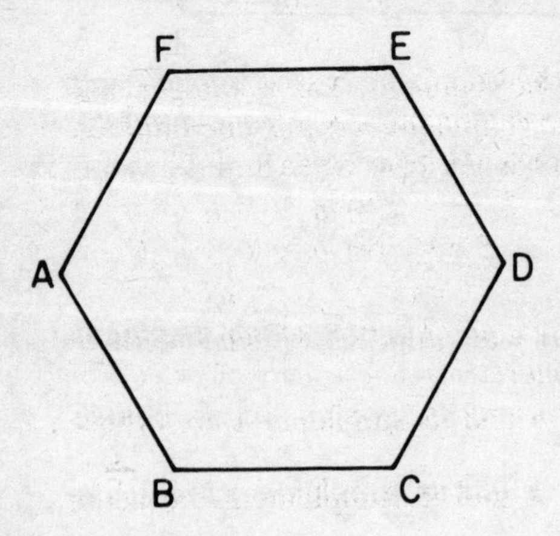

(1) point symmetry, only
(2) line symmetry, only
(3) point and line symmetry
(4) no symmetry

23_____

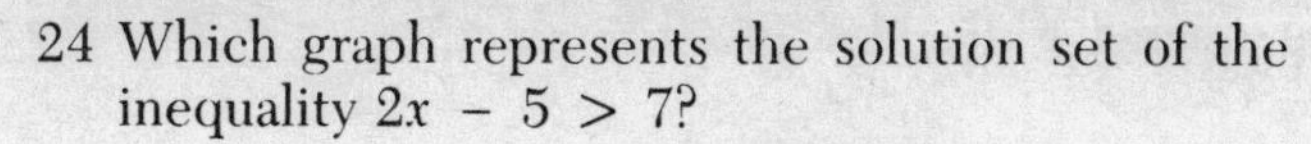

24 Which graph represents the solution set of the inequality $2x - 5 > 7$?

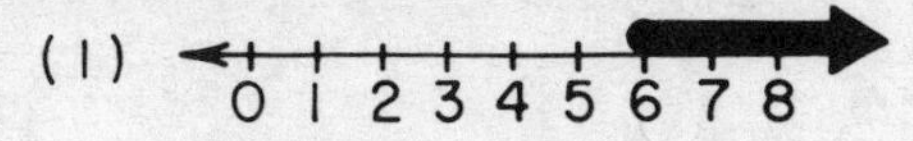

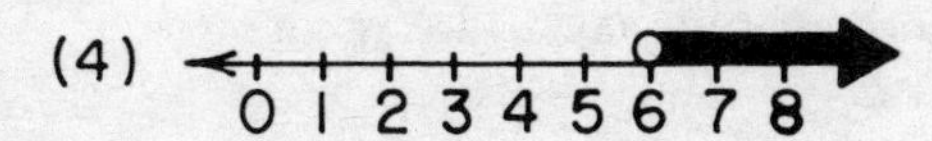

24_____

25 The vertices of rectangle $ABCD$ are $A(3,1)$, $B(-5,1)$, $C(-5,-3)$, and D. What are the coordinates of D?

(1) (3,–3)
(2) (–3,3)
(3) (3,3)
(4) (–3,–3)

25_____

26 If p represents the statement "x is an integer" and q represents the statement "x is a prime number," which statement is *not* true when $x = 7$?

(1) $p \rightarrow q$
(2) $p \leftrightarrow q$
(3) $p \wedge \sim q$
(4) $p \vee q$

26_____

27 If the measure of angle A is 60°, which statement must also be true?

(1) Both angle A and its supplement are obtuse angles.
(2) Both angle A and its supplement are acute angles.
(3) Both angle A and its complement are obtuse angles.
(4) Both angle A and its complement are acute angles.

27_____

28 In the accompanying diagram, which is an equation of line ℓ?

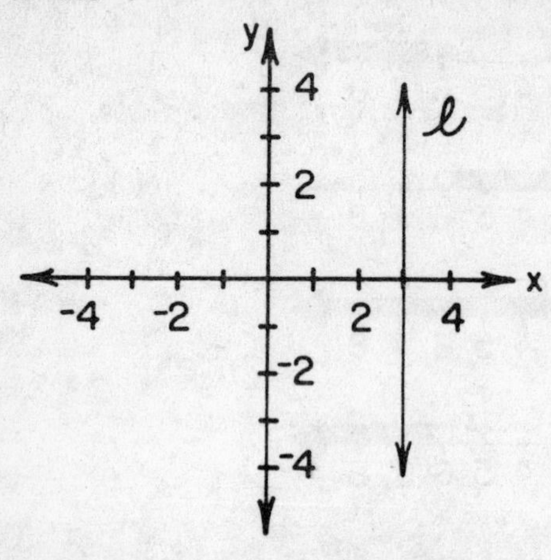

(1) $y = 3$ (3) $x + y = 3$
(2) $x = 3$ (4) $x - y = 3$ 28____

29 Written in factored form, the binomial $a^2b - ab^2$ is equivalent to

(1) $ab(a - b)$ (3) $a^2(b - b^2)$
(2) $(a - b)(a + b)$ (4) $a^2b^2(b - a)$ 29____

30 The solution set of the equation $x^2 - x - 6 = 0$ is

(1) $\{6,-1\}$ (3) $\{2,-3\}$
(2) $\{3,-2\}$ (4) $\{-6,1\}$ 30____

31 The statement "If p, then q" is false if and only if

(1) p is true and q is false
(2) p is true and q is true
(3) p is false and q is false
(4) p is false and q is true 31____

32 What is the converse of the statement "If an animal is a black cat, then it has four legs"?

(1) If an animal has four legs, then it is a black cat.
(2) If an animal does not have four legs, then it is not a black cat.
(3) If an animal has four legs, then it is not a black cat.
(4) If an animal is not a black cat, then it does not have four legs.

32_____

33 Based on the data in the table below, which interval contains the median?

Interval	Frequency
0–5	1
6–10	2
11–15	2
16–20	4

(1) 0–5
(2) 6–10
(3) 11–15
(4) 16–20

33_____

34 Which does *not* represent a rational number?

(1) $\frac{3}{2}$
(2) $\sqrt{7}$
(3) $\sqrt{16}$
(4) $0.\overline{29}$

34_____

35 If the edge of a cube is doubled, the volume is multiplied by

(1) 6
(2) 2
(3) 3
(4) 8

35_____

PART TWO

DIRECTIONS *Answer four questions from this part. Show all work unless otherwise directed.*

36 *a* On the same set of coordinate axes, graph the following system of inequalities:

$$y > 2x + 5$$
$$x + y \leq 4 \quad [8]$$

b Based on the graphs drawn in part *a*, write the coordinates of a point in the solution set of the system of inequalities. [2]

37 *a* Solve algebraically and check:

$$2x + 3y = -5$$
$$3x - 2y = 12 \quad [7,2]$$

b If the system of equations in part *a* were graphed, in which quadrant would the solution lie? [1]

38 In the accompanying diagram, *ABCD* is a rectangle with $\overline{AB}$ as a diameter of semicircle *O*. Diagonal $\overline{BD}$ is drawn in rectangle *ABCD*. The length of $\overline{BD}$ exceeds the length of $\overline{AB}$ by 1, and $AD = 5$.

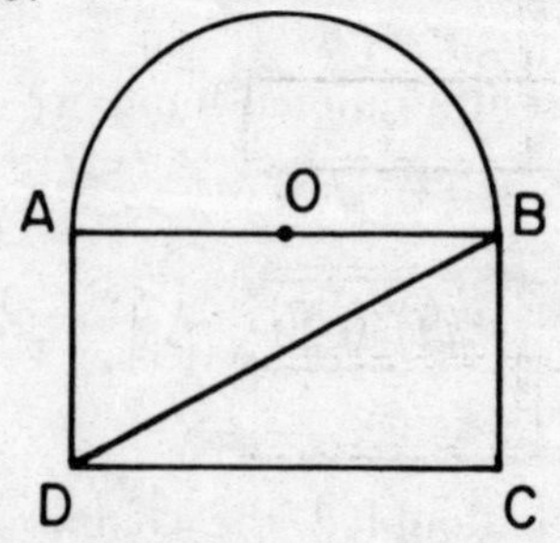

a Find *AB*. [6]

b Find, to the *nearest integer*, the area of the semicircle. [Use $\pi = 3.14$.] [4]

39 The larger of two positive integers is three more than the smaller. If twice the square of the smaller integer is increased by three times the larger integer, the result is 74. Find the integers. [*Only an algebraic solution will be accepted.*] [5,5]

40 Each part below consists of three statements. The truth value for two of the statements is given. *On your answer paper,* write the letters *a* through *e*, and next to each letter, write the missing truth value (TRUE or FALSE). If the truth value cannot be determined from the information given, write "CANNOT BE DETERMINED."

a

p	q	$p \leftrightarrow q$
	T	F

[2]

b

p	q	$q \rightarrow p$
	T	F

[2]

c

p	q	$p \vee q$
T		T

[2]

d

p	q	$\sim(p \wedge q)$
	T	F

[2]

e

p	q	$p \vee \sim q$
	T	F

[2]

41 *a* In $\triangle ABC$, the measures of angles A and B are in the ratio 1:3, and the measure of angle C is twice the measure of angle B. Write an equation or a system of equations that could be used to find the measure of each angle. State what the variable or variables represent. [*Solution of the equation(s) is not required.*] [5]

b Write an equation that could be used to find three consecutive integers such that the sum of the first two integers is nine more than the third integer. State what the variable represents. [*Solution of the equation is not required.*] [5]

42 A jar contains five disks numbered 1, 2, 3, 4, and 5, respectively. A two-digit number is formed by drawing a disk, not replacing it, and then drawing a second disk.

a Draw a tree diagram or list the sample space showing all possible outcomes of two-digit numbers. [4]

b Find the probability that the two-digit number

(1) is greater than 40 [2]
(2) is divisible by 5 [2]
(3) has the same digit in both places [2]

Answers January 1990

Three-Year Sequence for High School Mathematics—Course I

ANSWER KEY

PART ONE

1.	$\frac{2}{5}$	**13.**	2	**25.**	(1)
2.	1	**14.**	$16x$	**26.**	(3)
3.	16	**15.**	6	**27.**	(4)
4.	5	**16.**	50	**28.**	(2)
5.	9	**17.**	8	**29.**	(1)
6.	100	**18.**	10	**30.**	(2)
7.	30	**19.**	3	**31.**	(1)
8.	65	**20.**	(2)	**32.**	(1)
9.	50	**21.**	(2)	**33.**	(3)
10.	$(x + 5)(x - 3)$	**22.**	(3)	**34.**	(2)
11.	6	**23.**	(3)	**35.**	(4)
12.	6	**24.**	(4)		

Part Two—*See* Answers Explained.

ANSWERS EXPLAINED

PART ONE

1.

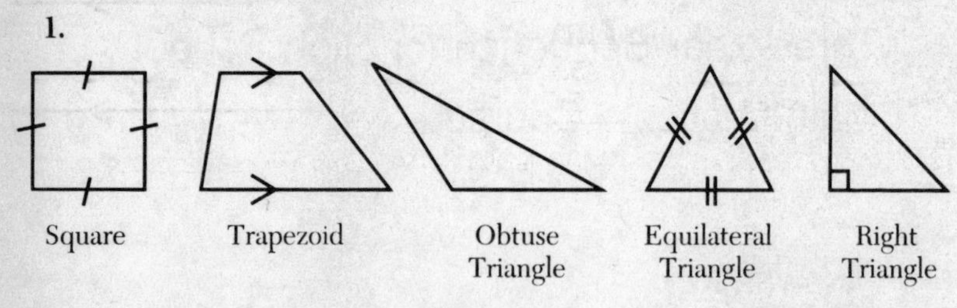

Both the square and the equilateral triangle have all their sides congruent. A trapezoid has one base longer than the other, so it cannot have all sides congruent. An obtuse triangle can have at most one obtuse angle, so it cannot be equiangular; only an equiangular triangle can have all its sides congruent. In a right triangle, the hypotenuse is always longer than either leg, so a right triangle cannot have all sides congruent.

$$\text{Probability of an event occurring} = \frac{\text{number of successful outcomes}}{\text{total possible number of outcomes}}$$

For the selection of a figure with all congruent sides, there are only 2 successful outcomes: the square and the equilateral triangle. Since any one of the 5 figures may be selected, the total possible number of outcomes is 5. The probability of choosing a figure with all sides congruent is $\frac{2}{5}$.

The probability is $\frac{2}{5}$.

2. The given equation contains parentheses: $5(x + 2) - 3x = 12$

Apply the distributive law to remove the parentheses by multiplying each term inside the parentheses by the coefficient, 5: $5x + 10 - 3x = 12$

Combine like terms: $2x + 10 = 12$

Add -10 (the additive inverse of 10) to both sides of the equation: $\underline{\quad -10 = -10}$

$2x = 2$

Divide both sides of the equation by 2: $\frac{2x}{2} = \frac{2}{2}$

$x = 1$

$x = 1$.

3. If two parallel lines are cut by a transversal, a pair of alternate interior angles are equal in measure:

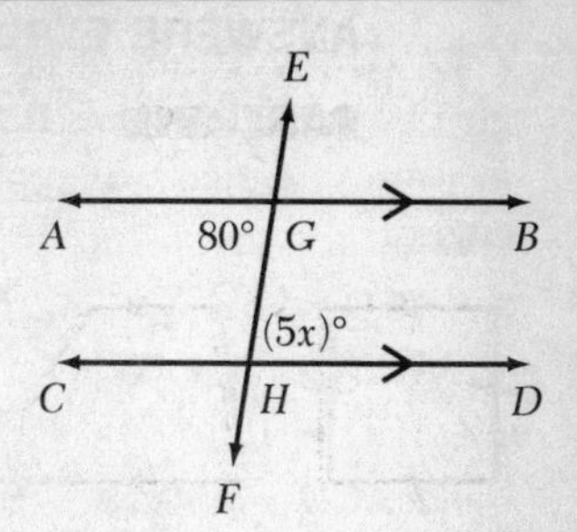

$$m\angle EHD = m\angle AGF$$
$$5x = 80$$

Divide both sides of the equation by 5:

$$\frac{5x}{5} = \frac{80}{5}$$
$$x = 16$$

The value of x is **16**.

4. If x varies directly as y, the relationship may be represented by: $x = ky$, where k is a constant

Since $x = 3$ when $y = 15$, let $x = 3$ and let $y = 15$:

$$3 = 15k$$

Divide both sides of the equation by 15:

$$\frac{3}{15} = \frac{15k}{15}$$
$$\frac{1}{5} = k$$

Since k is a constant, the relationship is always:

$$x = \frac{1}{5}y$$

To find x when $y = 25$, substitute 25 for y:

$$x = \frac{1}{5}(25)$$

The value of x is **5**.

$$x = 5$$

5. The given equation contains decimals:

$$0.5x - 4 = 0.5$$

Clear decimals by multiplying each term on both sides of the equation by 10:

$$10(0.5x) - 10(4) = 10(0.5)$$
$$5x - 40 = 5$$

Add 40 (the additive inverse of -40) to both sides of the equation:

$$+40 = +40$$
$$5x = 45$$

Divide both sides of the equation by 5:

$$\frac{5x}{5} = \frac{45}{5}$$
$$x = 9$$

$x =$ **9**.

6. The measure of an exterior angle of a triangle equals the sum of the measures of the two remote interior angles:

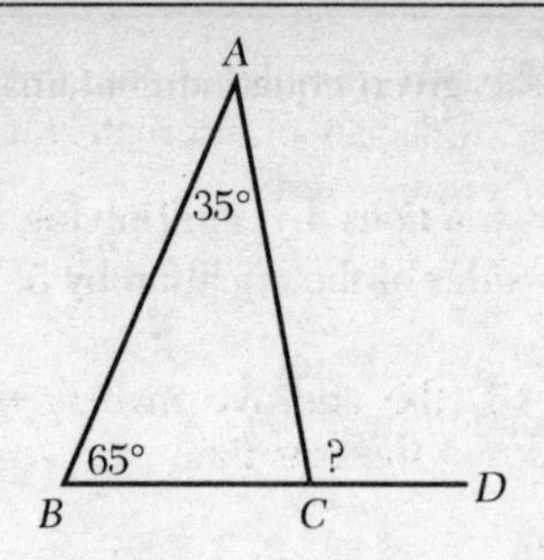

$m\angle ACD = m\angle A + m\angle B$

$m\angle ACD = 35 + 65$

$m\angle ACD = 100$

$m\angle ACD =$ **100.**

7. If two lines intersect, the measures of a pair of vertical angles are equal:

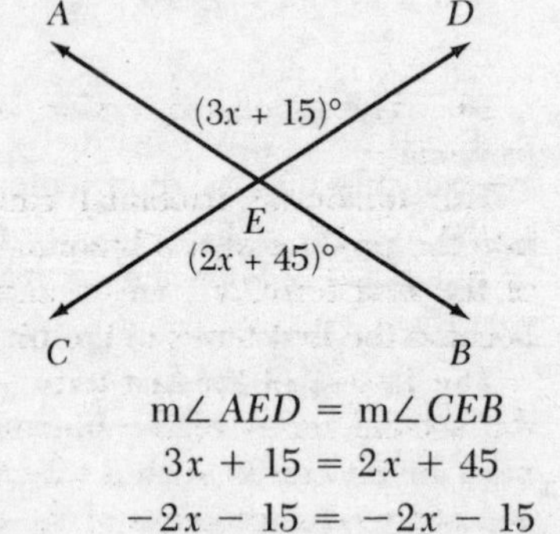

$m\angle AED = m\angle CEB$

$3x + 15 = 2x + 45$

Add $-2x$ (the additive inverse of $2x$) and also add -15 (the additive inverse of 15) to both sides of the equation:

$$\begin{aligned} -2x - 15 &= -2x - 15 \\ \hline x &= 30 \end{aligned}$$

The value of x is **30.**

8. Let $x = m\angle A$.

The base angles of an isosceles triangle are equal in measure: $m\angle C = m\angle A = x$

The sum of the measures of the three angles of a triangle is 180°: $m\angle A + m\angle B + m\angle C = 180$

$x + 50 + x = 180$

Combine like terms: $2x + 50 = 180$

Add -50 (the additive inverse of 50) to both sides of the equation:

$$\begin{aligned} -50 &= -50 \\ \hline 2x &= 130 \end{aligned}$$

Divide both sides of the equation by 2: $\dfrac{2x}{2} = \dfrac{130}{2}$

The measure of a base angle is **65.** $x = 65$

9. The given equation contains a fraction: $\frac{x}{5} - 3 = 7$

Clear fractions by multiplying each term on both sides of the equation by 5:

$$5\left(\frac{x}{5}\right) - 5(3) = 5(7)$$
$$x - 15 = 35$$

Add 15 (the additive inverse of -15) to both sides of the equation:

$$\underline{\quad 15 = 15}$$
$$x \quad = 50$$

$x =$ **50.**

10. The given expression is a *quadratic trinomial*: $x^2 + 2x - 15$

The quadratic trinomial can be factored into the product of two binomials. The factors of the first term, x^2, are x and x, and they become the first terms of the binomials: $(x \quad)(x \quad)$

The factors of the last term, -15, become the second terms of the binomials, but they must be chosen in such a way that the inner and outer cross-products of the binomials add up to the middle term, $+2x$, of the original trinomial. Try $+5$ and -3 as the factors of -15:

$+5x =$ inner product

$(x + 5)(x - 3)$

$-3x =$ outer product

Since $(+5x)+(-3x) = +2x$, these are the correct factors: $(x + 5)(x - 3)$

The product of two binomials is $\mathbf{(x+5)(x-3)}$.

11. The given expression is: $-2x^2y$

Let $x = -1$ and let $y = -3$: $-2(-1)^2(-3)$

Since $(-1)^2 = 1$: $-2(1)(-3)$

6

The value is **6.**

12. The given equation is: $x + 2y = 15$

If $(3, k)$ is on the graph of the equation, its coordinates must satisfy the equation when substituted for x and y. Let $x = 3$ and let $y = k$:

$$3 + 2k = 15$$

Add -3 (the additive inverse of 3) to both sides of the equation

$$\underline{-3 \quad = -3}$$

$$2k = 12$$

Divide both sides of the equation by 2:

$$\frac{2k}{2} = \frac{12}{2}$$

The value of k is **6**.

$$k = 6$$

13. If $x = 2$, the expression $\frac{1}{x-2}$ will become $\frac{1}{2-2}$ or $\frac{1}{0}$. But division by 0 is undefined, so $\frac{1}{x-2}$ is undefined for $x = 2$.

It is undefined for $x =$ **2**.

14. The perimeter, P, of a rectangle is the sum of the lengths of all four sides:

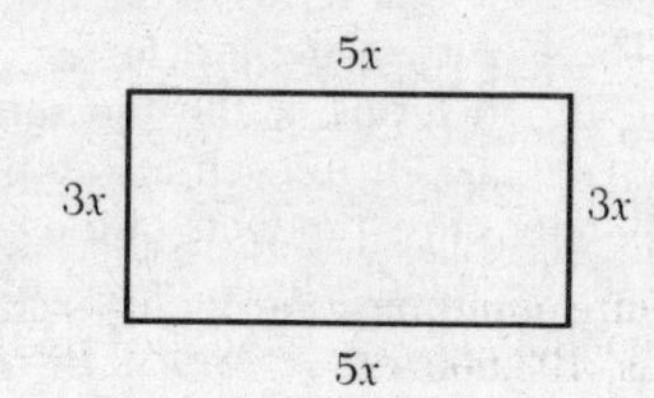

$$P = 5x + 3x + 5x + 3x$$

Combine like terms:

$$P = 16x$$

The perimeter is $\mathbf{16x}$.

15. Let $RT = x$.

$\angle B \cong \angle S$ since both are right angles and all right angles are congruent.

$\triangle ABC \sim \triangle RST$. In similar triangles, the corresponding sides (which are opposite congruent angles) are in proportion:

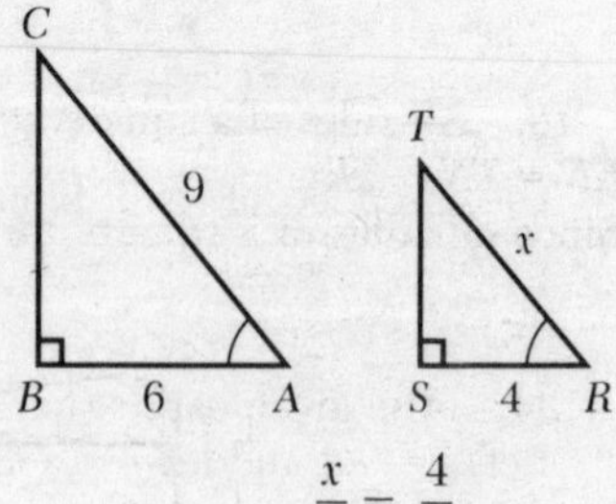

$$\frac{x}{9} = \frac{4}{6}$$

In a proportion, the product of the means equals the product of the extremes (cross-multiply):

$$6x = 4(9)$$

$$6x = 36$$

Divide both sides of the equation by 6:

$$\frac{6x}{6} = \frac{36}{6}$$

$RT =$ **6**.

$$x = 6$$

16. Let x = the number of defective bulbs expected out of each 1,000 bulbs produced.

$$\text{Probability of an event occurring} = \frac{\text{number of successful outcomes}}{\text{total possible number of outcomes}}$$

For the probability of producing a defective bulb, the number of defective bulbs, x, is the number of "successful" outcomes. We are told that 1,000 bulbs are produced, so this is the total possible number of outcomes.

$$\text{Probability of producing a defective bulb} = \frac{x}{1{,}000} = 0.05$$

Multiply both sides of the equation by 1,000:

$$1{,}000\left(\frac{x}{1{,}000}\right) = 1{,}000(0.05)$$
$$x = 50.00$$

50 defective light bulbs may be expected.

17. Let $198{,}000{,}000 = 1.98 \times 10^n$.

To convert 1.98 to 198,000,000, the decimal point in 1.98 must be moved eight places to the right. Multiplication by 10 moves the decimal point one place to the right. Moving it eight places to the right will require multiplication by 10 eight times, that is, multiplication by 10^8. Thus, $198{,}000{,}000 = 1.98 \times 10^8$ or $n = 8$.

The value of n is **8**.

18. All sides of a square are congruent: $TK = MT = 6$

$KN = TN - TK$: $KN = 14 - 6 = 8$

Since all sides of a square are congruent: $KP = TM = 6$

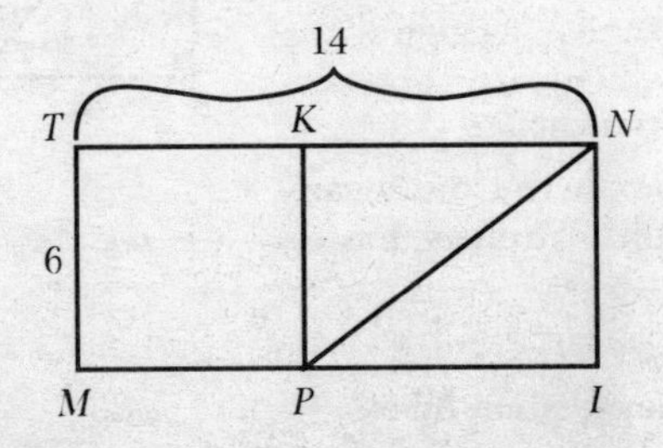

Right triangle NKP is a 3-4-5 right triangle with $KP = 6 = 2(3)$, and $KN = 8 = 2(4)$. Therefore, $PN = 2(5)$ or 10.

PN = **10**.

19. The given statement is indicated as an equation:

$$\sqrt{50} + x\sqrt{2} = 8\sqrt{2}$$

Factor out the perfect square factor in the first radicand:

$$\sqrt{25(2)} + x\sqrt{2} = 8\sqrt{2}$$

Remove the perfect square factor from under the radical sign by taking its square root and writing it as a coefficient of the radical:

$$5\sqrt{2} + x\sqrt{2} = 8\sqrt{2}$$

All the terms in the equation are now *similar radicals* since they all have the same radicand, 2, and they are all of the same index (all are square roots, that is, of index 2). Similar radicals are combined by combining their coefficients:

$$5 + x = 8$$

Add -5 (the additive inverse of 5) to both sides of the equation:

$$\begin{array}{r} -5 \quad = -5 \\ \hline x = 3 \end{array}$$

The value of x is **3.**

20. The product of a number and its reciprocal is 1. Use this fact to test each choice for the reciprocal of $\frac{1}{x}$:

(1) $-\frac{1}{x}$: $\left(-\frac{1}{x}\right)\left(\frac{1}{x}\right) \stackrel{?}{=} 1$

$-\frac{1}{x^2} \neq 1$ $\quad -\frac{1}{x}$ is *not* the reciprocal of $\frac{1}{x}$.

(2) x: $x\left(\frac{1}{x}\right) \stackrel{?}{=} 1$

$1 = 1\,\checkmark$ $\quad x$ is the reciprocal of $\frac{1}{x}$.

(3) $-x$: $(-x)\left(\frac{1}{x}\right) \stackrel{?}{=} 1$

$-1 \neq 1$ $\quad -x$ is *not* the reciprocal of $\frac{1}{x}$.

(4) $1 - \frac{1}{x}$: $\left(1 - \frac{1}{x}\right)\left(\frac{1}{x}\right) \stackrel{?}{=} 1$

$\frac{1}{x} - \frac{1}{x^2} \neq 1$ $\quad 1 - \frac{1}{x}$ is *not* the reciprocal of $\frac{1}{x}$.

The correct choice is **(2).**

21. The product can be indicated as: $\left(\frac{1}{2}x^3\right)\left(\frac{1}{4}x^5\right)$

The numerical coefficient of the product of two monomials is the product of their numerical coefficients: $\left(\frac{1}{2}\right)\left(\frac{1}{4}\right) = \frac{1}{8}$

The literal factor of the product of two monomials is the product of their literal factors. The product of two powers of the same base is the power of that base obtained by adding the two exponents: $(x^3)(x^5) = x^8$

Combine the two results above: $\left(\frac{1}{2}x^3\right)\left(\frac{1}{4}x^5\right) = \frac{1}{8}x^8$

The correct choice is **(2)**.

22. The 75th percentile is the score at or below which 75 percent of all the scores fall. If the 75th percentile is 80, then 75 percent of all the scores are at or below 80.

The correct choice is **(3)**.

23. If point X is taken at the center of the regular hexagon, any line drawn through X will intersect the hexagon in two points, P and P', such that $PX = XP'$. This property fulfills the requirement for point symmetry.

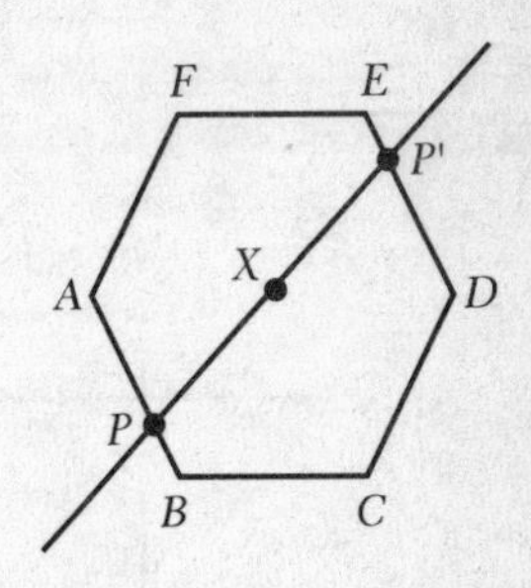

If the hexagon is folded along the dashed line shown in the second drawing, each point on the right side of the hexagon will fall on a point on the left side, and each point on the left side will correspond to a point on the right side. This property indicates that the hexagon fulfills the requirement for line symmetry. Note that there are two other possible locations for folds with the same property as the one shown, and three more locations for folds joining a pair of opposite vertices of the hexagon that could also serve as axes for line symmetry—six axes of line symmetry in all. The regular hexagon has both point and line symmetry.

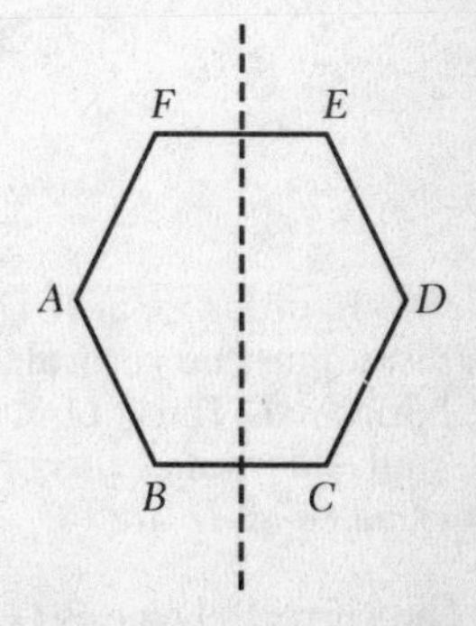

The correct choice is **(3)**.

24. The given inequality is: $2x - 5 > 7$

Add 5 (the additive inverse of -5) to both sides of the inequality: $5 = 5$

$$2x > 12$$

Divide both sides of the inequality by 2:

$$\frac{2x}{2} > \frac{12}{2}$$

$$x > 6$$

(1)

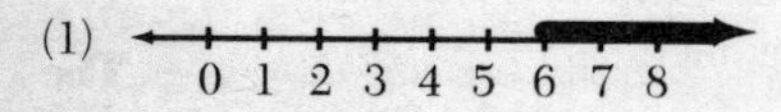

(3) 0 1 2 3 4 5 6 7 8

(2) 0 1 2 3 4 5 6 7 8

(4)

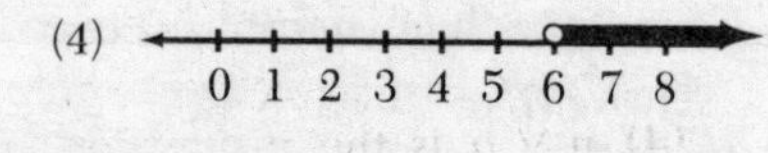

The graph of $x > 6$ must show a solid line extending indefinitely to the right from 6 to take in all real numbers greater than 6. Since $x > 6$ does not include 6 as a solution, the graph must show that 6 is excluded by indicating an open, unshaded small circle at 6.

The correct choice is **(4)**.

25.

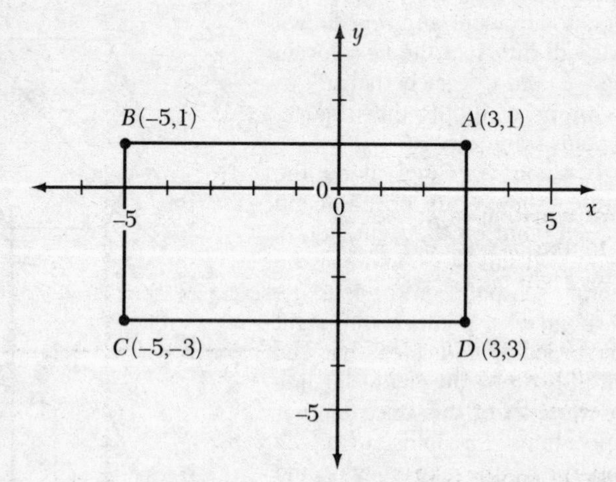

In order for $ABCD$ to be a rectangle, $\overline{AD}$ must be parallel to $\overline{BC}$, that is, it must be vertical, and $\overline{CD}$ must be parallel to $\overline{BA}$, that is, it must be horizontal. Thus, D must have the same x-coordinate as A (namely, 3), and the same y-coordinate as C (namely, -3). Therefore, the coordinates of D are $(3, -3)$.

The correct choice is **(1)**.

26. p represents "x is an integer." If $x = 7$, p is true.
q represents "x is a prime number." If $x = 7$, q is true.

Consider each choice when $x = 7$:

(1) $p \rightarrow q$ is the *conditional* or *implication*, "If p is true, then q is true." Since p and q are both true, $p \rightarrow q$ is true.

(2) $p \leftrightarrow q$ is the *equivalence relation* or *biconditional* between p and q. The biconditional is true if p and q have the same truth value; both are true, so $p \leftrightarrow q$ is true.

(3) $p \wedge \sim q$ is the *conjunction* between p and the *negation* of q. The negation of q is "7 is not a prime number," and it is false. For the conjunction $p \wedge \sim q$ to be true, both p and $\sim q$ must be true. Since $\sim q$ is false, the conjunction $p \wedge \sim q$ is false.

(4) $p \vee q$ is the *disjunction* between p and q. If either p or q or both are true, the disjunction $p \vee q$ will be true. Since both p and q are true, $p \vee q$ is true.

The correct choice is **(3)**.

27. The sum of the measures of two supplementary angles is 180°. Therefore, the supplement of 60° is 120°. Angle A (60°) is an acute angle, but its supplement is obtuse.

The sum of the measures of two complementary angles is 90°. Therefore, the complement of 60° is 30°. Angle A (60°) is an acute angle, and its complement is also acute.

The correct choice is **(4)**.

28. An equation of a line parallel to the y-axis and a units to the right of it is of the form $x = a$.

The line ℓ is parallel to the y-axis and 3 units to the right of it. An equation of this line is $x = 3$.

The correct choice is **(2)**.

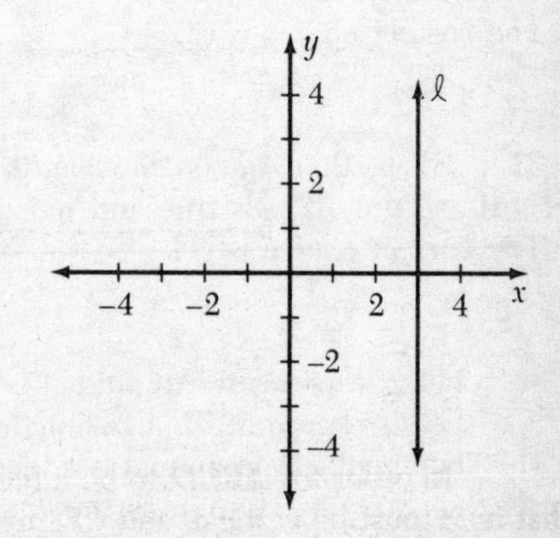

29. The given expression is: $a^2b - ab^2$

Both terms contain a *common monomial factor* of ab. The other factor is obtained by dividing each of the terms by ab: $ab(a - b)$

The correct choice is **(1)**.

30. The given equation is a *quadratic equation*: $x^2 - x - 6 = 0$

The left side is a *quadratic trinomial* that can be factored into the product of two binomials. The factors of the first term, x^2, are x and x, and they become the first terms of the binomials: $(x \quad)(x \quad) = 0$

The factors of the last term, -6, become the second terms of the binomials, but they must be chosen in such a way that the inner and outer cross-products of the binomials add up to the middle term, $-x$, of the original trinomial. Try -3 and $+2$ as the factors of -6:

$-3x$ = inner product

$(x - 3)(x + 2) = 0$

$+2x$ = outer product

Since $(-3x)+(+2x) = -x$, these are the correct factors: $(x - 3)(x + 2) = 0$

If the product of two factors is zero, either factor may equal zero: $x - 3 = 0 \lor x + 2 = 0$

Add the appropriate additive inverse to both sides of the equations, 3 for the left equation, and -2 for the right:

$$\begin{array}{rr} 3 = 3 & -2 = -2 \\ \hline x \quad = 3 & x \quad = -2 \end{array}$$

The solution set is $\{3, -2\}$.

The correct choice is **(2)**.

31. "If p, then q" is the *conditional* or *implication*, $p \to q$. It is false if and only if p is true and q is false.

The correct choice is **(1)**.

32. Let p represent "An animal is a black cat."

Let q represent "It has four legs."

The statement "If an animal is a black cat, then it has four legs" is the *implication* $p \to q$. Here p is the *hypothesis* or *antecedent* of $p \to q$, and q is the *conclusion* or *consequent*.

The *converse* of $p \to q$ is formed by interchanging its hypothesis and conclusion. The converse of $p \to q$ is therefore $q \to p$, or "If an animal has four legs, then it is a black cat."

The correct choice is **(1)**.

33.

Interval	Frequency
0–5	1
6–10	2
11–15	2
16–20	4
	9

The median is the middle value if the data are arranged in order of magnitude.

Find the total of the frequency column. The total shows that there are 9 items of data. The 5th one is the median; 4 items are less than it and 4 are greater than it.

Count down from the top of the table to reach the 5th item. The first three intervals have respectively 1, 2, and 2 items—a total of 5 items of data. Thus the 5th item lies within the 11–15 interval. The median is in the 11–15 interval.

The correct choice is (3).

34. A rational number is a number that can be represented as the quotient of two integers.

Consider each choice in turn:

(1) $\frac{3}{2}$ is in the form of the quotient of two integers, so it is a rational number.

(2) $\sqrt{7}$ cannot be expressed as the quotient of two integers. It does not have a decimal equivalent that is either a terminating or a repeating decimal (such decimals are the only decimals that can be expressed as quotients of two integers). Therefore $\sqrt{7}$ is an irrational number.

(3) $\sqrt{16} = 4$, which can be expressed as $\frac{4}{1}$. Since $\frac{4}{1}$ is the quotient of two integers, $\sqrt{16}$ is a rational number.

(4) $0.\overline{29}$ represents a nonterminating repeating decimal, 0.292929 . . . . Note that, if $x = 0.292929\ldots$, then $0.01x = 0.00292929\ldots$.

Therefore: $x = 0.29 + 0.01x$

Clear decimals by multiplying all terms in the equation by 100:

$$100x = 100(0.29) + 100(0.01x)$$
$$100x = 29 + x$$

Add $-x$ (the additive inverse of x) to both sides of the equation:

$$-x = \quad -x$$
$$99x = 29$$

Divide both sides of the equation by 99:
$$\frac{99x}{99} = \frac{29}{99}$$
$$x = \frac{29}{99}$$

Since $0.\overline{29} = \frac{29}{99}$, we have expressed it as the quotient of two integers; hence $0.\overline{29}$ is a rational number.

The correct choice is **(2)**.

35. A cube is a rectangular solid whose length, width, and height are all equal. Its volume is the product of its length, width, and height. If the volume of a cube is V and the length of one edge is x, then:
$$V = x^3$$

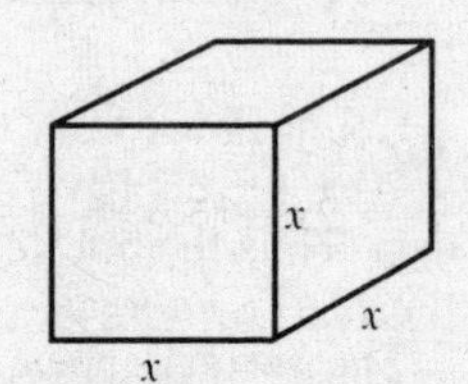

If the edge of the cube is doubled, its length will now be $2x$. Call the new volume V':
$$V' = (2x)^3$$
$$V' = 8x^3$$

The old volume V, or x^3, has been multiplied by 8 to give the new volume, $8x^3$.

The correct choice is **(4)**.

PART TWO

36. a. To draw the graph of the inequality $y > 2x + 5$, we first draw the graph of the equation $y = 2x + 5$. Prepare a table of pairs of values of x and y by selecting any three convenient values for x and substituting them in the equation to calculate the corresponding values of y:

x	$2x + 5$	$= y$
-3	$2(-3) + 5 = -6 + 5$	$= -1$
0	$2(0) + 5 = 0 + 5$	$= 5$
2	$2(2) + 5 = 4 + 5$	$= 9$

Plot the points $(-3, -1)$, $(0, 5)$, and $(2, 9)$, and draw a *broken* line through them. The line is the graph of the equation $y = 2x + 5$. The *broken* line indicates that points on it are *not* part of the solution set of $y > 2x + 5$.

The solution set of the inequality $y > 2x + 5$ lies on one side of the line $y = 2x + 5$. To find out on which side of the line the solution set lies, choose a convenient point, say (0, 0), and substitute its coordinates in the inequality:

$$y > 2x + 5$$
$$0 \overset{?}{>} 2(0) + 5$$
$$0 \overset{?}{>} 0 + 5$$
$$0 \not> 5$$

Since (0, 0) does *not* satisfy the inequality, it does *not* lie in the solution set of $y > 2x + 5$. Therefore shade the *opposite* side of the line $y = 2x + 5$ with cross-hatching extending up and to the left. The shaded part represents the graph of $y > 2x + 5$.

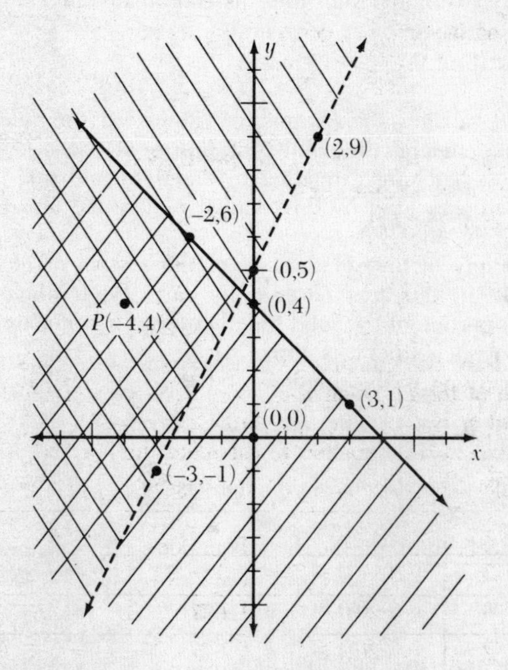

To graph $x + y \leq 4$, rearrange it so that it is solved for y by adding $-x$ (the additive inverse of x) to both sides:

$$\begin{array}{r} x + y \leq 4 \\ \underline{-x \quad\ = \ -x} \\ y \leq 4 - x \end{array}$$

Graph the equation $y = 4 - x$ by preparing a table of pairs of values of x and y after choosing three convenient values for x:

x	$4 - x$	$= y$
-2	$4 - (-2) = 4 + 2$	$= 6$
0	$4 - 0$	$= 4$
3	$4 - 3$	$= 1$

Plot the points $(-2, 6)$, $(0, 4)$, and $(3, 1)$, and draw a *solid* line through them. The line is the graph of $x + y = 4$. It is shown as a *solid* line to indicate that points on it are part of the solution set of $x + y \leq 4$.

The part of the solution set represented by $x + y < 4$ lies on one side of the line $x + y = 4$. To find out on which side, choose a convenient test point, say $(0, 0)$, and substitute its coordinates in $x + y < 4$ to see whether they satisfy it:

$$x + y < 4$$
$$0 + 0 \overset{?}{<} 4$$
$$0 < 4 \checkmark$$

Since $(0, 0)$ satisfies the inequality, it lies on the side of the line $x + y = 4$ that contains the points that represent $x + y < 4$. Shade this side with cross-hatching extending to the right and up. The graph of $x + y \leq 4$ is represented by this shaded region and also the solid-line boundary on one side of it.

b. The points in the solution set of the system of inequalities are those that lie in the area covered by *both* types of cross-hatching, including the portion of the solid line that forms a boundary of that area. An example of such a point is $P(-4, 4)$.

$(-4, 4)$ is a point in the solution set.

37. a. The given system of equations is: $\begin{cases} 2x + 3y = -5 \\ 3x - 2y = 12 \end{cases}$

Multiply each term of the first equation by 2: $2(2x) + 2(3y) = 2(-5)$

Multiply each term of the second equation by 3: $3(3x) - 3(2y) = 3(12)$

Perform the indicated multiplications:

$$4x + 6y = -10$$
$$9x - 6y = 36$$

Add the last two equations to eliminate y: $13x = 26$

Divide both sides of the equation by 13: $\frac{13x}{13} = \frac{26}{13}$

$$x = 2$$

Substitute 2 for x in the first equation:

$$2(2) + 3y = -5$$
$$4 + 3y = -5$$

Add -4 (the additive inverse of 4) to both sides of the equation:

$$\underline{-4 \qquad = -4}$$
$$3y = -9$$

Divide both sides of the equation by 3:

$$\frac{3y}{3} = \frac{-9}{3}$$
$$y = -3$$

The solution is **(2, −3)** or $\boldsymbol{x = 2, y = -3}$.

CHECK: The solution must satisfy *both original* equations. Substitute 2 for x and -3 for y in *both* of the *original* equations to see whether they are satisfied:

$$\underline{2x + 3y = -5} \qquad \underline{3x - 2y = 12}$$
$$2(2) + 3(-3) \stackrel{?}{=} -5 \qquad 3(2) - 2(-3) \stackrel{?}{=} 12$$
$$4 - 9 \stackrel{?}{=} -5 \qquad 6 + 6 \stackrel{?}{=} 12$$
$$-5 = -5 \checkmark \qquad 12 = 12 \checkmark$$

b. The point (2, −3) lies in Quadrant IV.
Quadrant IV.

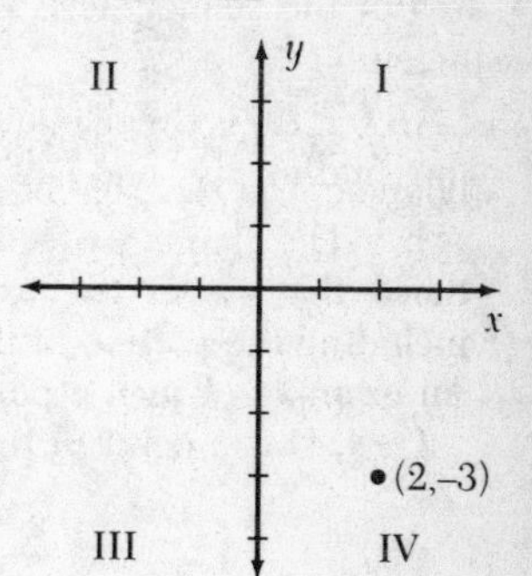

38. a. Let x = the length of $\overline{AB}$.
Then $x + 1$ = the length of $\overline{BD}$.

Since $ABCD$ is a rectangle and all angles of a rectangle are right angles, $\angle A$ is a right angle.

Apply the Pythagorean Theorem to right triangle BAD: In a right triangle, the square of the length of the hypotenuse is equal to the sum of the squares of the lengths of the legs:

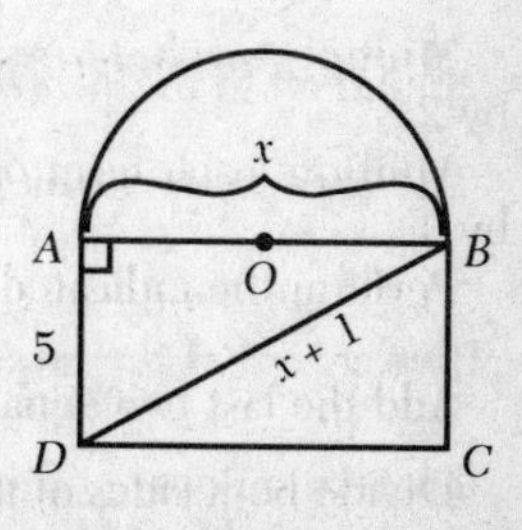

$$(x + 1)^2 = x^2 + 5^2$$

$$\begin{array}{r} x + 1 \\ x + 1 \\ \hline x^2 + x \\ x + 1 \\ \hline x^2 + 2x + 1 \end{array}$$

$$x^2 + 2x + 1 = x^2 + 25$$

Add $-x^2$ (the additive inverse of x^2) and also add -1 (the additive inverse of 1) to both sides of the equation:

$$\begin{array}{rcl} -x^2 \quad -1 & = & -x^2 - 1 \\ \hline 2x & = & 24 \end{array}$$

Divide both sides of the equation by 2:

$$\frac{2x}{2} = \frac{24}{2}$$

$$x = 12$$

AB = **12.**

b. The area, A, of a circle whose radius is r is given by the formula:

$$A = \pi r^2$$

Let S = the area of the semicircle. The area of the semicircle is one-half the area of the circle:

$$S = \frac{\pi r^2}{2}$$

AB (or 12) is the diameter of the semicircle. The radius is one-half the diameter; $r = \frac{1}{2}(12) = 6$:

$$S = \frac{\pi(6)^2}{2}$$

Let $\pi = 3.14$:

$$S = \frac{3.14(6)^2}{2}$$

$$S = 1.57(36)$$

$$\begin{array}{r} 1.57 \\ \times \ 36 \\ \hline 9\,42 \\ 47\,1 \\ \hline 56.52 \end{array}$$

$$S = 56.52$$

Round off to the *nearest integer*:

$$S = 57$$

The area of the semicircle to the *nearest integer* is **57**.

39. Let x = the smaller positive integer.
Then $x + 3$ = the larger integer.

Twice	the square of the smaller	increased by	3	times the larger	is	74
↓		↓	↓		↓	↓
2	x^2	+	3	$(x + 3)$	=	74

The equation to use is: $2x^2 + 3(x + 3) = 74$

Apply the distributive law to remove parentheses by multiplying each term inside the parentheses by 3: $2x^2 + 3x + 9 = 74$

Add -74 (the additive inverse of 74) to both sides of the equation:

$$\begin{array}{r} 2x^2 + 3x + 9 = 74 \\ -74 = -74 \\ \hline 2x^2 + 3x - 65 = 0 \end{array}$$

This is a *quadratic equation*. The left side is a *quadratic trinomial* that can be factored into the product of two binomials. The factors of the first term, $2x^2$, are $2x$ and x, and they become the first terms of the binomials: $(2x \quad)(x \quad) = 0$

The factors of the last term, -65 become the second terms of the binomials, but they must be chosen in such a way that the sum of the inner and outer cross-products of the binomials equals the middle term, $+3x$, of the original trinomial.

Try $+13$ and -5 as the factors of -65:

$+13x$ = inner product

$(2x + 13)(x - 5) = 0$

$-10x$ = outer product

Since $(+13x) + (-10x) = +3x$, these are the correct factors: $(2x + 13)(x - 5) = 0$

If the product of two factors is zero, either factor may equal zero: $2x + 13 = 0 \lor x - 5 = 0$

Add the appropriate additive inverse to both sides of each equation, -13 for the left equation, and $+5$ for the right:

$$\begin{array}{rr} -13 = -13 & +5 = +5 \\ \hline 2x = -13 & x = 5 \end{array}$$

Divide both sides of the left equation by 2:

$$\frac{2x}{2} = \frac{-13}{2}$$

$$x = -\frac{13}{2}$$

Reject $-\frac{13}{2}$ as a solution to the problem since it is neither positive nor an integer:

$$x = 5$$

$$x + 3 = 8$$

The positive integers are **5** and **8**.

40. a.

p	q	$p \leftrightarrow q$
	T	F

$p \leftrightarrow q$ is the *equivalence relation* or *biconditional* between p and q. The biconditional has the truth value T when p and q have the same truth value, either both T or both F. Since $p \leftrightarrow q$ is F and q is T, p must be F.

p is **false.**

b.

p	q	$q \rightarrow p$
	T	F

$q \rightarrow p$ is the *conditional*, q implies p. The conditional has the truth value F if p is F and q is T. Therefore p must be F.

p is **false.**

c.

p	q	$p \vee q$
T		T

$p \vee q$ is the *disjunction* between p and q. The disjunction will have the truth value T if either p or q, or both, have the truth value T. Here p has the value T, so if q were either T or F, $p \vee q$ would still be T.

The truth value of q **cannot be determined.**

d.

p	q	$\sim(p \wedge q)$
	T	F

$\sim (q \wedge q)$ is the *negation* of the *conjunction* of p and q. Since $\sim (p \wedge q)$ has the truth value F, $(p \wedge q)$ must have the truth value T. But the conjunction $(p \wedge q)$ has the truth value T only if p and q *both* have the truth value T. Thus p must have the value T.

p is **true.**

e.

p	q	$p \vee \sim q$
	T	F

$p \vee \sim q$ is the *disjunction* between $\sim q$ and p; $\sim q$ is the *negation* of q. Since q is T, $\sim q$ must be F. The disjunction $p \vee \sim q$ has the truth value T if either p, or $\sim q$, or both, have the truth value T. We have shown that $\sim q$ is F. Since p cannot be T, p must be F.

p is **false.**

41. a. **Let x = the measure of $\angle A$.**

Then $3x$ = the measure of $\angle B$.

And $2(3x)$ or $6x$ = the measure of $\angle C$.

The sum of the measures of the angles of a triangle is 180°.
The equation is $\mathbf{x + 3x + 6x = 180}$.

b. Let x = the first consecutive integer.
Then $x + 1$ = the second consecutive integer.
And $x + 2$ = the third consecutive integer.

The first integer plus the second is the third integer plus 9

$$\begin{array}{ccccccc} \downarrow & \downarrow & \downarrow & \downarrow & \downarrow & \downarrow & \downarrow \\ x & + & (x+1) & = & (x+2) & + & 9 \end{array}$$

The equation is $\mathbf{x + (x + 1) = (x + 2) + 9}$.

42. a.

Tree Diagram:

First Drawing	Second Drawing
1	2, 3, 4, 5
2	1, 3, 4, 5
3	1, 2, 4, 5
4	1, 2, 3, 5
5	1, 2, 3, 4

(START branches to each first drawing.)

Sample Space:

First Drawing	Second Drawing
1	2
1	3
1	4
1	5
2	3
2	4
2	5
3	1
3	2
3	4
3	5
4	1
4	2
4	3
4	5
5	1
5	2
5	3
5	4

b. Probability of an event occurring

$$= \frac{\text{number of successful outcomes}}{\text{total possible number of outcomes}}$$

(1) The successful outcomes are the numbers greater than 40. In the tree diagram, they are all on the paths that lead to either a 4 or a 5 on the first drawing. Each of these paths has 4 second branches, so there are 8 numbers greater than 40. In the sample space, the numbers greater than 40 are on the last 8 lines, the ones that have either a 4 or a 5 listed under "First Drawing."

The total possible number of outcomes is the total number of lines in the sample space, 20, or the total number of complete paths on the tree diagram from "Start" through "Second Drawing"; there are 5 × 4, or 20, complete paths.

The probability that the two-digit number is greater than 40 is $\frac{8}{20}$.

The probability is $\mathbf{\frac{8}{20}}$.

(2) For the probability that the two-digit number is divisible by 5, the successful outcomes are the numbers that end in 5. On the tree diagram, one branch in each cluster except the last leads to a 5 on the "Second Drawing"; thus, there are 4 successful outcomes. In the sample space, 4 of the lines have numbers ending in 5: 15, 25, 35, and 45.

The probability that the two-digit number is divisible by 5 is $\frac{4}{20}$.

The probability is $\mathbf{\frac{4}{20}}$.

(3) Since the first disk is not replaced before the second drawing is made, any number selected on the "First Drawing" is not available for selection on the "Second Drawing." Thus, there is no successful outcome in which the two-digit number has the same digit in both places.

The probability of the same digit in both places is $\frac{0}{20}$ or 0.

The probability is **0**.

Topic	Question Numbers	Number of Points	Your Points	Your Percentage
1. Numbers (rat'l, irrat'l); Percent	34	2		
2. Properties of No. Systems	13, 20	2 + 2 = 4		
3. Operations on Rat'l Nos. and Monomials	21	2		
4. Operations on Multinomials	—	0		
5. Square Root; Operations involving Radicals	19	2		
6. Evaluating Formulas and Expressions	11	2		
7. Linear Equations (simple cases incl. parentheses)	2	2		
8. Linear Equations Containing Decimals or Fractions	5, 9	2 + 2 = 4		
9. Graphs of Linear Functions (slope)	12, 28	2 + 2 = 4		
10. Inequalities	24	2		
11. Systems of Eqs. & Inequalities (alg. & graphic solutions)	36a, b, 37a, b	8 + 2 + 9 + 1 = 20		
12. Factoring	10, 29	2 + 2 = 4		
13. Quadratic Equations	30	2		
14. Verbal Problems	39	10		
15. Variation	4, 35	2 + 2 = 4		
16. Literal Eqs.; Expressing Relations Algebraically	14, 41a, b	2 + 5 + 5 = 12		
17. Factorial *n*	—	0		
18. Areas, Perims., Circums., Vols. of Common Figures	38b	2		
19. Geometry (≅, ∡ meas., ∥ lines, compls., suppls.)	3, 7, 27	2 + 2 + 2 = 6		
20. Ratio & Proportion (incl. similar triangles & polygons)	15	2		
21. Pythagorean Theorem	38a	6		
22. Logic (symbolic rep., logical forms, truth tables)	26, 31, 32, 40a, b, c, d, e	2 + 2 + 2 + 2 + 2 + 2 + 2 + 2 = 16		
23. Probability (incl. tree diagrams & sample spaces)	16, 42b	2 + 6 = 8		
24. Combinatorics (arrangements, permutations)	—	0		
25. Statistics (central tend., freq. dist., histograms, quartiles, percentiles)	22, 33, 42a	2 + 2 + 4 = 8		

Topic	Question Numbers	Number of Points	Your Points	Your Percentage
26. Properties of Triangles and Quadrilaterals	1, 6, 8, 18, 25	2 + 2 + 2 + 2 + 2 = 10		
27. Transformations (reflect., translat., rotat., dilat.)	—	0		
28. Symmetry	23	2		
29. Area from Coord. Geom.	—	0		
30. Dimensional Analysis	—	0		
31. Scientific Notation; Neg. & Zero Exponents	17	2		

Examination June 1990

Three-Year Sequence for High School Mathematics—Course I

PART ONE

DIRECTIONS: *Answer 30 questions from this part. Each correct answer will receive 2 credits. No partial credit will be allowed. Write your answers in the spaces provided. Where applicable, answers may be left in terms of* π *or in radical form.*

1 A 50-foot tree casts a shadow of 40 feet. At the same time, a boy casts a shadow of 4 feet. Expressed in feet, how tall is the boy? 1_____

2 In the accompanying diagram of $\triangle ABC$, $\overline{AC} \cong \overline{BC}$ and $m\angle A = 70$. Find the measure of the vertex angle.

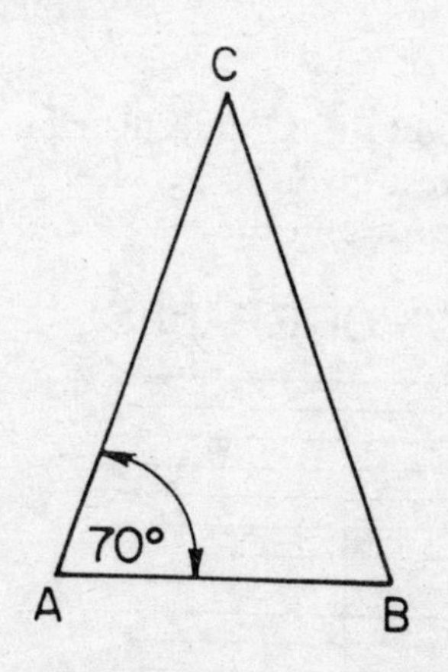

2_____

3 In the accompanying diagram, $\overleftrightarrow{AB}$ and $\overleftrightarrow{CD}$ intersect at E. If $m\angle AED = 9x + 10$ and $m\angle BEC = 2x + 52$, find the value of x.

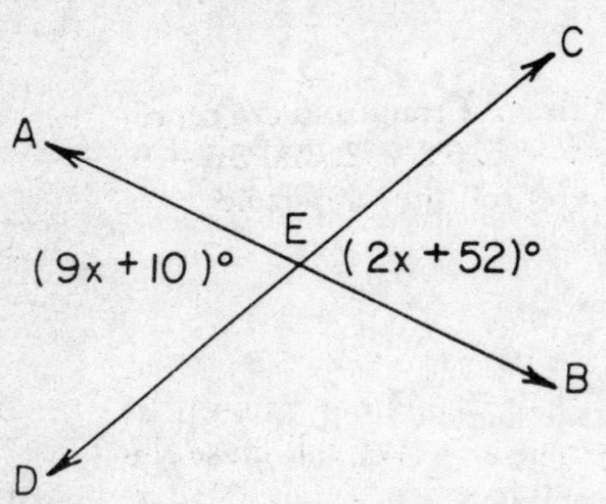

3_____

4 Let p represent the statement "I will win," and let q represent the statement "I practice." Write in symbolic form: "If I do not practice, then I will not win." 4_____

5 Solve for x: $\frac{3x}{4} - 1 = 2$ 5_____

6 The histogram below shows the grade distribution for a mathematics test given to Ms. Keith's class. How many students are in the class?

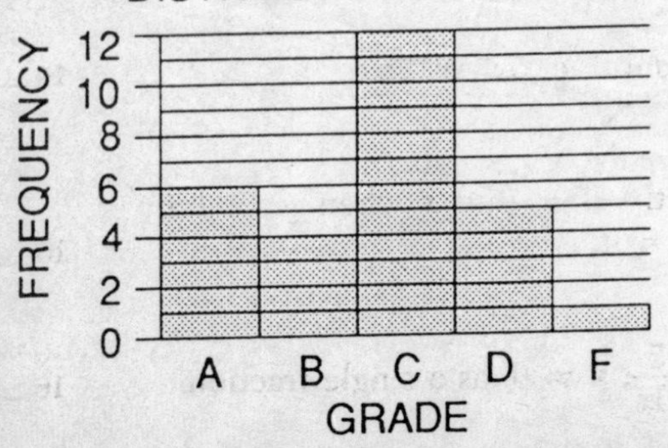

6_____

7 Solve for x: $0.03x - 2.1 = 0.3$ 7____

8 Solve for x: $\frac{3}{x + 2} = \frac{1}{x}$, $x \neq 0, x \neq -2$ 8____

9 The lengths of the sides of a trapezoid are represented by $2x + 3$, $4x - 5$, $3x + 2$, and $5x - 9$. Express the perimeter of the trapezoid as a binomial in terms of x. 9____

10 If the replacement set for x is $\{2,3,4,5,6\}$, what is the probability that a number chosen at random from the replacement set will make the sentence $3x + 2 \leq 20$ true? 10____

11 Solve for x: $4(2x - 1) = 2x + 35$ 11____

12 In rectangle *ABCD*, *AB* is represented by $2x + 1$ and *BC* is represented by $x + 3$. Express the area of rectangle *ABCD* as a trinomial in terms of x. 12____

13 The measures of two supplementary angles are in the ratio 4:5. Find the number of degrees in the measure of the *smaller* angle. 13____

14 From $7x^2 - 4x$ subtract $5x^2 + 2x$. 14____

15 If x varies directly as y, find x when $y = 1$ if $x = 12$ when $y = 4$. 15____

16 Express $\frac{3}{2x} + \frac{5}{3x}$, $x \neq 0$, as a single fraction. 16____

17 In the accompanying diagram, $\angle ACD$ is an exterior angle of $\triangle ABC$. If $m\angle B = 40$, $m\angle A = 2x$, and $m\angle ACD = 3x$, what is the value of x? 17_____

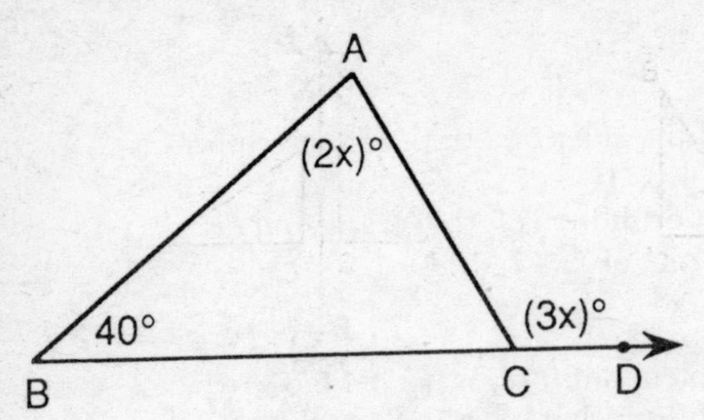

18 The volume of a rectangular solid is 180 cubic centimeters. The length is 10 centimeters and the width is 4 centimeters. Using the formula $V = lwh$, find the number of centimeters in the height. 18_____

DIRECTIONS (19–35): *For each question chosen, write the* numeral *preceding the word or expression that best completes the statement or answers the question.*

19 The expression $(x - 4)^2$ is equivalent to

(1) $x^2 - 16$
(2) $x^2 + 16$
(3) $x^2 - 8x + 16$
(4) $x^2 + 8x + 16$

19_____

20 What is the product of $3x^4y^2$ and $2xy^3$?

(1) $6x^5y^5$
(2) $6x^4y^5$
(3) $6x^4y^6$
(4) $6x^5y^6$

20_____

21 Which property is *not* true for *all* parallelograms?

(1) Opposite angles are congruent.
(2) Consecutive angles are supplementary.
(3) Opposite sides are congruent.
(4) Diagonals are congruent.

21_____

22 In which figure is $\triangle A'B'C'$ a reflection of $\triangle ABC$ in line ℓ?

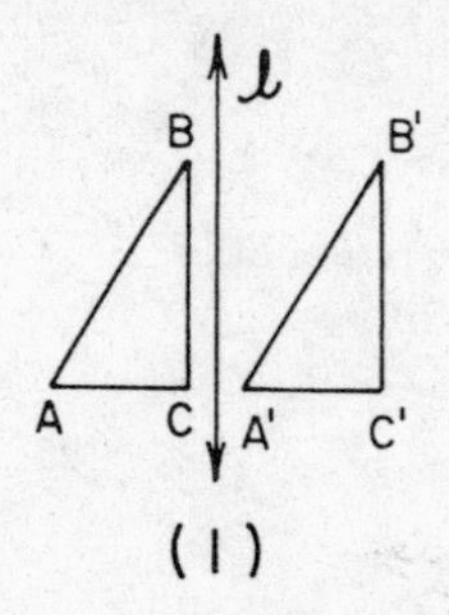

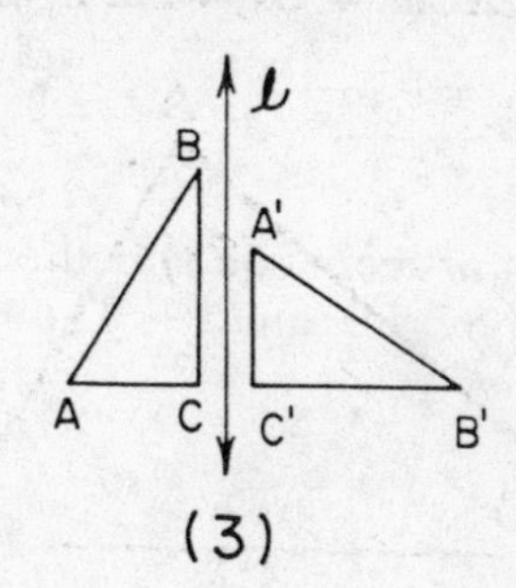

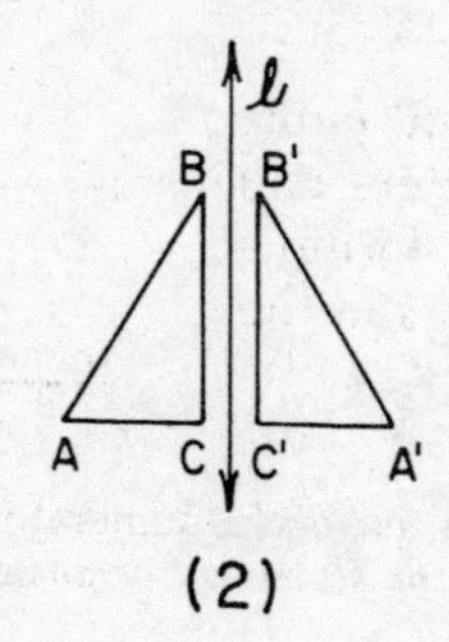

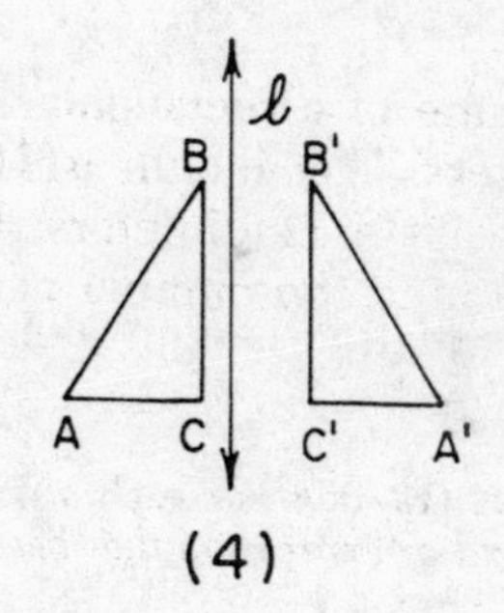

22_____

23 If the length of a rectangle is 3 and the width is 2, the length of the diagonal is

(1) $\sqrt{5}$ (3) 5
(2) $\sqrt{13}$ (4) 13

23_____

24 Which phrase describes the graph of $y = -1$ on the coordinate plane?

(1) a line parallel to the y-axis and 1 unit to the right of it
(2) a line parallel to the y-axis and 1 unit to the left of it
(3) a line parallel to the x-axis and 1 unit below it
(4) a line parallel to the x-axis and 1 unit above it

24_____

25 What is the value of $\frac{6!}{3!}$?

(1) 6 (3) 120
(2) 2 (4) 720

25_____

26 The set of scores on a mathematics test is 72, 80, 80, 82, 87, 89, and 91. The mean score is

(1) 84 (3) 82
(2) 83 (4) 80

26_____

27 If x is an integer, which is the solution set of $-1 \leq x < 2$?

(1) $\{0,1\}$ (3) $\{0,1,2\}$
(2) $\{-1,0,1,2\}$ (4) $\{-1,0,1\}$

27_____

28 Which equation is equivalent to $x + 2y = 6$?

(1) $y = -x + 6$ (3) $y = -x + 3$
(2) $y = -\frac{1}{2}x + 6$ (4) $y = -\frac{1}{2}x + 3$

28_____

29 Let p represent "$x > 5$" and let q represent "x is a multiple of 3." If $x = 12$, which statement is false?

(1) $p \vee q$ (3) $p \rightarrow q$
(2) $\sim q \wedge p$ (4) $p \leftrightarrow q$

29_____

30 What is the solution set of $x^2 - x - 20 = 0$?

(1) $\{5,-4\}$ (3) $\{-10,2\}$
(2) $\{-5,4\}$ (4) $\{10,-2\}$

30_____

31 Which number is equal to 3.6×10^5?

(1) 360,000 (3) 0.000036
(2) 3,600,000 (4) 0.0000036

31_____

32 Which value of x will make the fraction $\frac{x - 3}{x + 6}$ undefined?

(1) 6
(2) −6
(3) 3
(4) −3

32_____

33 If the coordinates of the vertices of $\triangle ABC$ are $A(3,-2)$, $B(7,-2)$, and $C(5,5)$, what is the area of the triangle?

(1) 10
(2) 14
(3) 20
(4) 28

33_____

34 The expression $2\sqrt{3} - \sqrt{27}$ is equivalent to

(1) $2\sqrt{24}$
(2) $5\sqrt{3}$
(3) $-5\sqrt{3}$
(4) $-\sqrt{3}$

34_____

35 Which figure does *not* have line symmetry?

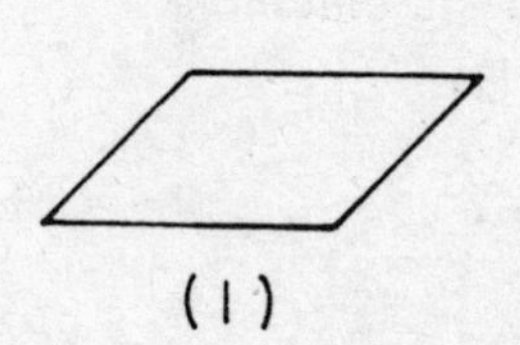
(1)

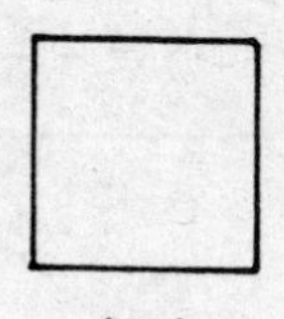
(3)

(2)

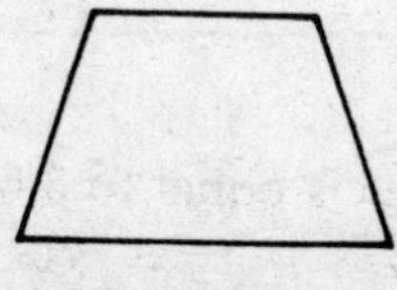
(4)

35_____

PART TWO

DIRECTIONS: *Answer* four *questions from this part. Show all work unless otherwise directed.*

36 *a* On the same set of coordinate axes, graph the following system of inequalities:

$$y \le -3x + 2$$
$$y - x > 0 \quad [8]$$

b Write the coordinates of a point *not* in the solution set of the inequalities graphed in part *a*. [2]

37 Twice the square of an integer is five less than eleven times the integer. Find the integer. [*Only an algebraic solution will be accepted.*] [4,6]

38 Solve the following system of equations algebraically and check:

$$2x + 3y = 11$$
$$5x - 2y = -20 \quad [8,2]$$

39 In rectangle $ABCD$, the ratio of $AB:BC$ is 4:3. The perimeter of the rectangle is 56 centimeters.

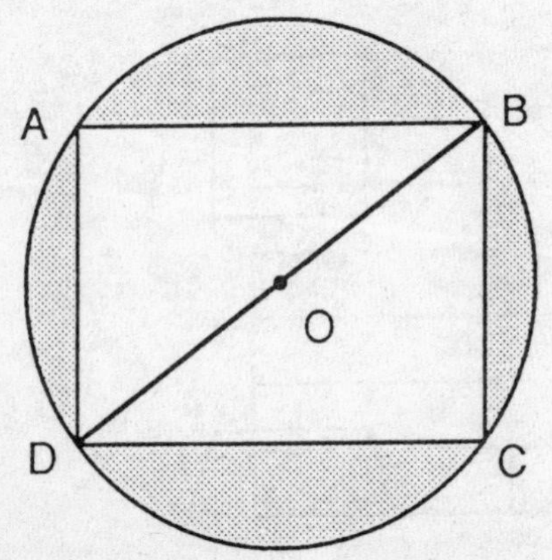

a Find AB. [2]
b Find BD. [3]

c Express, in terms of π, the area of circle O. [2]

d Express, in terms of π, the area of the shaded region. [3]

40 The table below shows the cumulative frequency of the ages of 35 people standing in a cafeteria line.

Interval	Cumulative Frequency
10–19	2
10–29	17
10–39	27
10–49	32
10–59	32
10–69	35

a *On your answer paper*, copy and complete the frequency table below, based on the data given in the cumulative frequency table above. [1]

Interval	Frequency
10–19	2
20–29	
30–39	
40–49	
50–59	
60–69	

b Construct a frequency histogram using the table completed in part *a*. [4]

c Using the frequency table in part *a*, in which interval does the median occur? [2]

d What is the probability that a person chosen at random from the line is at least 40 years old? [2]

e What is the probability that a person chosen at random from the line is between 50 and 59 years old? [1]

41 Adam has a bag containing four yellow gumdrops and one red gumdrop. He will eat one of the gumdrops, and a few minutes later, he will eat a second gumdrop.

a What is the probability Adam will eat a yellow gumdrop first and a red gumdrop second? [3]

b What is the probability Adam will eat two yellow gumdrops? [3]

c What is the probability Adam will eat two gumdrops having different colors? [2]

d What is the probability Adam will eat two red gumdrops? [2]

42 Let p represent: "The flowers are not in bloom."
Let q represent: "It is raining."
Let r represent: "The grass is green."

a Write, in symbolic form, the converse of "If the flowers are in bloom, then the grass is green." [2]

b Write, in symbolic form, the inverse of "If it is raining, the grass is green." [2]

c Write in sentence form: $p \wedge \sim q$ [2]

d Write in sentence form: $\sim r \vee \sim p$ [2]

e Which of these four statements must have the same truth value as $\sim q \rightarrow r$? [2]

(1) $q \rightarrow \sim r$
(2) $\sim r \rightarrow q$
(3) $\sim q \rightarrow \sim r$
(4) $r \rightarrow q$

Answers June 1990

Three-Year Sequence for High School Mathematics—Course I

ANSWER KEY

PART ONE

1. 5
2. 40
3. 6
4. $\sim q \rightarrow \sim p$
5. 4
6. 28
7. 80
8. 1
9. $14x - 9$
10. 1
11. 6.5
12. $2x^2 + 7x + 3$
13. 80
14. $2x^2 - 6x$
15. 3
16. $\frac{19}{6x}$
17. 40
18. 4.5
19. (3)
20. (1)
21. (4)
22. (2)
23. (2)
24. (3)
25. (3)
26. (2)
27. (4)
28. (4)
29. (2)
30. (1)
31. (1)
32. (2)
33. (2)
34. (4)
35. (1)

Part Two—*See* Answers Explained.

ANSWERS EXPLAINED

PART ONE

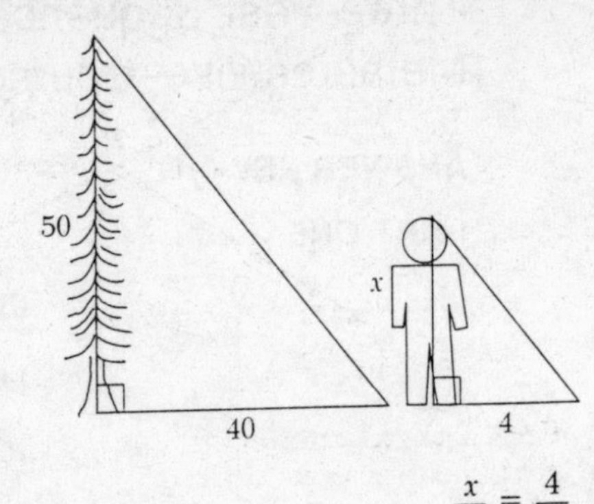

1. Let x = the number of feet in the boy's height.

The tree and its shadow and the boy and his shadow form the legs of two similar right triangles. In similar triangles, the corresponding sides are in proportion:

$$\frac{x}{50} = \frac{4}{40}$$

In a proportion, the product of the means equals the product of the extremes (cross-multiply):

$$40x = 4(50)$$
$$40x = 200$$

Divide both sides of the equation by 40:

$$\frac{40x}{40} = \frac{200}{40}$$
$$x = 5$$

The boy is **5** feet tall.

2. 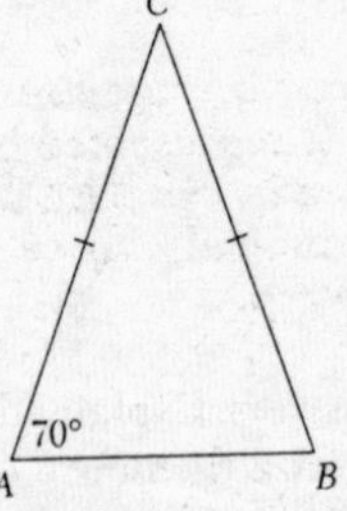

Since $\overline{AC} \cong \overline{BC}$, ΔABC is isosceles. In an isosceles triangle, the base angles are equal in measure:

$$m\angle B = m\angle A = 70$$

The sum of the measures of the three angles of a triangle is 180:

$$m\angle A + m\angle B + m\angle C = 180$$
$$70 + 70 + m\angle C = 180$$

Combine like terms:

$$140 + m\angle C = 180$$

Add −140 (the additive inverse of 140) to both sides of the equation:

$$\underline{-140 \qquad\quad = -140}$$
$$m\angle C = 40$$

The measure of the vertex angle is **40.**

3.

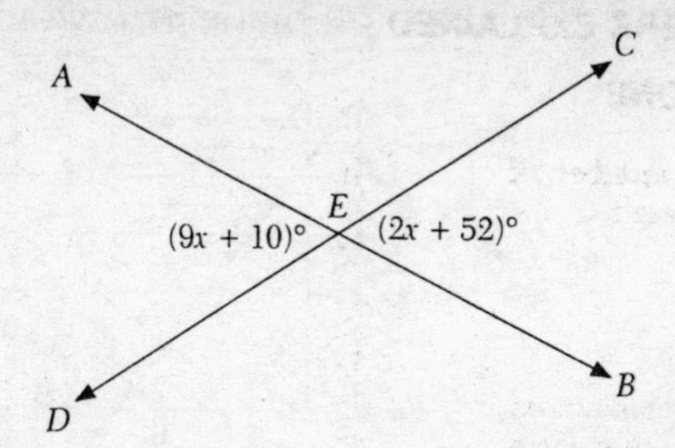

Vertical angles are congruent, that is, equal in measure:

$$m\angle AED = m\angle CEB$$
$$9x + 10 = 2x + 52$$

Add -10 (the additive inverse of 10) and also add $-2x$ (the additive inverse of $2x$) to both sides of the equation:

$$\underline{-2x - 10 = -2x - 10}$$
$$7x = 42$$

Divide both sides of the equation by 7:

$$\frac{7x}{7} = \frac{42}{7}$$
$$x = 6$$

$x = 6$.

4. p represents "I will win" and q represents "I practice." "I do not practice" is the *negation* of q, and is represented by $\sim q$. "I will not win" is the *negation* of p, and is represented by $\sim p$. "If I do not practice, then I will not win" is the *implication* or *conditional*, $\sim q \rightarrow \sim p$.

The symbolic form is $\boldsymbol{\sim q \rightarrow \sim p}$.

5. The given equation contains a fraction:

$$\frac{3x}{4} - 1 = 2$$

Clear fractions by multiplying all terms on both sides of the equation by 4:

$$4\left(\frac{3x}{4}\right) - 4(1) = 4(2)$$
$$3x - 4 = 8$$

Add 4 (the additive inverse of -4) to both sides of the equation:

$$\underline{4 = 4}$$
$$3x = 12$$

Divide both sides of the equation by 3:

$$\frac{3x}{3} = \frac{12}{3}$$
$$x = 4$$

$x = 4$.

6.

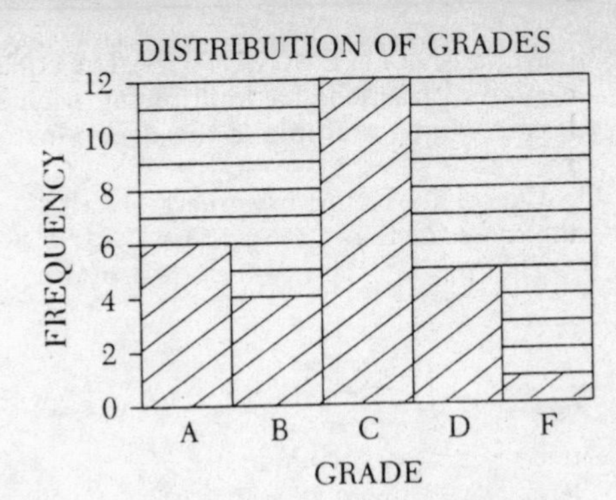

The frequency shown on the graph for each grade is the number of students who achieved that grade:

A	6
B	4
C	12
D	5
F	1
	28

Add all the frequencies to determine the number of students in the class:

There are 28 students in the class.

7. The given equation contains decimals: $0.03x - 2.1 = 0.3$

Clear decimals by multiplying each term on both sides of the equation by 100:

$$100(0.03x) - 100(2.1) = 100(0.3)$$
$$3x - 210 = 30$$

Add 210 (the additive inverse of -210) to both sides of the equation:

$$\underline{\quad 210 = 210}$$
$$3x \quad = 240$$

Divide both sides of the equation by 3:

$$\frac{3x}{3} = \frac{240}{3}$$
$$x = 80$$

$x = 80$.

8. The given equation is in the form of a proportion: $\frac{3}{x+2} = \frac{1}{x}, x \neq -2$

In a proportion, the product of the means equals the product of the extremes (cross-multiply):

$$3x = x + 2$$

Add $-x$ (the additive inverse of x) to both sides of the equation:

$$\underline{-x = -x}$$
$$2x = \quad 2$$

Divide both sides of the equation by 2:

$$\frac{2x}{2} = \frac{2}{2}$$
$$x = 1$$

$x = 1$.

ALTERNATIVE SOLUTION: The equation can be cleared of fractions by multiplying each side by the least common multiple of the denominators which is $x(x + 2)$:

$$\frac{3x(x+2)}{x+2} = \frac{x(x+2)}{x}$$

Cancel like factors in numerator and denominator of the same fraction:

$$3x = x + 2$$

The solution then continues as above.

9. The perimeter of the trapezoid is the sum of the lengths of all four sides: perimeter $= 2x + 3 + 4x - 5 + 3x + 2 + 5x - 9$

Combine like terms: perimeter $= 14x - 9$

The expression for the perimeter is $\mathbf{14x - 9}$.

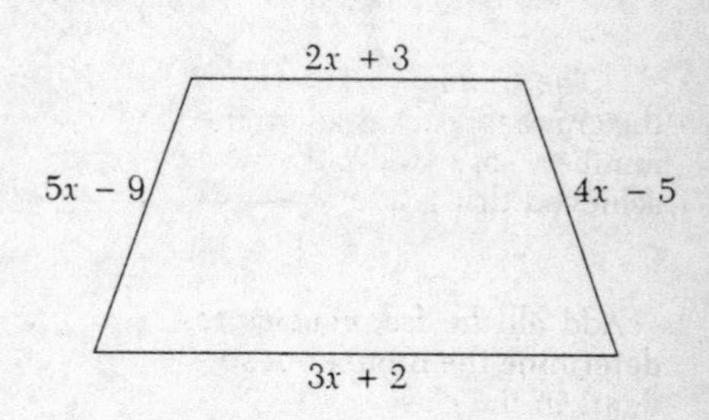

10. First solve the given inequality:

$$3x + 2 \leq 20$$

Add -2 (the additive inverse of 2) to both sides of the inequality:

$$\frac{\quad -2 = -2}{3x \quad \leq 18}$$

Divide both sides of the inequality by 3:

$$\frac{3x}{3} \leq \frac{18}{3}$$

$$x \leq 6$$

$$\text{The probability of an event occurring} = \frac{\text{number of successful outcomes}}{\text{total possible number of outcomes}}.$$

The solution shows that all numbers less than or equal to 6 will make the sentence true. The replacement set for x is $\{2, 3, 4, 5, 6\}$ which contains 5 numbers less than or equal to 6. Therefore, there can be 5 successful outcomes if a number is chosen at random from the replacement set. The total possible number of outcomes is also 5 since there are exactly 5 numbers in the replacement set from which to choose. The probability of choosing a number that will make the sentence true is $\frac{5}{5}$ or 1.

The probability is **1**.

ALTERNATIVE SOLUTION: Note that since all the numbers in the replacement set will make the sentence true, it is certain that no matter what number is chosen it will satisfy the sentence. Certainty is expressed as a probability of 1.

11. The given equation contains parentheses: $4(2x - 1) = 2x + 35$

Remove the parentheses by applying the distributive law, multiplying each term within the parentheses by the factor in front of it:

$$4(2x) - 4(1) = 2x + 35$$
$$8x - 4 = 2x + 35$$

Add +4 (the additive inverse of −4) and also add $-2x$ (the additive inverse of $2x$) to both sides of the equation:

$$\begin{array}{r} -2x + 4 = -2x + 4 \\ \hline 6x \quad = \quad 39 \end{array}$$

Divide both sides of the equation by 6:

$$\frac{6x}{6} = \frac{39}{6}$$
$$x = 6\frac{1}{2} \text{ or } 6.5$$

$x = 6\frac{1}{2}$ **or 6.5.**

12.

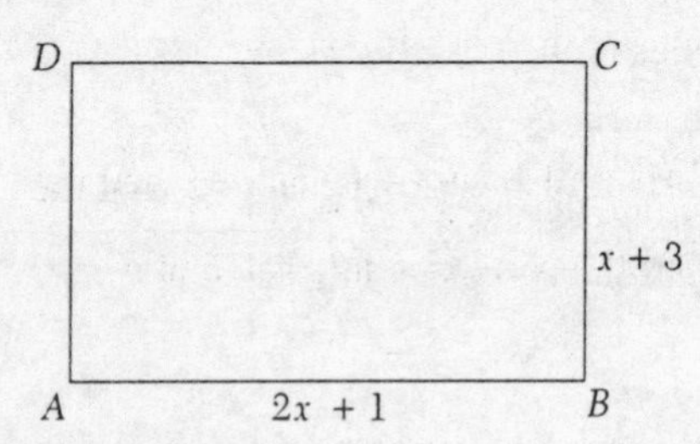

The area, A, of a rectangle is equal to the product of its length and width: $A = (2x + 1)(x + 3)$

Multiply the two binomials together:

$$\begin{array}{r} 2x + 1 \\ x + 3 \\ \hline 2x^2 + x \quad\; \\ 6x + 3 \\ \hline 2x^2 + 7x + 3 \end{array}$$

$A = 2x^2 + 7x + 3$

The trinomial expressing the area is $\mathbf{2x^2 + 7x + 3}$.

13. Let $4x$ = the measure of the smaller angle.

Then $5x$ = the measure of the larger angle.

The sum of the measures of two supplementary angles is 180: $4x + 5x = 180$

Combine like terms: $9x = 180$

Divide both sides of the equation by 9:

$$\frac{9x}{9} = \frac{180}{9}$$
$$x = 20$$

The *smaller* angle is represented by $4x$: $4x = 4(20) = 80$

The measure of the *smaller* angle is **80.**

14. Write the binomial to be subtracted beneath the other one with like terms in the same column:

$$\begin{array}{r} 7x^2 - 4x \\ \underline{5x^2 + 2x} \end{array}$$

Change the signs of the terms in the binomial to be subtracted:

$$\begin{array}{r} 7x^2 - 4x \\ \underline{-5x^2 - 2x} \end{array}$$

Combine the coefficients in each column:

$$2x^2 - 6x$$

The difference is $2x^2 - 6x$.

15. The relationship, x varies directly as y, may be represented by:

$$x = ky$$

where k is a constant.

It is given that $x = 12$ when $y = 4$:

$$12 = 4k$$

Divide both sides of the equation by 4:

$$\frac{12}{4} = \frac{4k}{4}$$

$$3 = k$$

Since k is a constant, it is always equal to 3, and the relationship can be written as:

$$x = 3y$$

Let $y = 1$ to find the corresponding value of x:

$$x = 3(1)$$

$$x = 3$$

$x = 3$ when $y = 1$.

16. In their given form, the fractions cannot be combined because they have different denominators:

$$\frac{3}{2x} + \frac{5}{3x}$$

Find the least common denominator (L.C.D.). The L.C.D. is the simplest expression into which all the denominators will divide evenly. The L.C.D. for $2x$ and $3x$ is $6x$.

Convert each fraction to an equivalent fraction having the L.C.D. Multiply the first fraction by 1 in the form $\frac{3}{3}$, and the second fraction by 1 in the form $\frac{2}{2}$:

$$\frac{3}{3}\left(\frac{3}{2x}\right) + \frac{2}{2}\left(\frac{5}{3x}\right)$$

$$\frac{9}{6x} + \frac{10}{6x}$$

Since the fractions now have the same denominator, they may be combined by combining their numerators:

$$\frac{9 + 10}{6x}$$

Combine like terms in the numerator:

$$\frac{19}{6x}$$

The single fraction is $\frac{19}{6x}$.

17.

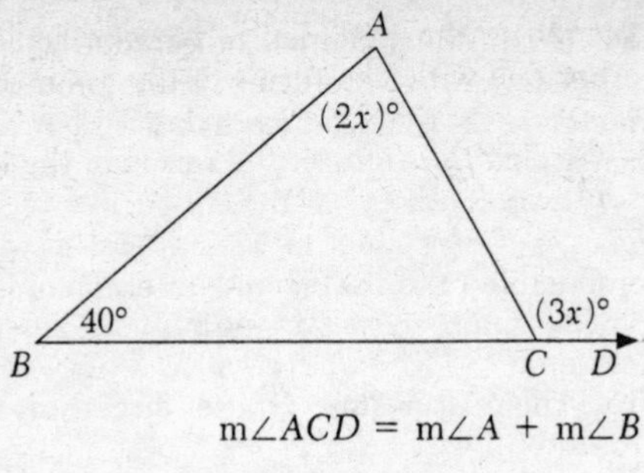

The measure of an exterior angle of a triangle is equal to the sum of the measures of the two remote interior angles: $m\angle ACD = m\angle A + m\angle B$

$m\angle ACD = 3x$, $m\angle A = 2x$, and $m\angle B = 40$: $3x = 2x + 40$

Add $-2x$ (the additive inverse of $2x$) to both sides of the equation:

$$\begin{aligned} -2x &= -2x \\ \hline x &= 40 \end{aligned}$$

The value of x is **40**.

18.

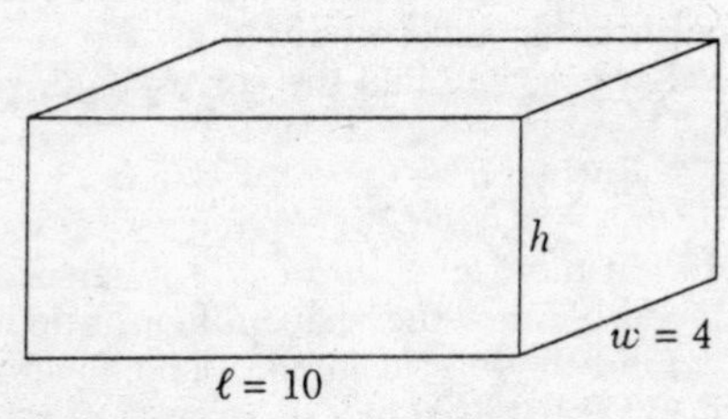

The volume, V, of a rectangular solid is given by the formula: $V = \ell wh$

where ℓ is the length, w is the width, and h is the height.

It is given that the volume is 180 cubic centimeters, the length is 10 centimeters and the width is 4 centimeters; substitute 180 for V, 10 for ℓ, and 4 for w:

$$180 = 10(4)h$$
$$180 = 40h$$

Divide both sides of the equation by 40: $\frac{180}{40} = \frac{40h}{40}$

Reduce the fraction on the left by dividing numerator and denominator by 20: $\frac{9}{2} = h$ or $h = 4.5$

The height is **4.5** centimeters.

19. $(x-4)^2$ means $(x-4)(x-4)$. Multiply out by multiplying each term of $x - 4$ by x and by -4:

$$\begin{array}{r} x - 4 \\ x - 4 \\ \hline x^2 - 4x \quad\quad \\ -4x + 16 \\ \hline x^2 - 8x + 16 \end{array}$$

$(x-4)^2$ is equivalent to: $x^2 - 8x + 16$

The correct choice is (3).

20. The product can be indicated as:

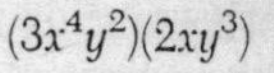

$(3x^4y^2)(2xy^3)$

The numerical coefficient of the product of two monomials is the product of their numerical coefficients: $(3)(2) = 6$

The literal factor of the product of two monomials is the product of their literal factors. The product of two powers of the same base is the power of that base obtained by adding the two exponents. Remember that x is understood to have the exponent, 1: $(x^4y^2)(xy^3) = x^5y^5$

Combine the two results above: $(3x^4y^2)(2xy^3) = 6x^5y^5$

The correct choice is **(1)**.

21. The opposite angles of all parallelograms are congruent and any two consecutive angles of a parallelogram are supplementary. The opposite sides of all parallelograms are congruent.

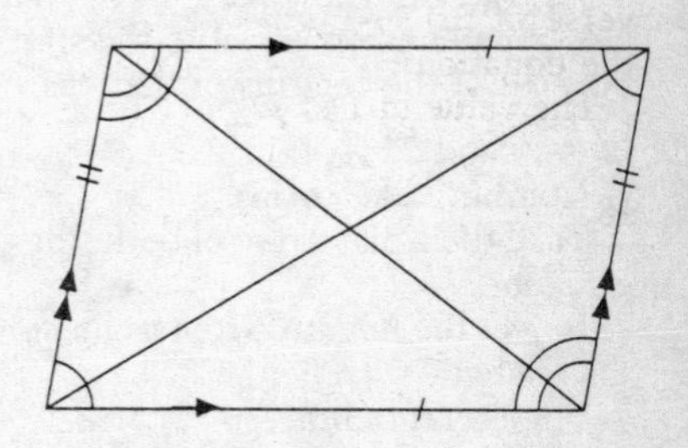

The diagonals of a parallelogram are *not* congruent except in the special case in which the parallelogram is a rectangle.

The correct choice is **(4)**.

22. In order for $\triangle A'B'C'$ to be a reflection of $\triangle ABC$ in line ℓ, line ℓ must be the perpendicular bisector of the line segments joining any two corresponding points of $\triangle A'B'C'$ and $\triangle ABC$. For example, line ℓ must be the perpen-

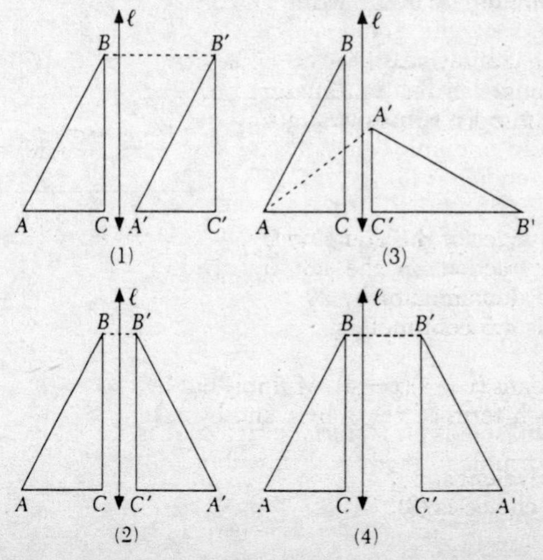

dicular bisector of $\overline{AA'}$, $\overline{BB'}$, and $\overline{CC'}$. All such line segments are either not perpendicular to ℓ or are not bisected by it except in figure (2).

The correct choice is (2).

23. Let x = the length of the diagonal.

Since the angles of a rectangle are right angles, the diagonal is the hypotenuse of a right triangle formed with one length and one width of the rectangle.

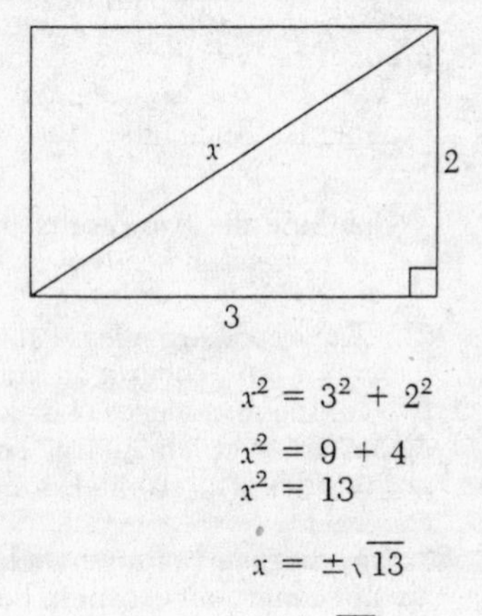

Use the Pythagorean Theorem: In a right triangle, the square of the length of the hypotenuse is equal to the sum of the squares of the lengths of the legs: $x^2 = 3^2 + 2^2$

$3^2 = 9$ and $2^2 = 4$: $x^2 = 9 + 4$

Combine like terms: $x^2 = 13$

Take the square root of both sides of the equation: $x = \pm\sqrt{13}$

Reject the negative root as meaningless for a length: $x = \sqrt{13}$

The correct choice is (2).

24. The graph of $y = -1$ consists of all points whose y-coordinate is -1. Since x does not appear in the equation, the points on the graph will include those with all possible values of x for the x-coordinate. Thus, the graph of $y = -1$ is a line parallel to the x-axis and one unit below it.

The correct choice is (3).

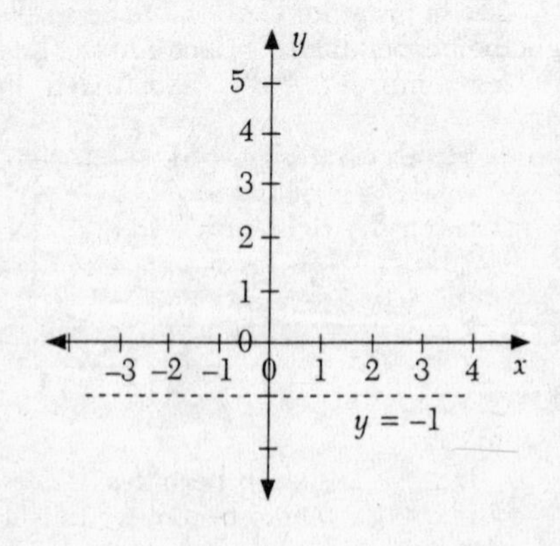

25. The given expression is: $\frac{6!}{3!}$

The symbol $n!$ stands for *factorial n*. Its value is given by the formula $n! = n(n-1)(n-2)(n-3) \ldots (3)(2)(1)$.

$$\frac{6!}{3!} = \frac{6(5)(4)(3)(2)(1)}{3(2)(1)}$$

Divide out like factors appearing in both numerator and denominator:

$$\frac{6!}{3!} = \frac{6(5)(4)\overset{1}{\cancel{(3)}}\overset{1}{\cancel{(2)}}(1)}{\underset{1}{\cancel{3}}\underset{1}{\cancel{(2)}}(1)}$$

Multiply the remaining factors:

$$\frac{6!}{3!} = \frac{120}{1} = 120$$

The correct choice is (3).

26. There are 7 scores: 72, 80, 80, 82, 87, 89, and 91

The mean score is the sum of all the scores divided by the number of them:

$$\text{mean} = \frac{72 + 80 + 80 + 82 + 87 + 89 + 91}{7}$$

$$\text{mean} = \frac{581}{7} = 83$$

The correct choice is (2).

27. The given statement is: $-1 \leq x < 2$, x is an integer

This means that x is any integer that is greater than or equal to -1 but less than 2. The integers, -1, 0, and 1, are the only ones that meet these conditions. The solution set is: $\{-1, 0, 1\}$

The correct choice is (4).

28. The given equation is: $x + 2y = 6$

Since all of the choices involve equations that are solved for y, the given equation must also be solved for y. Add $-x$ (the additive inverse of x) to both sides of the equation:

$$\begin{array}{r} -x \quad = -x \\ \hline 2y = 6 - x \end{array}$$

Divide both sides of the equation by 2:

$$\frac{2y}{2} = \frac{6 - x}{2}$$

Apply the distributive law by dividing each term on the right side by 2:

$$y = 3 - \frac{1}{2}x$$

To correspond to the form of the choices, this can be rewritten as:

$$y = -\frac{1}{2}x + 3$$

The correct choice is (4).

29. It is given that p represents "$x > 5$" and q represents "x is a multiple of 3."

If $x = 12$, then p becomes "$12 > 5$" which is true.

If $x = 12$, then q becomes "12 is a multiple of 3" which is true since $12 = 4 \times 3$.

Find the truth value of each of the choices:

(1) $p \vee q$: $p \vee q$ is the *disjunction* of p and q. The disjunction is true if either p or q or both are true. Since both p and q are true, $p \vee q$ is true.

(2) $\sim q \wedge p$: $\sim q$ is the *negation* of q. Since q is true, its negation, $\sim q$, is false. $\sim q \wedge p$ is the *conjunction* of $\sim q$ and p. The conjunction is true if $\sim q$

and p are both true, but is false if either $\sim q$ or p or both are false. Since $\sim q$ is false, $\sim q \wedge p$ is false.

(3) $p \rightarrow q$: $p \rightarrow q$ is the *implication* or *conditional* that if p is true, then q is true. The implication, $p \rightarrow q$, is true whenever q is true and also when p and q are both false. Since q is true for $x = 12$, the implication, $p \rightarrow q$, is true.

(4) $p \leftrightarrow q$: $p \leftrightarrow q$ is the *equivalence relation* or *biconditional* between p and q. The biconditional is true whenever p and q have the same truth value, either both true or both false, but it is false if p and q have different truth values. Since for $x = 12$, p and q are both true, $p \leftrightarrow q$ is true.

The only false statement is (2).

The correct choice is **(2)**.

30. The given equation is a *quadratic equation*: $x^2 - x - 20 = 0$

The left side is a *quadratic trinomial* that can be factored into the product of two binomials. The factors of the first term, x^2, are x and x, and they become the first terms of the binomials: $(x \quad)(x \quad) = 0$

The factors of the last term, -20, become the second terms of the binomials, but they must be chosen in such a way that the inner and outer cross-products of the binomials add up to the middle term, $-x$, of the original trinomial. Try -5 and $+4$ as the factors of -20:

$$-5x = \text{inner product}$$
$$(x - 5)(x + 4) = 0$$
$$+4x = \text{outer product}$$

Since $(-5x) + (+4x) = -x$, these are the correct factors: $(x - 5)(x + 4) = 0$

If the product of two factors is 0, either factor may equal 0: $x - 5 = 0 \vee x + 4 = 0$

Add the appropriate additive inverse to both sides of the equation, $+5$ for the left equation and -4 for the right equation:

$$\begin{array}{rr} 5 = 5 & -4 = -4 \\ \hline x \quad = 5 & x \quad = -4 \end{array}$$

The solution set is $\{5, -4\}$.

The correct choice is **(1)**.

31. The given number is written in scientific notation: 3.6×10^5

The notation indicates that 3.6 is to be multiplied by 10 five times. Each multiplication by 10 moves the decimal point one place to the right. Multiplication by 10^5 will move the decimal point 5 places to the right: $3.6 \times 10^5 = 360{,}000$

The correct choice is **(1)**.

32. The given fractional expression is: $\frac{x-3}{x+6}$

Test each of the choices:

(1) If $x = 6$, the fraction becomes $\frac{6-3}{6+6}$ or $\frac{3}{12}$ or $\frac{1}{4}$.

(2) If $x = -6$, the fraction becomes $\frac{-6-3}{-6+6}$ or $\frac{-9}{0}$. But division by 0 is undefined, so $\frac{-9}{0}$ is undefined.

(3) If $x = 3$, the fraction becomes $\frac{3-3}{3+6}$ or $\frac{0}{9}$ or 0.

(4) If $x = -3$, the fraction becomes $\frac{-3-3}{-3+6}$ or $\frac{-6}{3}$ or -2.

The correct choice is (2).

33. Consider $\overline{AB}$ to be the base of the triangle. The altitude to $\overline{AB}$ will be a perpendicular segment dropped from C to $\overline{AB}$.

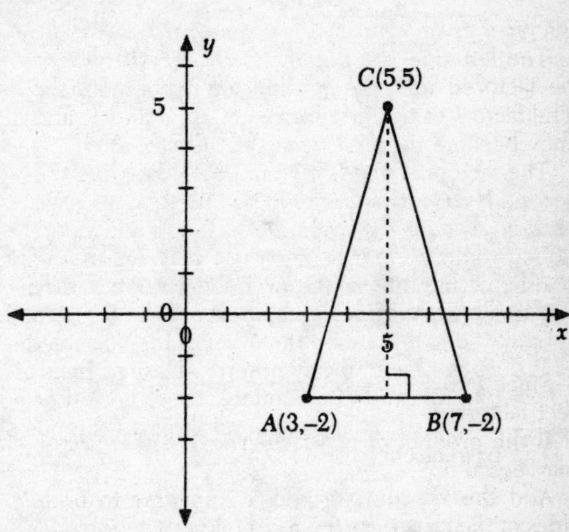

The area, A, of a triangle whose base is b and whose altitude is h is given by the formula: $A = \frac{1}{2}bh$

Here, $b = AB = 4$ and $h = 7$: $A = \frac{1}{2}(4)(7)$

$A = 2(7)$

$A = 14$

The correct choice is (2).

34. The given expression is: $2\sqrt{3} - \sqrt{27}$

Factor out the perfect square factor in the second radicand: $2\sqrt{3} - \sqrt{9(3)}$

Remove the perfect square factor from under the radical sign by taking its square root and writing it as a coefficient of the radical: $2\sqrt{3} - 3\sqrt{3}$

Both terms are now similar radicals since they have the same radicand, 3, and they are both of the same index (all are square roots, that is, of index 2). Similar radicals can be combined by combining their coefficients: $-\sqrt{3}$

The correct choice is (4).

35.

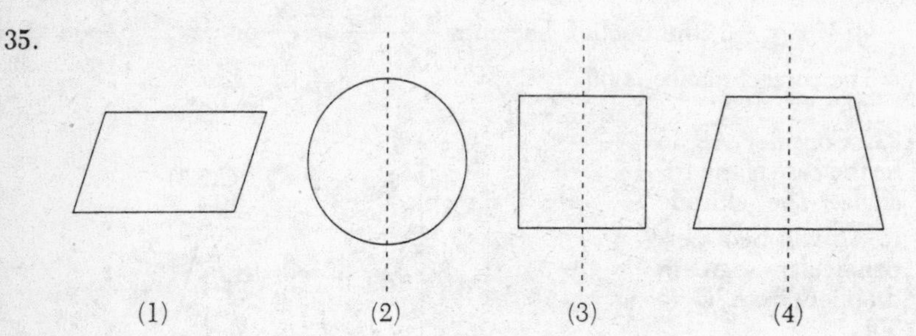

A figure has line symmetry if there exists a line such that when the figure is folded along that line, every point of the figure on each side of the line will correspond to a point on the other side. No such line is possible for figure (1) so it does *not* have line symmetry. Dashed lines of symmetry are shown for figures (2), (3), and (4). Actually, there are other positions for lines of symmetry in figures (2) and (3) in addition to the one shown in each.

The correct choice is (1).

PART TWO

36. a. To draw the graph of the inequality $y \leq -3x + 2$, we first draw the graph of the equation $y = -3x + 2$. Prepare a table of pairs of values of x and y by selecting any three convenient values for x and substituting them in the equation to calculate the corresponding values of y:

x	$-3x + 2$	$= y$
-2	$-3(-2) + 2 = 6 + 2$	$= 8$
0	$-3(0) + 2 = 0 + 2$	$= 2$
3	$-3(3) + 2 = -9 + 2$	$= -7$

Plot the points $(-2,8)$, $(0,2)$, and $(3,-7)$ and draw a *solid* line through them. This line is the graph of the equation $y = -3x + 2$. It is solid to signify that it is part of the solution set of $y \leq -3x + 2$.

The solution set of the inequality $y < -3x + 2$ lies on one side of the line $y = -3x + 2$. To find out on which side, choose a convenient point, say (0,0), and substitute its coordinates in the inequality:

$y < -3x + 2$
$0 \overset{?}{<} -3(0) + 2$
$0 \overset{?}{<} 0 + 2$
$0 < 2$ ✓ Since (0,0) satisfies the inequality, it lies on the same side as the solution set of $y < -3x + 2$. Shade this side with cross-hatching extending down and to the left.

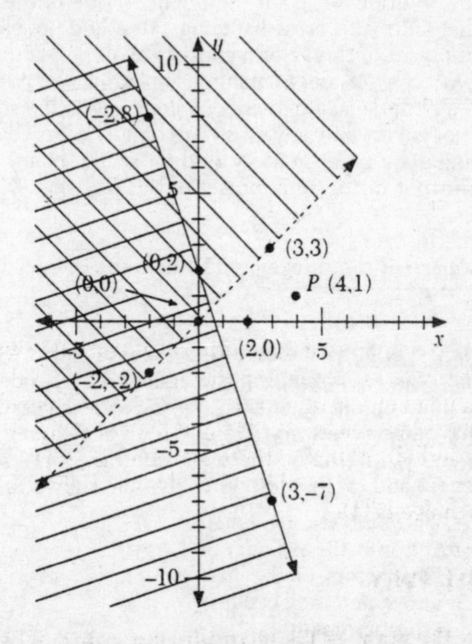

To graph $y - x > 0$, first rearrange it so that it is solved for y by adding x (the additive inverse of $-x$) to both sides of the inequality:

$$\begin{array}{r} y - x > 0 \\ x = x \\ \hline y > x \end{array}$$

Graph the equation $y = x$ by preparing a table of pairs of corresponding values of x and y after choosing three convenient values for x:

x	y
−2	= −2
0	= 0
3	= 3

Plot the points (−2,−2), (0,0), and (3,3), and draw a

broken line through them. Showing the graph of $y = x$ as a broken line indicates that the equation $y = x$ is *not* part of the solution set of $y - x > 0$.

To find out on which side of the line $y = x$ the solution set of $y - x > 0$ lies, choose a convenient point, say (2,0), and substitute its coordinates in $y - x > 0$ to see whether they satisfy it:

$$y - x > 0$$
$$0 - 2 \overset{?}{>} 0$$
$$-2 \not> 0$$

Since (2,0) does *not* satisfy $y - x > 0$, it does *not* lie in the solution set of $y - x > 0$. The solution set is on the opposite side of the line $y = x$ from (2,0). Shade that side with cross-hatching extending up and to the left.

b. The solution set of the system consists of all points that lie in the area covered by *both* types of cross-hatching, including the portion of the solid line that forms a boundary of that area. A point *not* in the solution set would be in the area not covered by any cross-hatching or covered by only one type of cross-hatching. The point $P(4,1)$ would be such a point.

(4,1) is a point *not* in the solution set of the system.

37. Let $x =$ the integer.

Twice the square of the integer is 11 times the integer less 5.

$$2 \quad x^2 \quad = 11 \quad \times \quad x \quad - 5$$

The equation to use is: $2x^2 = 11x - 5$

This is a *quadratic equation*. Rearrange it so that all terms are on one side equal to zero by adding $-11x$ and $+5$ to both sides of the equation:

$$\underline{-11x + 5 = -11x + 5}$$
$$2x^2 - 11x + 5 = 0$$

The left side is a *quadratic trinomial* that can be factored into the product of two binomials. The factors of the first term, $2x^2$, are $2x$ and x, and they become the first terms of the binomials: $(2x \quad)(x \quad) = 0$

The factors of the last term, $+5$, become the second terms of the binomials but they must be chosen in such a way that the inner and outer cross-products of the binomials add up to the middle term, $-11x$, of the original trinomial. Try -1 and -5 as the factors of $+5$:

$-x$ = inner product

$$(2x - 1)(x - 5) = 0$$

$-10x$ = outer product

Since $(-x) + (-10x) = -11x$, these are the correct factors: $(2x - 1)(x - 5) = 0$

If the product of two factors is 0, either factor may equal 0: $2x - 1 = 0 \vee x - 5 = 0$

$$2x - 1 = 0 \vee x - 5 = 0$$

Add the appropriate additive inverse to both sides of the equation, 1 for the left equation and 5 for the right equation:

$$\begin{array}{rr} \underline{\quad 1 = 1} & \underline{\quad 5 = 5} \\ 2x \quad = 1 & x \quad = 5 \end{array}$$

Divide both sides of the left equation by 2:

$$\frac{2x}{2} = \frac{1}{2}$$

$$x = \frac{1}{2}$$

Reject the fractional root since the question calls for an integer:

$$x = 5$$

The integer is **5**.

38. The given system of equations is:

$$\begin{cases} 2x + 3y = 11 \\ 5x - 2y = -20 \end{cases}$$

Multiply each term of the first equation by 2, and each term of the second equation by 3:

$$\begin{cases} 2(2x) + 2(3y) = 2(11) \\ 3(5x) + 3(-2y) = 3(-20) \end{cases}$$

$$\begin{cases} 4x + 6y = 22 \\ 15x - 6y = -60 \end{cases}$$

Adding the equations will eliminate y:

$$19x = -38$$

Divide both sides of the equation by 19:

$$\frac{19x}{19} = \frac{-38}{19}$$

$$x = -2$$

Substitute -2 for x in the first equation:

$$2(-2) + 3y = 11$$

$$-4 + 3y = 11$$

Add 4 (the additive inverse of -4) to both sides of the equation:

$$\underline{4 \quad = 4}$$

$$3y = 15$$

Divide both sides of the equation by 3:

$$\frac{3y}{3} = \frac{15}{3}$$

$$y = 5$$

The solution is **(−2,5) or $x = -2$, $y = 5$.**

CHECK: The solution must satisfy *both original* equations. Substitute -2 for x and 5 for y in *both* of the *original* equations to see whether they are satisfied:

$$\underline{2x + 3y = 11} \qquad \underline{5x - 2y = -20}$$

$$2(-2) + 3(5) \stackrel{?}{=} 11 \qquad 5(-2) - 2(5) \stackrel{?}{=} -20$$

$$-4 + 15 \stackrel{?}{=} 11 \qquad -10 - 10 \stackrel{?}{=} -20$$

$$11 = 11 \checkmark \qquad -20 = -20 \checkmark$$

39. a. Let $AB = 4x$
Then $BC = 3x$

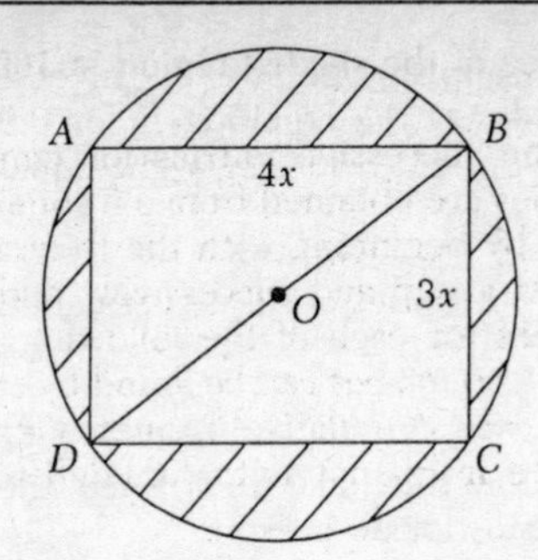

The opposite sides of a rectangle are congruent: $DC = AB = 4x$
$AD = BC = 3x$

The perimeter of a rectangle is the sum of the lengths of the four sides: Perimeter $= AB + BC + CD + AD$
The perimeter is 56 centimeters: $56 = 4x + 3x + 4x + 3x$
Combine like terms: $56 = 14x$
Divide both sides of the equation by 14: $\frac{56}{14} = \frac{14x}{14}$
$4 = x$
$AB = 4x$: $4x = 4(4) = 16$
$AB =$ **16** centimeters.

b. Since all the angles of a rectangle are right angles, ΔBDC is a right triangle. In right triangle BDC, $DC = AB = 16$ and $BC = 3x = 3(4) = 12$.

$\overline{BD}$ is the hypotenuse of a right triangle whose legs are 12 and 16. Since $BC = 3 \times 4$ and $AB = 4 \times 4$, ΔBDC is a 3-4-5 right triangle, and the hypotenuse BD must be 5×4 or 20.

$BD =$ **20** centimeters.

NOTE: BD can also be found by using the Pythagorean Theorem which gives $(BD)^2 = (DC)^2 + (BC)^2$ or $(BD)^2 = 16^2 + 12^2$.

c. $\overline{BD}$ is a diameter of circle O. The radius, r, of the circle is one-half the length of the diameter. Since $BD = 20$, $r = \frac{1}{2}(20)$, or $r = 10$.

The area, A, of a circle whose radius is r is given by the formula: $A = \pi r^2$
Since $r = 10$: $A = \pi(10)^2$
$A = 100\pi$

The area of the circle is **100π** square centimeters.

d. The area of the shaded region is the area of the circle minus the area of the rectangle.

The area, A, of a rectangle whose length is ℓ and whose width is w is given by the formula: $A = \ell w$
$\ell = AB = 16$, and $w = BC = 12$: $A = 16(12)$
$A = 192$

The area of the shaded region = **100π − 192** square centimeters.

40. a. The successive entries for cumulative frequencies are obtained from a frequency distribution by beginning with the frequency for the lowest group and successively adding the frequencies for each of the following groups. Thus the frequencies can be found by subtracting from each cumulative frequency entry the cumulative frequency entry immediately preceding it:

Interval	Cumulative Frequency
10–19	2
10–29	17
10–39	27
10–49	32
10–59	32
10–69	35

Interval	Frequency
10–19	2
20–29	15
30–39	10
40–49	5
50–59	0
60–69	3

b.

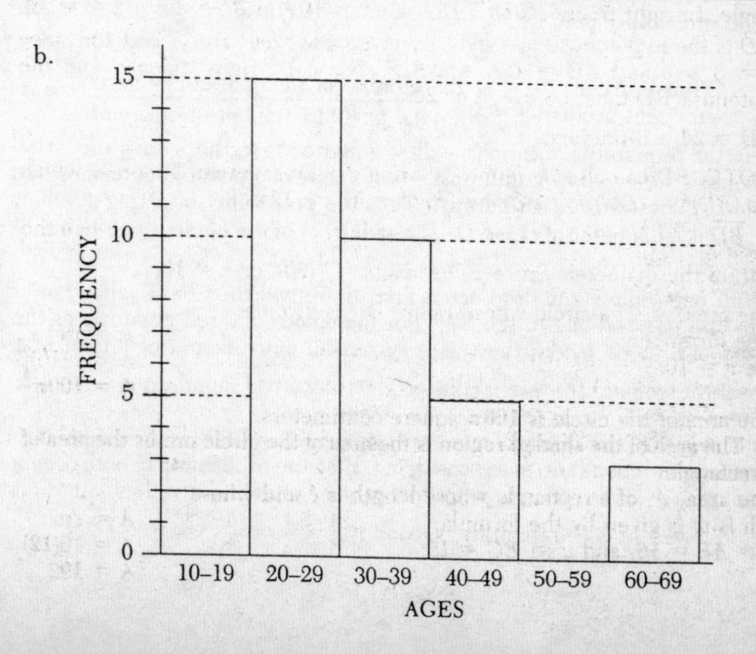

c. The median is the age of the middle person when all persons are arranged in the order of their ages. Since there are 35 people, the median is the age of the 18th person when arranged in order of age; there will be 17 people younger than this person and 17 people older than him or her. Counting down from the top of the frequency table, the first two intervals account for 2 + 15, or the 17 lowest ages. The 18th age in order of size therefore lies in the next interval, which is the 30–39 group.

The median lies in the **30–39** interval.

d. $$\text{Probability of an event occurring} = \frac{\text{the number of successful outcomes}}{\text{the total possible number of outcomes}}.$$

The successful outcomes for the random choice of a person at least 40 years old are those in the 40–45, 50–59, and 60–69 intervals. These intervals contain 5 + 0 + 3 persons or a total of 8 who constitute successful outcomes.

The total possible number of outcomes is 35, the total number of people from which the choice is made.

The probability of choosing a person at least 40 years old is $\frac{8}{35}$.

The probability is $\mathbf{\frac{8}{35}}$.

e. There are no people in the line between 50 and 59 years old. Thus the number of successful outcomes for the choice of a person in this age range is 0, and the probability of choosing such a person is $\frac{0}{35}$ or 0.

The probability is **0**.

41. a. $$\text{Probability of an event occurring} = \frac{\text{the number of successful outcomes}}{\text{the total possible number of outcomes}}.$$

For the probability of eating a yellow gumdrop first, there are 4 successful outcomes (the 4 yellow gumdrops) out of a total of 5 possible outcomes (the total of yellow and red gumdrops). Thus the probability of eating a yellow gumdrop first is $\frac{4}{5}$.

With one yellow gumdrop eaten first, there will then be 3 yellow gumdrops and one red one in the bag. For the choice of a red gumdrop as the second one to be eaten, there is 1 successful outcome out of a total of 4 possible outcomes. The probability of choosing a red gumdrop second is $\frac{1}{4}$.

If $P(A)$ is the probability of one event occurring and $P(B)$ is the probability of another (independent) event occurring, then the probability of both occurring is $P(A) \times P(B)$. Therefore the probability of eating a yellow gumdrop first, followed by a red gumdrop second is $\frac{4}{5} \times \frac{1}{4}$ or $\frac{4}{20}$.

The probability of a yellow followed by a red is $\frac{4}{20}$.

b. From part a, the probability of selecting a yellow gumdrop first is $\frac{4}{5}$. For a second selection, there are then 3 yellow gumdrops and 1 red gumdrop in the bag. The probability of selecting a second yellow gumdrop involves 3 successful outcomes out of a possible total of 4; the probability of selecting a second yellow gumdrop is therefore $\frac{3}{4}$. The probability of selecting two yellow gumdrops in succession is the product of their separate probabilites, or $\frac{4}{5} \times \frac{3}{4}$ or $\frac{12}{20}$.

c. Eating two gumdrops having different colors can occur in two ways. Either Adam can eat a yellow followed by a red OR he can eat a red followed by a yellow.

Part a shows the probability of a yellow followed by a red to be $\frac{4}{20}$. But suppose he eats the red first. The probability of a red first involves 1 successful outcome (the one red gumdrop) out of 5 possible outcomes; hence the probability of a red first is $\frac{1}{5}$. If he then eats a yellow, he has 4 successful outcomes (only the 4 yellows are left) out of the remaining 4 possible outcomes; the probability of a yellow on the second choice is $\frac{4}{4}$. The probability of a red followed by a yellow is $\frac{1}{5} \times \frac{4}{4}$ or $\frac{4}{20}$.

If $P(A)$ is the probability of one event occurring and $P(B)$ is the probability of another event, the probability of one OR the other occurring is $P(A) + P(B)$. Therefore, the probability of eating a yellow followed by red OR a red followed by yellow is $\frac{4}{20} + \frac{4}{20}$ or $\frac{8}{20}$.

The probability of eating two gumdrops of different colors is $\frac{8}{20}$.

d. Since there is only one red gumdrop in the bag, the probability that Adam will eat two red gumdrops is 0. There can be no successful outcomes for the choice of a second red gumdrop after one is eaten.

The probability of eating two red gumdrops is **0**.

42. p represents "The flowers are not in bloom."
q represents "It is raining."
r represents "The grass is green."

a. "The flowers are in bloom" is the *negation* of p, and it is represented by $\sim p$.

"If the flowers are in bloom, then the grass is green" is the *implication* or *conditional*, $\sim p \rightarrow r$. Its *hypothesis* or *antecedent* (the if clause) is $\sim p$, and its *conclusion* or *consequent* (the then clause) is r. The converse of a conditional is another conditional formed by interchanging the hypothesis and conclusion of the original conditional. Therefore, the converse of $\sim p \rightarrow r$ is $r \rightarrow \sim p$.

The converse is $r \rightarrow \sim p$.

b. "If it is raining, the grass is green" is the *implication* or *conditional*, $q \rightarrow r$. Its *hypothesis* is q and its *conclusion* is r. The *inverse* of a conditional is another conditional formed by negating both the hypothesis and conclusion of the original conditional. The negation of q is $\sim q$, and the negation of r is $\sim r$. Therefore, the inverse of $q \rightarrow r$ is $\sim q \rightarrow \sim r$.

The inverse is $\sim q \rightarrow \sim r$.

c. $p \wedge \sim q$ is the conjunction of p and the inverse of q. In sentence form it stands for **"The flowers are not in bloom and it is not raining."**

d. $\sim r \vee \sim p$ is the disjunction of the negation of r and the negation of p. In sentence form, it stands for **"The grass is not green or the flowers are in bloom."**

e. The implication, $\sim q \rightarrow r$, will have the same truth value as its contrapositive. The contrapositive of an implication is another implication formed by negating the hypothesis and conclusion of the original implication and then interchanging them. The negation of $\sim q$ is q, and the negation of r is $\sim r$. Therefore, the contrapositive of $\sim q \rightarrow r$ is $\sim r \rightarrow q$.

The correct choice is **(2)**.

Topic	Question Numbers	Number of Points	Your Points	Your Percentage
1. Numbers (rat'l, irrat'l); Percent	—	0		
2. Properties of No. Systems	32	2		
3. Operations on Rat'l Nos. and Monomials	16, 20	2 + 2 = 4		
4. Operations on Multinomials	14, 19	2 + 2 = 4		
5. Square Root; Operations involving Radicals	34	2		
6. Evaluating Formulas and Expressions	—	0		
7. Linear Equations (simple cases incl. parentheses)	11, 28	2 + 2 = 4		
8. Linear Equations Containing Decimals or Fractions	5, 7, 8	2 + 2 + 2 = 6		
9. Graphs of Linear Functions (slope)	24	2		
10. Inequalities	10, 27	2 + 2 = 4		
11. Systems of Eqs. & Inequalities (alg. & graphic solutions)	36a, b, 38	8 + 2 + 10 = 20		
12. Factoring	—	0		
13. Quadratic Equations	30	2		
14. Verbal Problems	37	10		
15. Variation	15	2		
16. Literal Eqs.; Expressing Relations Algebraically	12	2		
17. Factorial *n*	25	2		
18. Areas, Perims., Circums., Vols. of Common Figures	9, 18, 39a, c, d	2 + 2 + 2 + 2 + 2 = 10		
19. Geometry (≅, ‖ lines, compls., suppls.)	3, 13	2 + 2 = 4		
20. Ratio & Proportion (incl. similar triangles & polygons)	1	2		
21. Pythagorean Theorem	23, 39b	2 + 2 = 4		
22. Logic (symbolic rep., logical forms, truth tables)	4, 29, 42a, b, c, d, e	2 + 2 + 2 + 2 + 2 + 2 + 2 = 14		

Topic	Question Numbers	Number of Points	Your Points	Your Percentage
23. Probability (incl. tree diagrams & sample spaces)	40d, e, 41a, b, c, d	2 + 1 + 3 + 3 + 2 + 2 = 13		
24. Combinations, Arrangements, Permutations	—	0		
25. Statistics (central tend., freq. dist., histogr., quartiles, percentiles)	6, 26, 40a, b, c	2 + 2 + 1 + 4 + 2 = 11		
26. Properties of Triangles and Quadrilaterals	2, 17, 21	2 + 2 + 2 = 6		
27. Transformations (reflect., translations, rotations, dilations)	22	2		
28. Symmetry	35	2		
29. Area from Coordinate Geometry	33	2		
30. Dimensional Analysis	—	0		
31. Scientific Notation; Negative & Zero Exponents	31	2		

Examination January 1991

Three-Year Sequence for High School Mathematics—Course I

PART ONE

DIRECTIONS: *Answer 30 questions from this part. Each correct answer will receive 2 credits. No partial credit will be allowed. Write your answers in the spaces provided. Where applicable, answers may be left in terms of π or in radical form.*

1 In 3 hours a car traveled 180 kilometers. At the same average rate, how many kilometers can the car travel in 5 hours? 1_____

2 In the accompanying diagram, parallel lines $\overleftrightarrow{AB}$ and $\overleftrightarrow{CD}$ are intersected by transversal $\overleftrightarrow{EF}$ at points G and H, respectively.
If $m\angle FGB = 2x + 25$ and $m\angle FHD = 3x - 5$, find x.

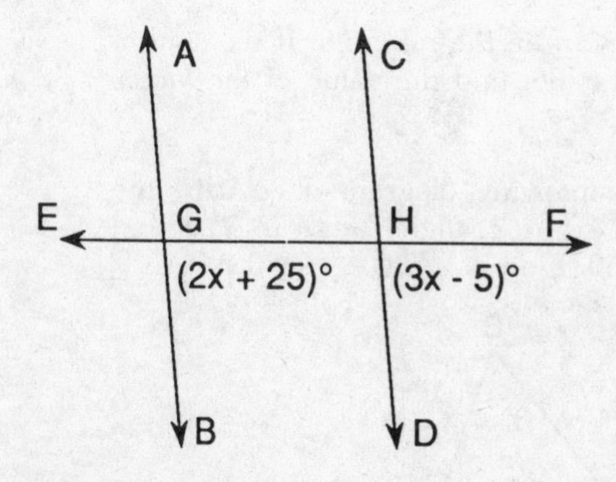

2_____

3 Solve for x: $0.03x - 1.2 = 0.24$ 3_____

4 If $a = -2$ and $b = 3$, what is the value of $-3a^2b$? 4_____

5 Solve for m: $\frac{2}{3}m - 4 = 14$ 5_____

6 In the accompanying diagram of rhombus $ABCD$, the lengths of the sides $\overline{AB}$ and $\overline{BC}$ are represented by $3x - 4$ and $2x + 1$, respectively. Find the value of x.

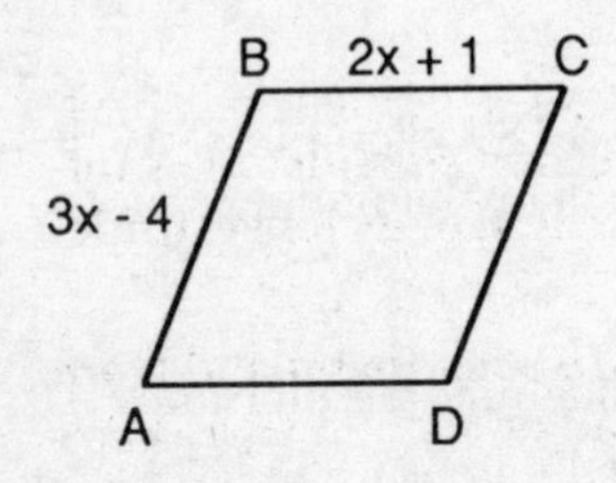

6_____

7 Perform the indicated operations and express as a trinomial:

$$(x + 4)(x - 2) + 3x$$

7_____

8 Two numbers are in the ratio 5:6. If the sum of the numbers is 66, find the value of the *larger* number. 8_____

9 In the accompanying diagram of $\triangle ABC$, the measure of exterior angle BCD is 110 and $m\angle BAC = 50$. Find $m\angle ABC$.

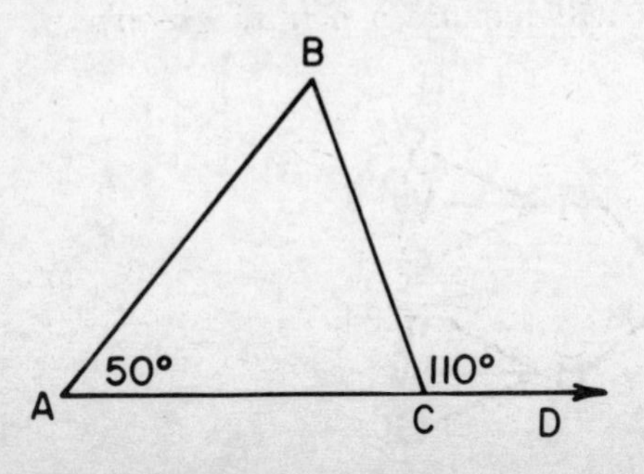

9_____

10 Express $9 - y^2$ as the product of two binomial factors. 10_____

11 The measure of the vertex angle of an isosceles triangle is three times the measure of a base angle. Find the number of degrees in the measure of a base angle. 11_____

12 Express, in terms of x, the mean of $(4x - 6)$, $(2x + 3)$, and $(3x + 3)$. 12_____

13 The lengths of the sides of a triangle are 3, 4, and 6. If the length of the shortest side of a similar triangle is 5, find the length of its longest side. 13_____

14 If $\sqrt{84}$ is simplified to $a\sqrt{b}$ such that a and b are integers, what is the value of a? 14_____

15 Find the numerical value of ${}_4P_3$. 15_____

16 For which value of x is the expression $\frac{3}{x + 1}$ undefined? 16_____

17 Mary has 2 blouses (1 red and 1 blue) and 3 pairs of slacks (1 yellow, 1 white, and 1 green). The tree diagram below represents the outfits she can wear. If Mary chooses 1 blouse and 1 pair of slacks at random, what is the probability that the outfit she chooses will include a pair of green slacks?

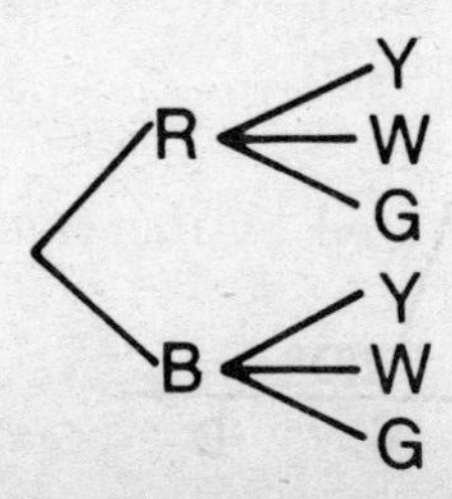

17_____

18 In the accompanying diagram, $ABCD$ is a parallelogram with vertices $A(2,0)$, $B(7,0)$, $C(10,3)$, and $D(5,3)$. What is the area of parallelogram $ABCD$?

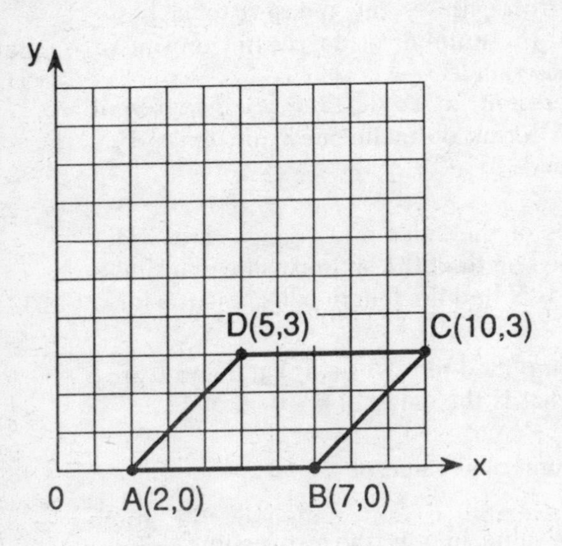

18_____

DIRECTIONS (**19–35**): *For each question chosen, write the* numeral *preceding the word or expression that best completes the statement or answers the question.*

19 The best description of a dilation of a figure is
(1) an enlargement or a reduction of the figure
(2) a slide of the figure
(3) a turning of the figure about some fixed point
(4) a mirror image of the figure

19_____

20 Which property is illustrated by the equation $3x - 6y = 3(x - 2y)$?

(1) associative (3) distributive
(2) commutative (4) closure

20_____

21 Which ordered pair is the solution to this system of equations?

$$y = x + 4$$
$$x + y = 2$$

(1) (1,5)
(2) (0,2)
(3) (−1,3)
(4) (−4,0)

21_____

22 Let p represent "x is odd" and let q represent "$x > 15$." Which statement is true if $x = 13$?

(1) $p \wedge \sim q$
(2) $\sim p \vee q$
(3) $p \rightarrow q$
(4) $p \wedge q$

22_____

23 Expressed as a single fraction in lowest terms, the sum of $\frac{3x}{4}$ and $\frac{2x}{3}$ is equivalent to

(1) $\frac{5x}{7}$
(2) $\frac{5x}{12}$
(3) $\frac{17x}{7}$
(4) $\frac{17x}{12}$

23_____

24 Which inequality is represented by the graph below?

(1) $-5 < x < 6$
(2) $-5 \leq x \leq 6$
(3) $-5 \leq x < 6$
(4) $-5 < x \leq 6$

24_____

25 In the following table, which interval contains the median?

Interval	Frequency
16–20	1
11–15	3
6–10	3
1–5	2

(1) 1–5
(2) 6–10
(3) 11–15
(4) 16–20

25_____

26 The converse of $m \rightarrow \sim r$ is

(1) $m \rightarrow r$ (3) $\sim m \rightarrow r$
(2) $\sim r \rightarrow m$ (4) $r \rightarrow \sim m$ 26_____

27 The expression $\frac{15k^3 - 9k^2 + 3k}{3k}$, $k \neq 0$, is equivalent to

(1) $5k^2 - 3k + 1$ (3) $15k^3 - 9k^2$
(2) $5k^2 - 3k$ (4) $3k$ 27_____

28 If the length of any rectangle is increased by 2 and the width is unchanged, the perimeter is

(1) increased by 2 (3) increased by 4
(2) multiplied by 2 (4) multiplied by 4 28_____

29 Which line ℓ has a slope of zero?

(1)
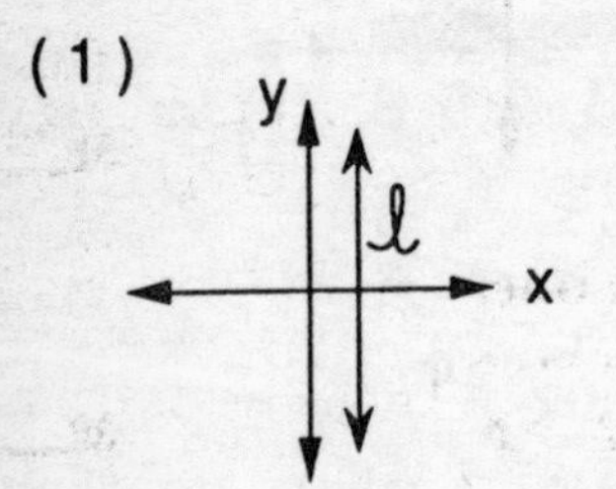

(3)
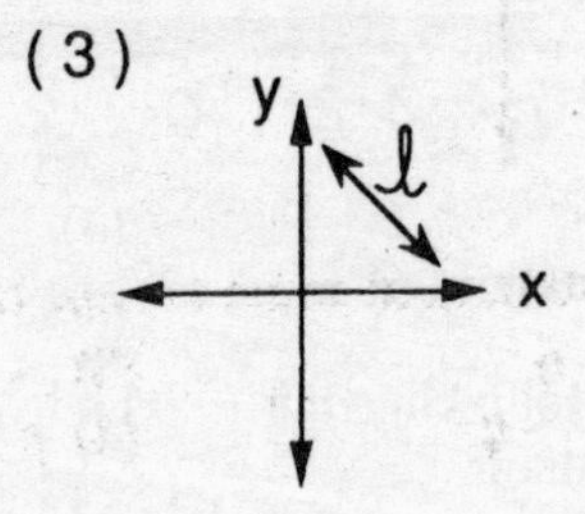

(2)
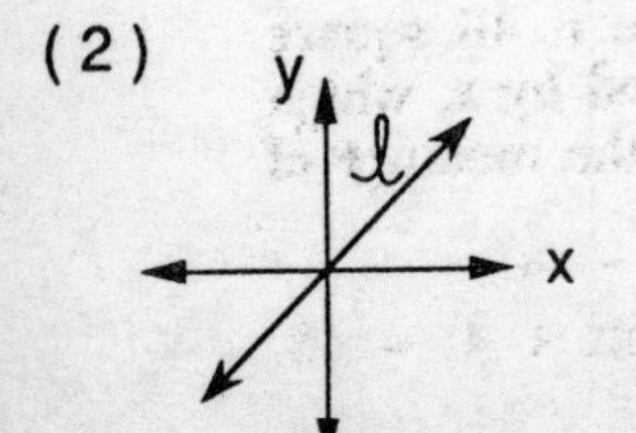

(4)
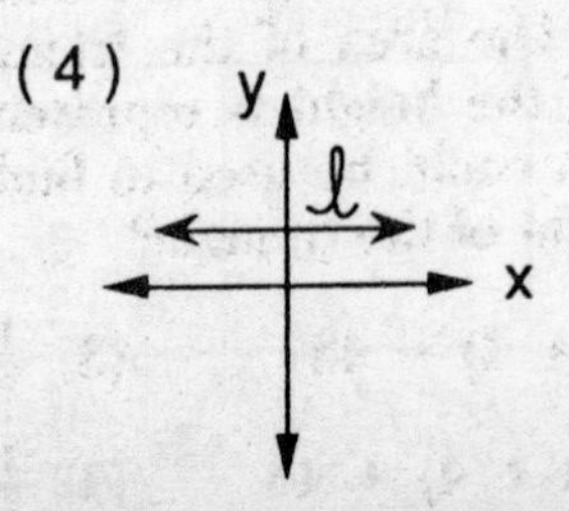

29_____

30 The measure of an angle is represented by x. The measure of the complement of this angle can be represented as

(1) $(90 - x)°$
(2) $(x - 90)°$
(3) $(180 - x)°$
(4) $(x - 180)°$

30_____

31 Which is a graphic representation of "y varies directly as x"?

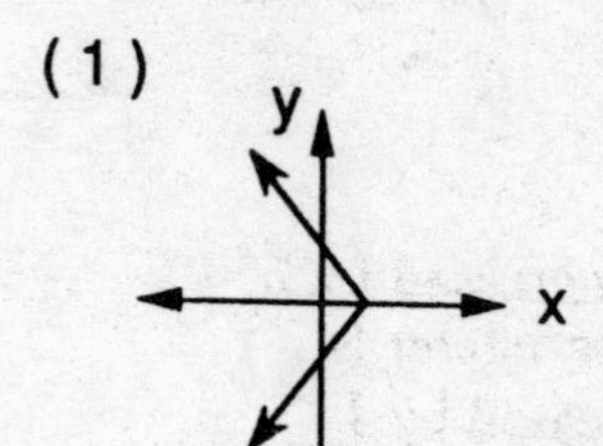

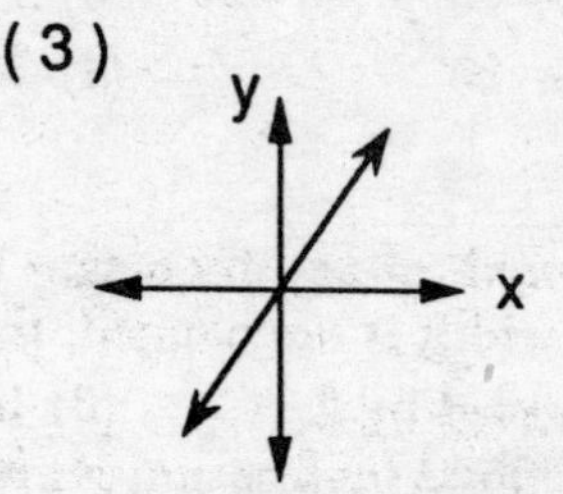

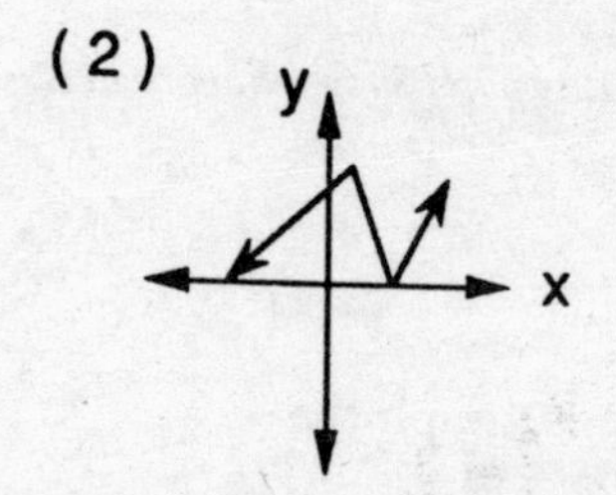

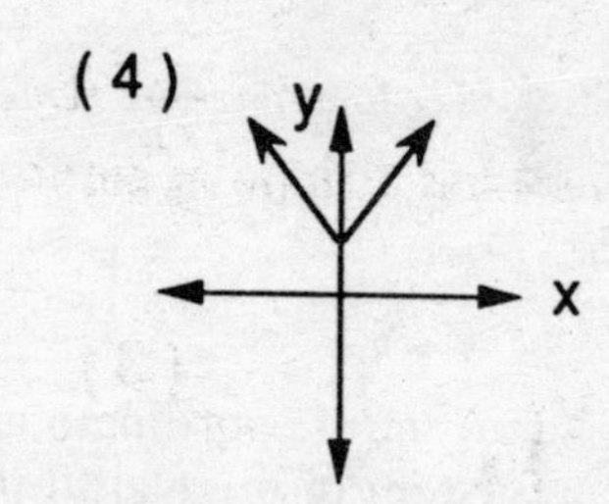

31_____

32 Which statement must *always* be true?

(1) $p \rightarrow q$
(2) $q \wedge \sim q$
(3) $\sim p \rightarrow \sim q$
(4) $p \vee \sim p$

32_____

33 The base of a triangle is 4 units more than the height. The area of the triangle is 48 square units. If the height is represented by x, which equation could be used to find the measure of the height of the triangle?

(1) $x(x + 4) = 48$
(2) $\frac{1}{2}x(x + 4) = 48$
(3) $\frac{1}{2}(2x + 4) = 48$
(4) $\frac{1}{2}x(x - 4) = 48$

33_____

34 The area of a circle is represented by 16π. What is the length of a diameter of the circle?

(1) 16
(2) 8
(3) $4\sqrt{2}$
(4) 4

34_____

35 Which number expresses 72 kilometers per hour as meters per hour?

(1) 7.2×10^{-2}
(2) 7.2×10^{2}
(3) 7.2×10^{-4}
(4) 7.2×10^{4}

35_____

PART TWO

DIRECTIONS *Answer* four *questions from this part. Show all work unless otherwise directed.*

36 *a* On the same set of coordinate axes, graph the following system of inequalities:

$$y + x \geq 5$$
$$y < 2x + 3$$

[8]

b Based on the graphs drawn in part *a*, write the coordinates of a point in the solution set. [2]

37 If the length of one side of a square is doubled and the length of an adjacent side is decreased by 3, the area of the resulting rectangle exceeds the area of the original square by 16. Find the length of a side of the original square. [*Only an algebraic solution will be accepted.*] [5,5]

38 In the accompanying diagram, $ABCD$ is a parallelogram, $\overline{DE} \perp \overline{BC}$, diagonal $\overline{BD}$ is drawn, and $AD = 24$.

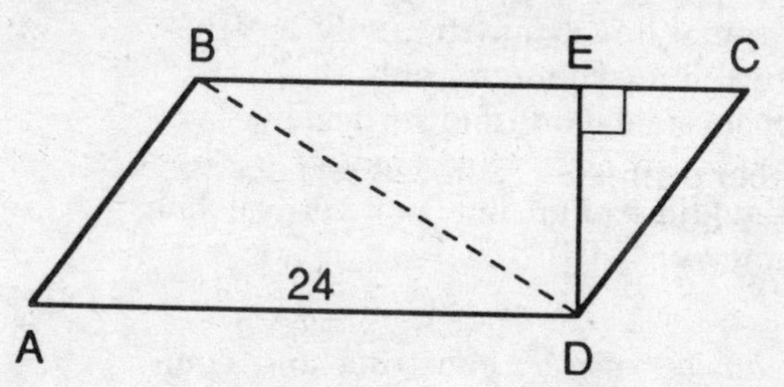

a If $AB = 2x + 4$ and $CD = x + 7$, find the length of $\overline{CD}$. [2]

b If the lengths of $\overline{EC}$ and $\overline{BE}$ are in the ratio 1:3, find the length of $\overline{EC}$. [2]

c Find the length of $\overline{ED}$. [2]

d Find the area of $\triangle ABD$. [2]

e Find the area of trapezoid $ABED$. [2]

39 The cost of a telephone call from Wilson, New York, to East Meadow, New York, is \$0.60 for the first three minutes plus \$0.17 for each *additional* minute. What is the greatest number of whole minutes of a telephone call if the cost cannot exceed \$2.50? [*Only an algebraic solution will be accepted.*] [4,6]

40 *a* How many different ways can the letters of the word "CHORD" be arranged? [2]

b How many of the arrangements in part *a* begin with either an "H" or an "O"? [2]

c If one of the arrangements in part *a* is selected at random, what is the probability it will begin with "C"? [2]

d If one letter is selected at random from the letters of the word "**CHORD**," find the probability that it will have
(1) horizontal line symmetry, only [1]
(2) vertical line symmetry, only [1]
(3) both horizontal line and vertical line symmetry [1]
(4) neither horizontal line nor vertical line symmetry [1]

41 *a* *On your answer paper*, construct and complete the truth table for the statement $(\sim p \rightarrow q) \leftrightarrow (p \vee q)$. [9]

b Based on the truth table constructed in part *a*, is $(\sim p \rightarrow q) \leftrightarrow (p \vee q)$ a tautology? [1]

42 In 20 games, a basketball player scored these points: 36, 32, 28, 30, 33, 36, 24, 33, 29, 30, 30, 25, 34, 36, 34, 31, 36, 29, 30, 34.

a *On your answer paper*, copy and complete the table below to find the frequency in each interval. [2]

Interval	Tally	Frequency
35–37		
32–34		
29–31		
26–28		
23–25		

b Construct a frequency histogram using the table completed in part *a*. [4]

c In what percent of the games played did the player score less than 29 points? [2]

d Which interval contains the 60th percentile? [2]

Answers January 1991

Three-Year Sequence for High School Mathematics—Course I

ANSWER KEY
PART ONE

1. 300	**13.** 10	**25.** (2)
2. 30	**14.** 2	**26.** (2)
3. 48	**15.** 24	**27.** (1)
4. –36	**16.** –1	**28.** (3)
5. 27	**17.** $\frac{2}{6}$	**29.** (4)
6. 5	**18.** 15	**30.** (1)
7. x^2+5x-8	**19.** (1)	**31.** (3)
8. 36	**20.** (3)	**32.** (4)
9. 60	**21.** (3)	**33.** (2)
10. $(3-y)(3+y)$	**22.** (1)	**34.** (2)
11. 36	**23.** (4)	**35.** (4)
12. $3x$	**24.** (4)	

PART TWO— *See **Answers Explained***

ANSWERS EXPLAINED

PART ONE

1. Let x = the number of kilometers the car can travel in 5 hours.

At the same rate of speed, the distance traveled is directly proportional to the time taken:

$$\frac{180}{3} = \frac{x}{5}$$

In a proportion, the product of the means equals the product of the extremes (cross-multiply):

$$3x = 5(180)$$
$$3x = 900$$

Divide both sides of the equation by 3:

$$\frac{3x}{3} = \frac{900}{3}$$
$$x = 300$$

The car can travel **300** kilometers in 5 hours.

2.

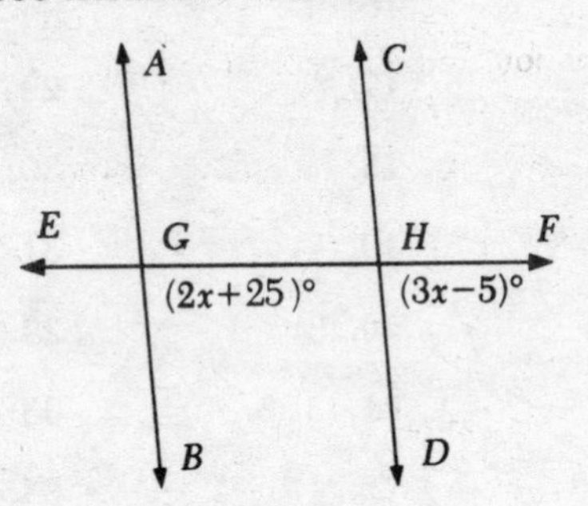

If two lines are parallel, a transversal to them makes a pair of corresponding angles congruent:

$$3x - 5 = 2x + 25$$

Add $-2x$ (the additive inverse of $2x$) to both sides of the equation; also add 5 (the additive inverse of -5) to both sides:

$$\underline{-2x + 5 = -2x + 5}$$
$$x = 30$$

x = **30.**

3. The given equation contains decimals:

$$0.03x - 1.2 = 0.24$$

To clear decimals, multiply each term on both sides of the equation by 100:

$$100(0.03x) - 100(1.2) = 100(0.24)$$
$$3x - 120 = 24$$

Add 120 (the additive inverse of -120) to both sides of the equation:

$$\underline{120 = 120}$$
$$3x = 144$$

Divide both sides of the equation by 3:

$$\frac{3x}{3} = \frac{144}{3}$$
$$x = 48$$

x = **48.**

4. The given expression is: $-3a^2b$

Substitute -2 for a and 3 for b: $-3(-2)^2(3)$

First evaluate the power; $(-2)^2 = 4$: $-3(4)(3)$

Multiply all the factors together: -36

The value is $\mathbf{-36}$**.**

5. The given equation contains a fraction:

$$\frac{2}{3}m - 4 = 14$$

To clear the fraction, multiply each term on both sides of the equation by 3:

$$3\left(\frac{2}{3}m\right) - 3(4) = 3(14)$$
$$2m - 12 = 42$$

Add 12 (the additive inverse of -12) to both sides of the equation:

$$\underline{ 12 = 12}$$
$$2m = 54$$

Divide both sides of the equation by 2:

$$\frac{2m}{2} = \frac{54}{2}$$
$$m = 27$$

m = **27.**

6.

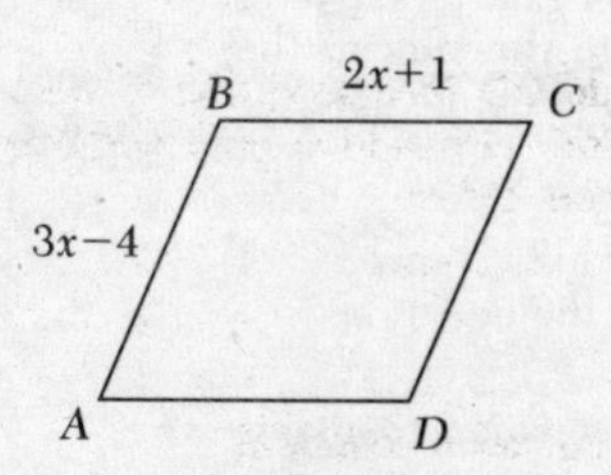

A rhombus is an equilateral parallelogram, so $AB = BC$: $3x - 4 = 2x + 1$

Add $-2x$ (the additive inverse of $2x$) and also add 4 (the additive inverse of -4) to both sides of the equation:

$$\underline{-2x + 4 = -2x + 4}$$
$$x = 5$$

x = **5.**

7. The given expression is: $(x + 4)(x - 2) + 3x$

Multiply the two binomials together:

$$\begin{array}{r} x + 4 \\ x - 2 \\ \hline x^2 + 4x \qquad \\ - 2x - 8 \\ \hline x^2 + 2x - 8 \end{array}$$

$x^2 + 2x - 8 + 3x$

Combine like terms: $x^2 + 5x - 8$

The trinomial is $\mathbf{x^2 + 5x - 8}$.

8. Let $5x$ = one number.
Then $6x$ = the other number.
The sum of the two numbers is 66: $5x + 6x = 66$
Combine like terms: $11x = 66$
Divide both sides of the equation by 11: $\frac{11x}{11} = \frac{66}{11}$

$x = 6$

The *larger* number is represented by $6x$: $6x = 6(6) = 36$
The *larger* number is **36.**

9.

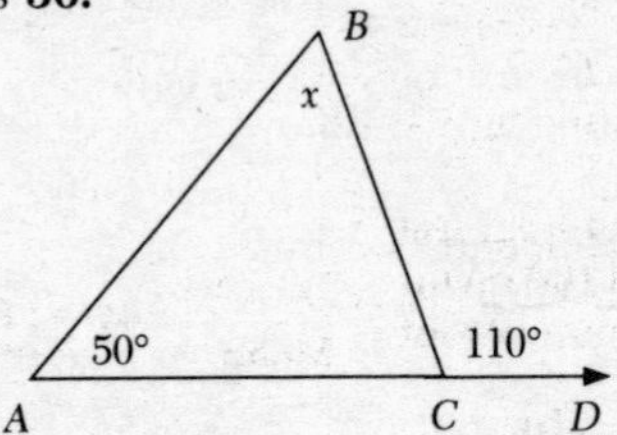

Let x = the measure of $\angle ABC$.
The measure of an exterior angle of a triangle is equal to the sum of the measures of the two remote interior angles:

$$m\angle BCD = m\angle BAC + m\angle ABC$$
$$110 = 50 + x$$

Add -50 (the additive inverse of 50) to both sides of the equation:

$$\begin{array}{rcl} -50 &=& -50 \\ \hline 60 &=& x \end{array}$$

$m\angle ABC$ = **60.**

10. The given expression is the difference between two perfect squares: $9 - y^2$

The difference between two perfect squares of the form $(A^2 - B^2)$ can be factored into the product of two binomials of the form $(A - B)(A + B)$; here $A^2 = 9$ and $B^2 = y^2$, so $A = 3$ and $B = y$: $(3 - y)(3 + y)$

The factored form is $\mathbf{(3 - y)(3 + y)}$.

11.

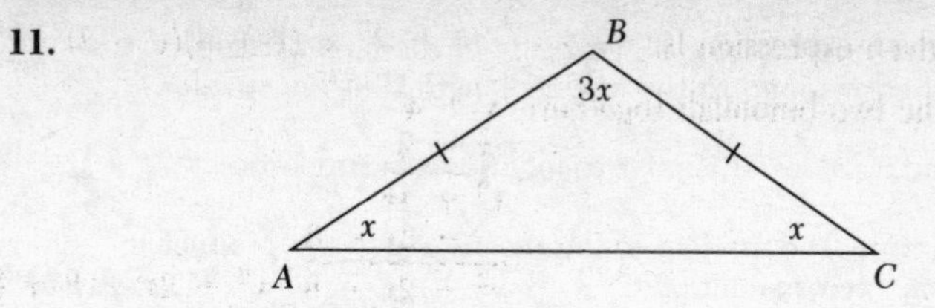

Let x = the measure of a base angle. Note that the measure of both base angles must be x since the base angles of an isosceles triangle are congruent.

Then $3x$ = the measure of the vertex angle.

The sum of the measures of the three angles of a triangle is 180°:

$$m\angle A + m\angle B + m\angle C = 180$$
$$x + 3x + x = 180$$

Combine like terms:

$$5x = 180$$

Divide both sides of the equation by 5:

$$\frac{5x}{5} = \frac{180}{5}$$
$$x = 36$$

The measure of a base angle is **36.**

12. The mean of several quantities is the sum of the quantities divided by the number of them:

$$\text{Mean} = \frac{4x - 6 + 2x + 3 + 3x + 3}{3}$$

Combine like terms in the numerator:

$$\text{Mean} = \frac{9x}{3}$$

Reduce the fraction by dividing numerator and denominator by 3:

$$\text{Mean} = 3x.$$

The mean is **3x.**

13.

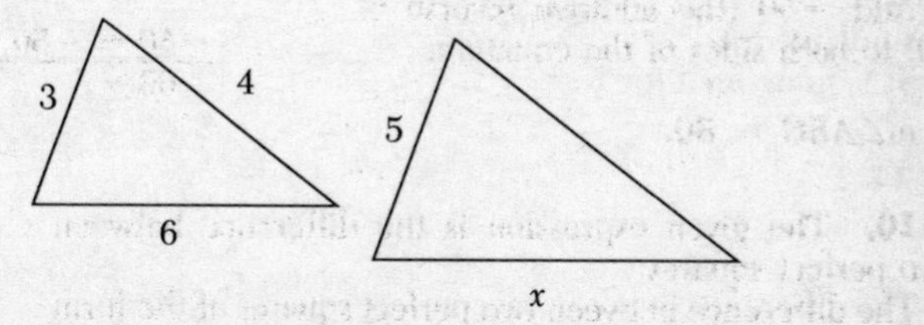

Since the triangles are similar, their shortest sides will correspond to one another; thus the sides whose respective lengths are 3 and 5 will correspond to one another.

Let x = the length of the longest side of the larger triangle; this side will correspond to the side of length 6 in the smaller triangle.

Corresponding sides of similar triangles are in proportion: $\frac{3}{5} = \frac{6}{x}$

In a proportion, the product of the means equals the product of the extremes (cross-multiply): $3x = 5(6)$

$3x = 30$

Divide both sides of the equation by 3: $\frac{3x}{3} = \frac{30}{3}$

$x = 10$

The length of the largest side is **10.**

14. The given radical is: $\sqrt{84}$

Factor out any perfect square factor in the radicand, 84: $\sqrt{4(21)}$

Remove the perfect square factor, 4, from under the radical sign by taking its square root and writing it as a coefficient of the radical: $2\sqrt{21}$

Compare $2\sqrt{21}$ with the form, $a\sqrt{b}$, in the question: $a = 2$

The value of a is **2.**

15. ${}_4P_3$ stands for the number of permutations of 4 things taken 3 at a time.

The vaue of ${}_nP_r$ is given by the formula: ${}_nP_r = n(n-1)(n-2)\ldots$ to r factors

For ${}_4P_3$, $n = 4$ and $r = 3$: ${}_4P_3 = 4(3)(2)$

${}_4P_3 = 24$

${}_4P_3 =$ **24.**

16. The expression $\frac{3}{x+1}$ will be undefined if its denominator equals 0, since division by 0 is undefined. If $x = -1$, the expression becomes $\frac{3}{0}$, which is undefined.

It is undefined for $x =$ **−1.**

17.

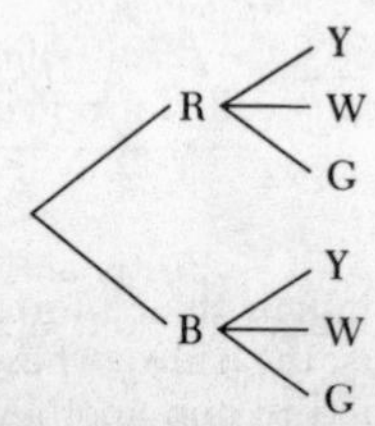

$$\text{The probability of an event occurring} = \frac{\text{the number of successful outcomes}}{\text{the total possible number of outcomes}}.$$

On the tree diagram, each outcome is represented by a path through a choice of blouse (R or B) to a choice of slacks (Y, W, or G). There is a total of 6 possible paths.

The successful outcomes are represented by paths that lead to a choice of G for slacks. There are 2 such paths.

The probability of Mary choosing an outfit that includes a pair of green slacks is $\frac{2}{6}$.

The probability is $\mathbf{\frac{2}{6}}$.

18.

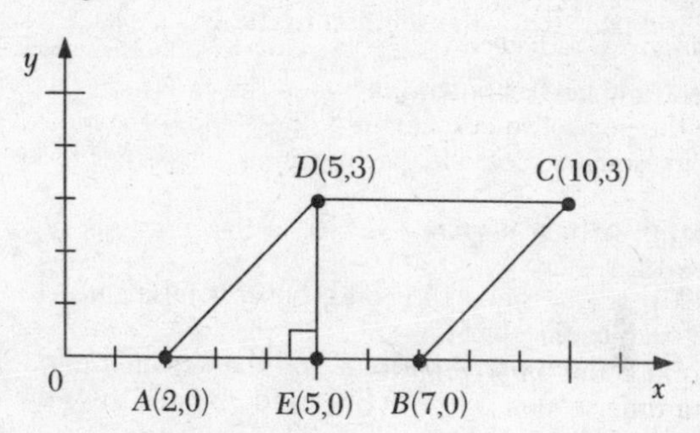

Draw $\overline{DE}$ from D perpendicular to the x-axis. $\overline{DE}$ is the altitude of parallelogram $ABCD$. The coordinates of E are (5,0).

The area of a parallelogram is equal to the product of its base and altitude:

$AB = 7 - 2 = 5$; $DE = 3 - 0 = 3$:

Area of $ABCD$ = $(AB)(DE)$
Area of $ABCD$ = $(5)(3)$
Area of $ABCD$ = 15

The area is **15.**

19. Consider each of the choices:

(1) An enlargement or a reduction of the figure is called a *dilation.* This is the correct answer.

(2) A slide of the figure is called a *translation.*

(3) A turning of the figure about some fixed point is called a *rotation.*

(4) A mirror image of the figure is called a *reflection.*

The correct choice is **(1).**

20. The equation $3x - 6y = 3(x - 2y)$ is an illustration of the distributive property. It shows that in multiplying out $3(x - 2y)$, the factor 3 is "distributed" to each of the terms, x and $-2y$.

The correct choice is **(3).**

21. The given system of equations is: $\begin{cases} y = x + 4 \\ x + y = 2 \end{cases}$

Substitute $x + 4$ for y in the second equation: $x + x + 4 = 2$

Combine like terms: $2x + 4 = 2$

Add -4 (the additive inverse of 4) to both sides of the equation: $-4 = -4$

$2x = -2$

Divide both sides of the equation by 2: $\frac{2x}{2} = \frac{-2}{2}$

$x = -1$

To obtain the value of y, substitute -1 for x in the first equation: $y = -1 + 4$

Combine like terms: $y = 3$

The ordered pair that is the solution to the system is $(-1,3)$.

The correct choice is **(3)**

22. p represents "x is odd."

q represents "$x > 15$."

If $x = 13$, p is true since 13 is odd, but q is false since $13 \not> 15$.

Examine each choice:

(1) $p \wedge \sim q$ is the *conjunction* of p and the *negation* of q. Since q is false, its negation, $\sim q$, is true. Since both p and $\sim q$ are true, the conjunction is true. This is the correct choice.

(2) $\sim p \vee q$ is the *disjunction* of q and the *negation* of p. Since p is true, its negation, $\sim p$, is false. Since $\sim p$ and q are both false, the disjunction is false.

(3) $p \rightarrow q$ is the *implication* that p implies q. Since the hypothesis p is true but the conclusion q is false, the implication is false.

(4) $p \wedge q$ is the *conjunction* of p and q. Since p is true and q is false, the conjunction is false; both p and q must be true in order for the conjunction to be true.

The correct choice is **(1)**.

23. The sum of the two functions can be represented as: $\frac{3x}{4} + \frac{2x}{3}$

In their present form, the fractions cannot be combined since they have different denominators. Convert each fraction to an equivalent fraction having the least common denominator (L.C.D.) of the denominators. The L.C.D. is the smallest number that is divisible by each of the denominators: The L.C.D. for 4 and 3 is 12.

Multiply the first fraction by 1 in the form $\frac{3}{3}$, and multiply the second fraction by 1 in the form $\frac{4}{4}$:

$$\frac{3(3x)}{3(4)} + \frac{4(2x)}{4(3)}$$

$$\frac{9x}{12} + \frac{8x}{12}$$

Since the fractions now have the same denominator, they may be combined by combining their numerators:

$$\frac{9x + 8x}{12}$$

Combine like terms in the numerator:

$$\frac{17x}{12}$$

The correct choice is **(4)**.

24.

−7 −6 −5 −4 −3 −2 −1 0 1 2 3 4 5 6 7

The heavy shaded line extends from -5 to 6, so it represents numbers greater than -5 and less than 6. It includes 6 since there is a heavy shaded dot at 6, but it does not include -5 since the empty circle at -5 indicates that it is not part of the set.

The inequality that is represented is $-5 < x \leq 6$.

The correct choice is **(4)**.

25.

Interval	Frequency
16-20	1
11-15	3
6-10	3
1-5	2
Total	9

Total the frequency column. There are 9 items covered in the table.

The median is the value of the middle item when they are arranged in order of size. Since $\frac{9}{2} = 4\frac{1}{2}$, the middle item is halfway between the fourth and fifth item, so the median lies midway between the values of these two items.

Counting up from the bottom of the table, there are 2 items in the 1-5 interval. We need 2 more to reach the fourth, and we need 3 more to reach the fifth item. Since the 6-10 interval contains 3 items, both the fourth and fifth items lie in this interval. The median, midway between them, must therefore lie in the 6-10 interval.

The correct choice is **(2)**.

26. In the *implication* $m \rightarrow \sim r$, the *hypothesis* or *antecedent* is m and the *conclusion* or *consequent* is $\sim r$.

The *converse* of an implication is another implication formed by interchanging the hypothesis and conclusion of the original implication. Thus, the converse of $m \rightarrow \sim r$ is $\sim r \rightarrow m$.

The correct choice is **(2)**.

27. The given expression is: $\dfrac{15k^3 - 9k^2 + 3k}{3k}$, $k \neq 0$

To simplify, divide each term in the numerator by $3k$, applying the distributive property. Divide the numerical coefficients to obtain the numerical coefficient of each quotient. Subtract the exponents to obtain the exponent of each power when dividing powers of the same base. Note that the provision that $k \neq 0$ insures that we are not dividing by 0; division by 0 is undefined: $5k^2 - 3k + 1$

The correct choice is **(1)**.

28.

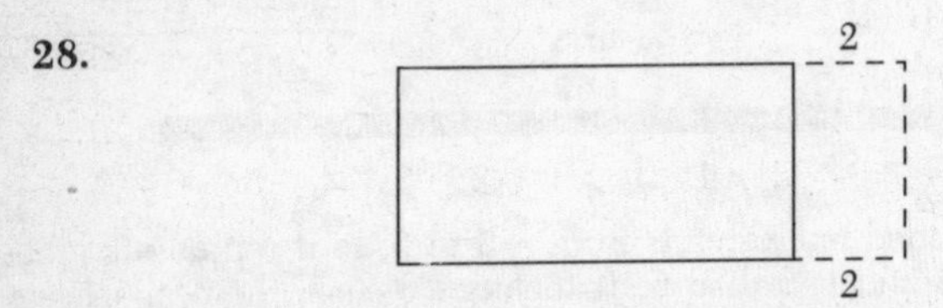

The perimeter of a rectangle is the sum of the measures of all four sides. If the length is increased by 2, the measures of two sides are each increased by 2, thus increasing the perimeter by 4.

The correct choice is **(3)**.

29.

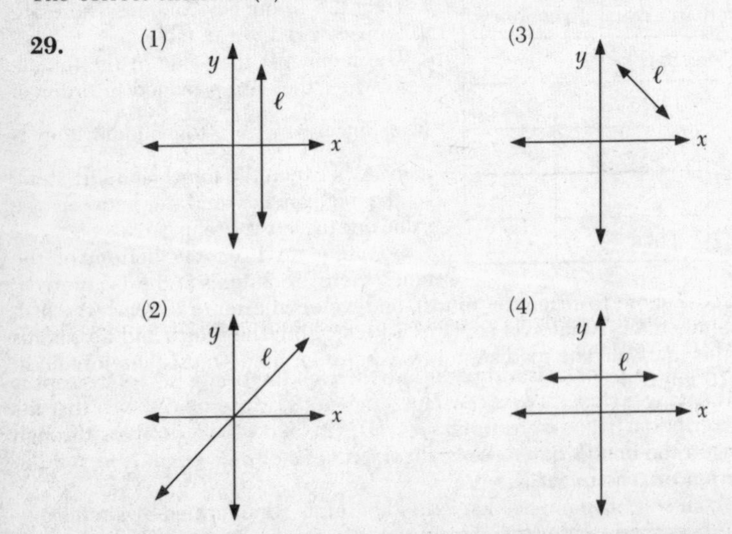

A line has a slope of zero if it is horizontal, that is, parallel to the x-axis.

The slope of a line through two points, (x_1,y_1) and (x_2,y_2), is $\dfrac{y_2 - y_1}{x_2 - x_1}$. In order for the slope to be zero, the numerator, $y_2 - y_1$, must be zero, or y_2 must equal y_1. This means that for a slope of zero, the y-coordinates of all points on the line must be the same.

The correct choice is **(4).**

30. Let y = the measure of the complement of the angle.

The sum of the measures of two complementary angles is 90°:

$$x + y = 90$$

Add $-x$ (the additive inverse of x) to both sides of the equations:

$$\begin{array}{rcr} -x & = & -x \\ \hline y & = & 90 - x \end{array}$$

The correct choice is **(1).**

31. (1)

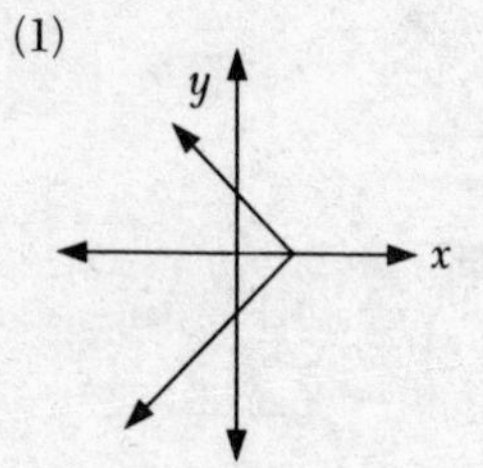

(3)

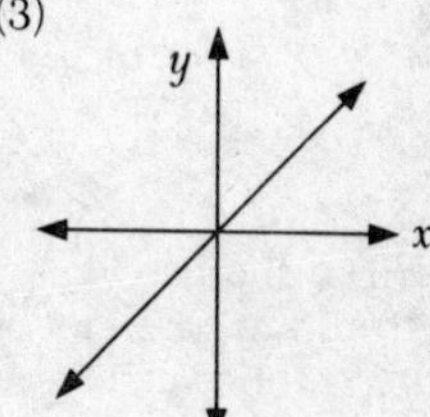

(2)

(4)

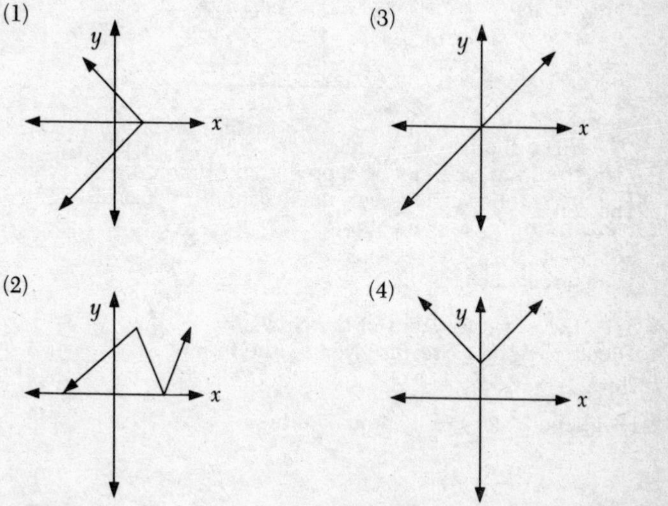

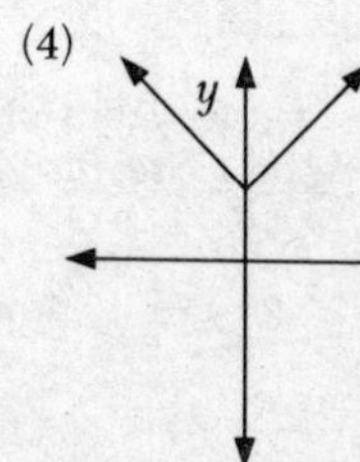

"y varies directly as x" can be represented algebraically as $y = kx$, where k is a constant.

The graph of $y = kx$ is a straight line through the origin (its y-intercept is 0), and having a slope of k, a constant. Choice (3) is the only choice that fits this description. The graphs in choices (1), (2), and (4) do not pass through the origin and do not have a constant slope.

The correct choice is **(3).**

32. Examine each of the choices:

(1): $p \rightarrow q$ is an *implication*. The implication is true if q is true or if p and q are both true or both false; it is not true if p is true and q is false.

(2) $q \wedge \sim q$ is the *conjunction* of q and the *negation* of q. The conjunction would be true only if both q and $\sim q$ are true. But q and its negation, $\sim q$, always have opposite truth values. Thus, the conjunction can *never* be true.

(3) $\sim p \rightarrow \sim q$ is an *implication*. It can be true or false depending on the truth values of $\sim p$ and $\sim q$ as explained for choice (1).

(4) $p \vee \sim p$ is the *disjunction* of p and the *negation* of p. If p is true, its negation, $\sim p$, is false; if p is false, its negation, $\sim p$, is true. The disjunction is true if p or $\sim p$ or both are true. Since either p or $\sim p$ must be true, the disjunction is *always* true.

The disjunction in (4) will *always* be true.

The correct choice is **(4).**

33.

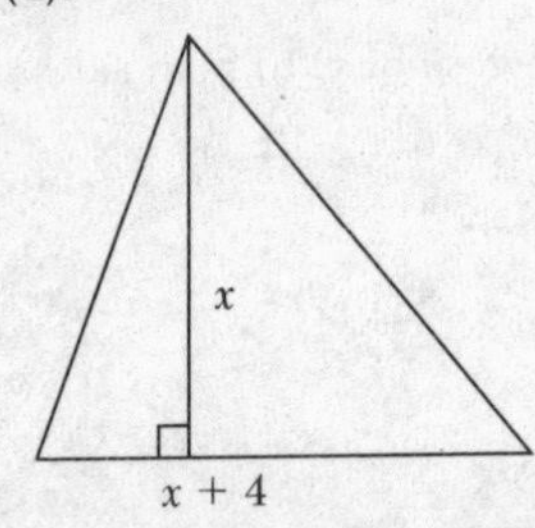

Since the base is 4 units more than the height, the length of the base can be represented by $x + 4$.

The area of a triangle equals one-half the product of the lengths of its base and altitude: $\frac{1}{2}x(x + 4) = 48$

The correct choice is **(2).**

34. Let r = the radius of the circle.

The area, A, of a circle is given by the formula: $A = \pi r^2$

Here, $A = 16\pi$: $16\pi = \pi r^2$

Divide both sides of the equation by π: $\frac{16\pi}{\pi} = \frac{\pi r^2}{\pi}$

$16 = r^2$

Take the square root of both sides of the equation: $\pm\sqrt{16} = r$

$\pm 4 = r$

Reject the negative root as meaningless for a length: $4 = r$

The length of the diameter, d, of a circle is twice the length of the radius: $d = 2(4) = 8$

The length of the diameter is 8.

The correct choice is **(2).**

35. 72 kilometers per hour may be written as

$\frac{72 \text{ kilometers}}{1 \text{ hour}}$

1 kilometer = 1,000 meters, or 1 kilometer = 10^3 meters.

Multiply $\frac{72 \text{ kilometers}}{1 \text{ hour}}$ by 1 in the form $\frac{10^3 \text{ meters}}{1 \text{ kilometer}}$:

$$\frac{72 \text{ kilometers}}{1 \text{ hour}} \times \frac{10^3 \text{ meters}}{1 \text{ kilometer}}$$

Cancel like "factors" (kilometer) in numerator and denominator:

$$\frac{72 \text{ \sout{kilometers}}}{1 \text{ hour}} \times \frac{10^3 \text{ meters}}{1 \text{ \sout{kilometer}}}$$

$$\frac{72 \times 10^3 \text{ meters}}{1 \text{ hour}}$$

Since the choices all contain 7.2 instead of 72 as the first factor, replace 72 by 7.2 × 10:

$$\frac{7.2 \times 10 \times 10^3 \text{ meters}}{1 \text{ hour}}$$

$10 \times 10^3 = 10^4$:

$$\frac{7.2 \times 10^4 \text{ meters}}{1 \text{ hour}}$$

The correct choice is **(4)**.

PART TWO

36. *a* To draw the graph of the inequality $y + x \geq 5$, first rearrange it so that it is solved for y by adding $-x$ (the additive inverse of x) to both sides:

$$\begin{aligned} y + x &\geq 5 \\ -x &= \quad -x \\ \hline y &\geq 5 - x \end{aligned}$$

Draw the graph of the equation $y = 5 - x$. Prepare a table of pairs of values of x and y by selecting three convenient values of x and substituting them in the equation to find the corresponding values of y:

x	$5 - x$	$= y$
-3	$5 - (-3) = 5 + 3$	$= 8$
0	$5 - 0$	$= 5$
3	$5 - 3$	$= 2$

Plot the points $(-3,8)$, $(0,5)$, and $(3,2)$ and draw a *solid* line through them. This line is the graph of the equation $y + x = 5$. It is shown as a *solid* line to indicate that the points on it are part of the solution set of $y + x \geq 5$.

The solution set of the inequality $y + x > 5$ lies on one side of the line $y + x = 5$. To find out on which side of the line the solution set lies, choose a convenient point, say $(0,0)$, and substitute its coordinates in the inequality:

$$\begin{aligned} y + x &> 5 \\ 0 + 0 &\overset{?}{>} 5 \\ 0 &\not> 5 \end{aligned}$$

Since (0,0) does *not* satisfy the inequality, it does *not* lie in the solution set of $y + x > 5$. Therefore shade the *opposite* side of the line $y + x = 5$ with cross-hatching extending up and to the right.

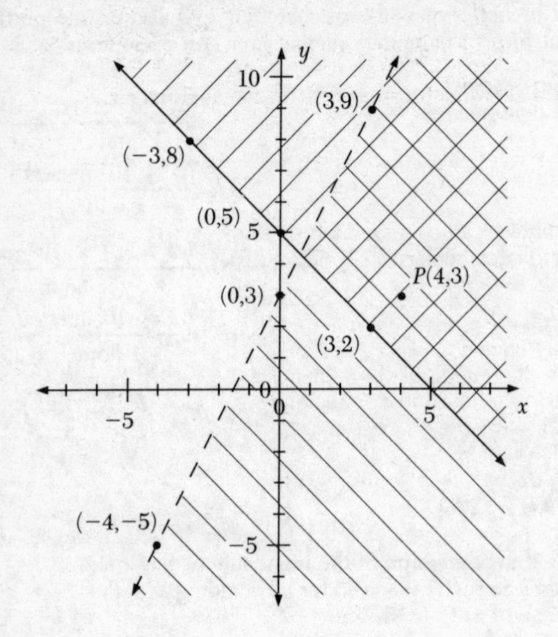

To draw the graph of the inequality $y < 2x + 3$, first draw the graph of the equation $y = 2x + 3$. Prepare a table of pairs of values of x and y after choosing three convenient values for x:

x	$2x + 3$	$= y$
-4	$2(-4) + 3 = -8 + 3$	$= -5$
0	$2(0) + 3 = 0 + 3$	$= 3$
3	$2(3) + 3 = 6 + 3$	$= 9$

Plot the points $(-4,-5)$, (0,3), and (3,9) and draw a *broken* line through them. The *broken* line indicates that the points on it are not part of the solution set of $y < 2x + 3$.

The solution set of $y < 2x + 3$ lies on one side of the line $y = 2x + 3$. To find out on which side, choose a convenient test point, say (0,0), and substitute its coordinates in the inequality to see if they satisfy it:

$$y < 2x + 3$$

$$0 \overset{?}{<} 2(0) + 3$$

$$0 \overset{?}{<} 0 + 3$$

$$0 < 3 \checkmark$$

Since (0,0) satisfies the inequality, it lies on the side of the line $y = 2x + 3$ that contains the points that represent $y < 2x + 3$. Shade this side with cross-hatching extending down and to the right.

b The points in the solution set of the inequalities are those that lie in the area covered by *both* types of cross-hatching and also on the portion of the solid line that forms a boundary of that area. An example of such a point is $P(4,3)$.

(4,3) are the coordinates of a point in the solution set.

37.

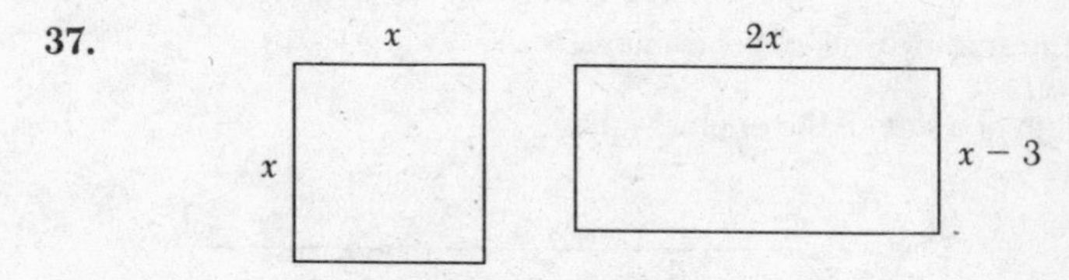

Let x = the length of a side of the square.

Then $2x$ = the length of the rectangle.

And $x - 3$ = the width of the rectangle

The area of a square = the square of the length of one side: Area of square $= x^2$

The area of a rectangle = the product of its length and width: Area of rectangle $= 2x(x - 3)$

The area of the rectangle exceeds the area of the square by 16 means that the area of the rectangle equals the area of the square plus 16.

$$2x(x - 3) \quad = \quad x^2 \quad + \; 16$$

The equation to use is: $2x(x - 3) = x^2 + 16$

Remove the parentheses by applying the distributive principle: $2x^2 - 6x = x^2 + 16$

This is a *quadratic equation.* Rearrange it so that all terms are on one side equal to 0 by adding $-x^2 - 16$ to both sides:

$$\begin{array}{r} 2x^2 - 6x = x^2 + 16 \\ -x^2 - 16 = -x^2 - 16 \\ \hline x^2 - 6x - 16 = 0 \end{array}$$

The left side is a *quadratic trinomial* that can be factored into the product of two binomials. The factors of the first term, x^2, are x and x, and they become the first terms of the binomials: $(x \quad)(x \quad) = 0$

The factors of the last term, -16, become the second terms of the binomials, but they must be chosen in such a way that the sum of the inner and outer cross-products equals the middle term, $-6x$, of the original trinomial. Try -8 and $+2$ as the factors of -16:

$-8x$ = inner product

$(x - 8)(x + 2) = 0$

$+2x$ = outer product

Since $(-8x) + (+2x) = -6x$, these are the correct factors: $(x - 8)(x + 2) = 0$

If the product of two factors equals 0, either factor may equal 0: $x - 8 = 0 \vee x + 2 = 0$

Add the appropriate additive inverse to both sides of each equation, 8 for the left equation, and -2 for the right one:

$$\begin{array}{rl} 8 = 8 & -2 = -2 \\ \hline x \quad = 8 & x \quad = -2 \end{array}$$

Reject the negative value as meaningless for a length: $x = 8$

The length of a side of the original square is **8**.

38.

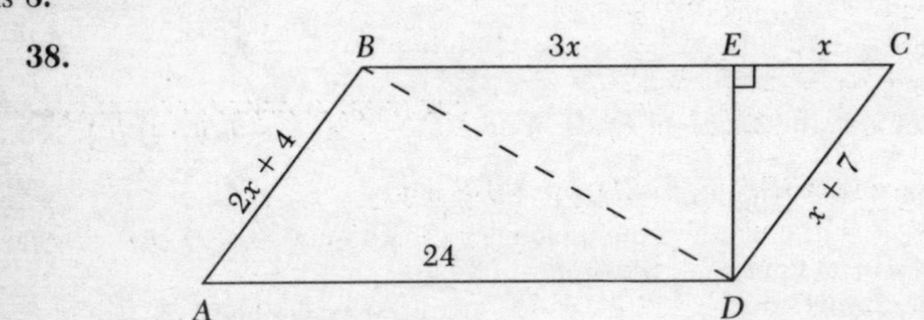

a The opposite sides of a parallelogram are congruent (equal in length): $AB = CD$

$AB = 2x + 4$; $CD = x + 7$: $2x + 4 = x + 7$

Add $-x$ (the additive inverse of x) and also add -4 (the additive inverse of 4) to both sides of the equation:

$$\begin{array}{rcl} -x - 4 & = & -x - 4 \\ \hline x & = & 3 \end{array}$$

$$CD = x + 7 = 3 + 7 = 10$$

The length of $\overline{CD}$ is **10.**

b Opposite sides of a parallelogram are congruent: $BC = AD$

$AD = 24$: $BC = 24$

$$EC + BE = BC$$

Let $EC = x$; then $BE = 3x$: $x + 3x = 24$

Combine like terms: $4x = 24$

Divide both sides of the equation by 4: $\frac{4x}{4} = \frac{24}{4}$

$$x = 6$$

The length of $\overline{EC}$ is **6.**

c $\triangle DEC$ is a right triangle since $\overline{DE} \perp \overline{BC}$.

From part *a*, $CD = 10$; from part *b*, $EC = 6$.

Right $\triangle DEC$ is a 3-4-5 right triangle with $EC = 2 \times 3$ and $CD = 2 \times 5$. Therefore, $ED = 2 \times 4$, or $ED = 8$.

$ED = $ **8**.

ALTERNATIVE SOLUTION: The Pythagorean Theorem may be used to find ED: $(ED)^2 + 6^2 = 10^2$

d The area, A, of a triangle whose base is b and whose altitude is h is given by the formula: $A = \frac{1}{2}bh$

For $\triangle ABD$, $b = AD$ and $h = ED$:

$$A = \frac{1}{2}(AD)(ED)$$
$$A = \frac{1}{2}(24)(8)$$
$$A = 12(8)$$
$$A = 96$$

The area of $\triangle ABD$ is **96.**

e The area, A, of a trapezoid whose altitude is h and whose bases are b_1 and b_2 is given by the formula: $A = \frac{1}{2}h(b_1 + b_2)$

For trapezoid $ABED$, $h = ED$, $b_1 = BE$, and $b_2 = AD$: $A = \frac{1}{2}(ED)(BE + AD)$

From part *c*, $BE = 3x = 3(6) = 18$:

$$A = \frac{1}{2}(8)(18 + 24)$$
$$A = 4(42)$$
$$A = 168$$

The area of trapezoid $ABED$ = **168.**

39. Let c = the cost of the telephone call.
Let m = the number of minutes the call takes.

The cost is \$0.60 for the first three minutes plus \$0.17 for each additional minute: $c = 0.60 + 0.17(m - 3)$

If the cost cannot exceed \$2.50, the total cost, c, must be less than \$2.50: $0.60 + 0.17(m - 3) < 2.50$

Remove the parentheses: $0.60 + 0.17m - 0.51 < 2.50$

Clear decimals by multiplying each term on both sides of the inequality by 100: $100(0.60) + 100(0.17m) - 100(0.51) < 100(2.50)$

$$60 + 17m - 51 < 250$$

Combine like terms: $17m + 9 < 250$

Add -9 (the additive inverse of 9) to both sides of the inequality:

$$\frac{\quad -9 = -9}{17m \qquad < 241}$$

Divide both sides of the inequality by 17:

$$\frac{17m}{17} < \frac{241}{17}$$

```
   14+
17)241
   17
    71
    68
```

$$m < 14+$$

The greatest number of whole minutes is **14.**

40. *a* The number of arrangements of the letters in "CHORD" is the number of permutations, ${}_5P_5$ of things taken 5 at a time.

$${}_nP_n = n! = n(n - 1)(n - 2) \ldots (3)(2)(1)$$

Here, $n = 5$: $\quad {}_5P_5 = 5! = 5(4)(3)(2)(1)$

$${}_5P_5 = 120$$

120 arrangements are possible.

b If an arrangement begins with "H," the remaining 4 letters may be arranged in any order. The number of arrangements is ${}_4P_4 = 4(3)(2)(1)$. The number of arrangements beginning with "H" is: $\quad {}_4P_4 = 24$

Similarly, if an arrangement begins with "O" the number of arrangements will again be: $\quad {}_4P_4 = 24$

The number of arrangements beginning with either "H" or "O" is 24 + 24, or 48.

The number of arrangements is **48.**

c Probability of an event occurring

$$= \frac{\text{the number of successful outcomes}}{\text{the total possible number of outcomes}}.$$

The successful outcomes are choices of arrangements beginning with "C." As explained in part *b* for "H" and "O," the number of arrangements beginning with "C" is 24. The total number of possible outcomes is the total number of possible arrangements, found in part *a* to be 120.

The probability that an arrangement selected at random will begin with "C" is: $\quad \frac{24}{120}$

Reduce the fraction by dividing numerator and denominator by 12: $\quad \frac{2}{10}$

Reduce again, dividing numerator and denominator by 2: $\quad \frac{1}{5}$

The probability is $\mathbf{\frac{1}{5}}$.

d

CHORD

C, H, O, and D all have horizontal line symmetry, that is, if folded along the dotted horizontal line shown, every point on the upper half will correspond to a point on the lower half and conversely.

H and O have vertical line symmetry. The fold to establish vertical line symmetry will be the vertical dotted line.

(1) For the probability of selecting a letter from "CHORD" having *only* horizontal line symmetry, there are two successful outcomes, D and C. The total possible number of outcomes is 5.

The probability of a letter with only horizontal line symmetry $= \frac{2}{5}$.

The probability is $\mathbf{\frac{2}{5}}$.

(2) There is no letter in "CHORD" that has vertical line symmetry only. Hence the number of successful outcomes for choosing such a letter is 0 and the probability is $\frac{0}{5}$ or 0.

The probability is **0.**

(3) For the selection of a letter having both horizontal line and vertical line symmetry, there are 2 successful outcomes, H and O.

The probability of selecting a letter with both types of symmetry $= \frac{2}{5}$.

The probability is $\mathbf{\frac{2}{5}}$.

(4) Only one letter, R, has neither horizontal line nor vertical line symmetry, so there is only 1 successful outcome for choosing such a letter.

The probability of selecting a letter with neither type of symmetry is $\frac{1}{5}$.

The probability is $\mathbf{\frac{1}{5}}$.

41. *a* Prepare a table with columns for p, q, $\sim p$, $(\sim p \rightarrow q)$, $(p \vee q)$, and $(\sim p \rightarrow q) \leftrightarrow (p \vee q)$:

p	q	$\sim p$	$\sim p \rightarrow q$	$p \vee q$	$(\sim p \rightarrow q) \leftrightarrow (p \vee q)$
T	T	F	T	T	T
T	F	F	T	T	T
F	T	T	T	T	T
F	F	T	F	F	T

Fill in the columns for p and q with all possible combinations of truth values, T and F, for p and q. There are 4 such combinations possible. For each of the other columns, the truth values, T or F, are determined in accordance with the values for p and q on the same line.

The statement $\sim p$ is the *negation* of p. The truth values for $\sim p$ are the opposite of those for p on the same line.

The statement $\sim p \rightarrow q$ is the *conditional* that $\sim p$ implies q. The conditional has the truth value T whenever q is T or when $\sim p$ and q are both F. It has the value F when $\sim p$ is T and q is F.

The statement $p \vee q$ is the *disjunction* of p and q. The disjunction has the truth value T when either p or q or both are T. The disjunction has the value F when p and q are both F.

The statement $(\sim p \rightarrow q) \leftrightarrow (p \vee q)$ is the *biconditional* or *equivalence relation* between $(\sim p \rightarrow q)$ and $(p \vee q)$. When the truth values of $(\sim p \rightarrow q)$ and $(p \vee q)$ are the same, either both T or both F, the bi-conditional has the value T; otherwise, it has the value F.

b A *tautology* is a compound statement that is always true regardless of the truth values of its component statements. Since $(\sim p \rightarrow q) \leftrightarrow (p \vee q)$ always has the truth value T for all possible combinations of values of p and q, it is a tautology.

The answer to the question is **yes.**

42. *a*

Interval	Tally	Frequency
35-37	////	4
32-34	~~////~~ /	6
29-31	~~////~~ //	7
26-28	/	1
23-25	//	2
Total		20

Tally the scores 36, 32, 28, 30, 33, 36, 24, 33, 29, 30, 30, 25, 34, 36, 34, 31, 36, 29, 30, and 34.

Enter the total for each interval in the frequency column.

b

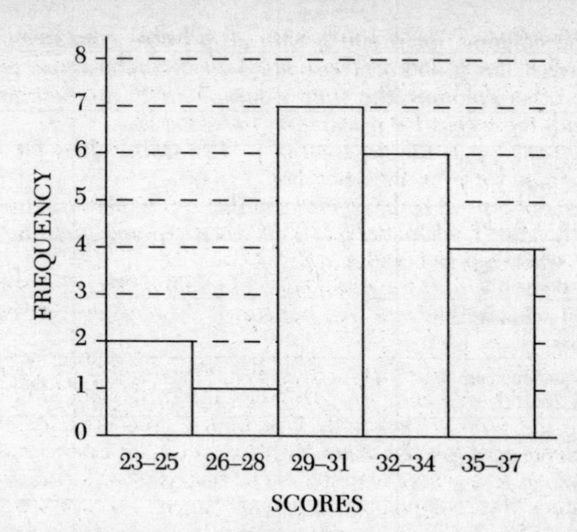

c The lowest two intervals contain the scores that are less than 29 points. These intervals cover 2 + 1, or 3, games.

20 games were played in all. The fraction in which the player scored less than 29 points is $\frac{3}{20}$. Since $\frac{3}{20} = \frac{15}{100}$, this represents 15% of the games played.

The player scored less than 29 points in **15%** of the games.

d The 60th percentile is the point at or below which 60% of the scores lie.

60%, or $\frac{6}{10}$ of the total of 20 scores is $\frac{6}{10}(20) = 12$. Thus, 12 scores are at or below the 60th percentile. Counting up from the bottom, we find 2 + 1 + 7, or 10, scores in the three lowest intervals. We need 2 more to reach the 12th score. Since the next interval, 32-34, contains 6 scores, it must contain the 2 needed to reach the 12th. Therefore, the 60th percentile lies in the 32-34 interval.

The **32-34** interval contains the 60th percentile.

Topic	Question Numbers	Number of Points	Your Points	Your Percentage
1. Numbers (rat'l, irrat'l); Percent	42c	2		
2. Properties of No. Systems	16, 20	2 + 2 = 4		
3. Operations on Rat'l Nos. and Monomials	23	2		
4. Operations on Multinomials	7, 27	2 + 2 = 4		
5. Square Root; Operations involving Radicals	14	2		
6. Evaluating Formulas and Expressions	4	2		
7. Linear Equations (simple cases incl. parentheses)	—	0		
8. Linear Equations Containing Decimals or Fractions	3, 5	2 + 2 = 4		
9. Graphs of Linear Functions (slope)	29	2		
10. Inequalities	24	2		
11. Systems of Eqs. & Inequal. (alg. & graphic solutions)	21, 36a,b	2 + 8 + 2 = 12		
12. Factoring	10	2		
13. Quadratic Equations	—	0		
14. Verbal Problems	8, 37, 39	2 + 10 + 10 = 22		
15. Variation	31	2		
16. Literal Eqs.; Expressing Relations Algebraically	38b	2		
17. Factorial n	—	2		
18. Areas, Perims., Circums., Volumes of Common Figures	28, 34, 38d,e	2 + 2 + 2 + 2 = 8		

Topic	Question Numbers	Number of Points	Your Points	Your Percentage
19. Geometry (congruence, parallel lines, compls., suppls.)	2, 30	2 + 2 = 4		
20. Ratio & Proportion (incl. similar triangles & polygons)	1, 13	2 + 2 = 4		
21. Pythagorean Theorem	38c	2		
22. Logic (symbolic rep., logical forms, truth tables)	22, 26, 32, 41a,b	2 + 2 + 2 + 9 + 1 = 16		
23. Probability (incl. tree diagrams & sample spaces)	17, 40c	2 + 2 = 4		
24. Combinations, Arrangements, Permutations	15, 40a,b	2 + 2 + 2 = 6		
25. Statistics (central tend., freq. dist., histogr., quartiles, percentiles)	12, 25, 42a,b,d	2 + 2 + 2 + 4 + 2 = 12		
26. Properties of Triangles and Quadrilaterals	6, 9, 11, 33, 38a	2 + 2 + 2 + 2 + 2 = 10		
27. Transformations (reflect., translations, rotations, dilations)	19	2		
28. Symmetry	40d	4		
29. Area from Coordinate Geometry	18	2		
30. Dimensional Analysis	35	2		
31. Scientific Notation; Negative & Zero Exponents	—	0		

Examination June 1991

Three-Year Sequence for High School Mathematics—Course I

PART ONE

DIRECTIONS: *Answer 30 questions from this part. Each correct answer will receive 2 credits. No partial credit will be allowed. Write your answers in the spaces provided. Where applicable, answers may be left in terms of π or in radical form.*

1 The mean of a set of 5 numbers is 10. If all the numbers are doubled, what is the mean of this new set of numbers? 1_____

2 Solve for x: $0.3x + 1.7 = 2$ 2_____

3 Find the value of $a^2 - b$ if $a = 3$ and $b = -4$. 3_____

4 Solve for x in terms of a and b:

$$2x + a = b$$

4_____

5 In the accompanying diagram, $\overleftrightarrow{AB}$ and $\overleftrightarrow{CD}$ intersect at E. If $m\angle AEC = 3x - 40$ and $m\angle BED = 2x + 10$, find the value of x.

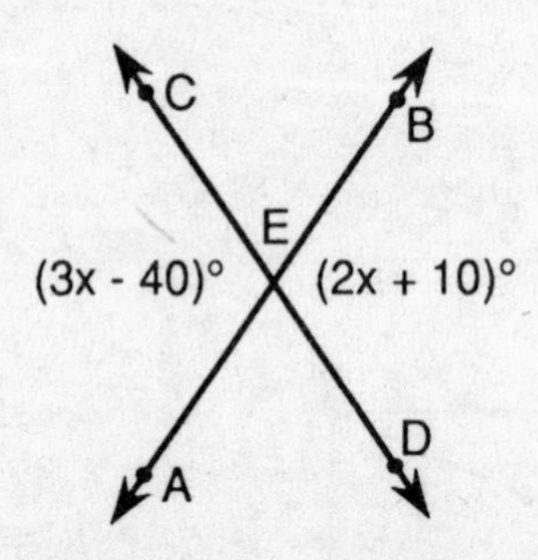

5_____

6 If the measures of the angles of a triangle are in the ratio 1:2:3, find the number of degrees in the *smallest* angle. 6_____

7 If x varies directly as y and $x = 60$ when $y = 5$, find the value of y when $x = 36$. 7_____

8 In the accompanying diagram, $\overleftrightarrow{AOB}$ is a straight line, $m\angle AOD = 3x - 8$, and $m\angle BOD = x$. Find x.

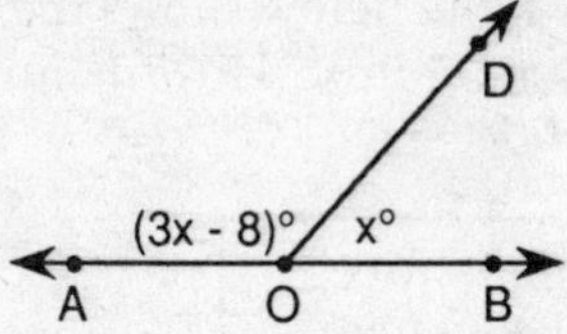

8_____

9 Write, in symbolic form, the inverse of $p \to \sim q$. 9_____

10 Find the area of the triangle whose vertices have coordinates (8,0), (0,10), and (0,0). 10_____

11 Solve for x: $\frac{2}{3}x - 2 = 10$ 11_____

12 Express $\frac{5a}{6} - \frac{4a}{9}$ as a single fraction in simplest form. 12_____

13 In the accompanying diagram, $ABCD$ is a rectangle. If $DB = 26$ and $DC = 24$, find BC.

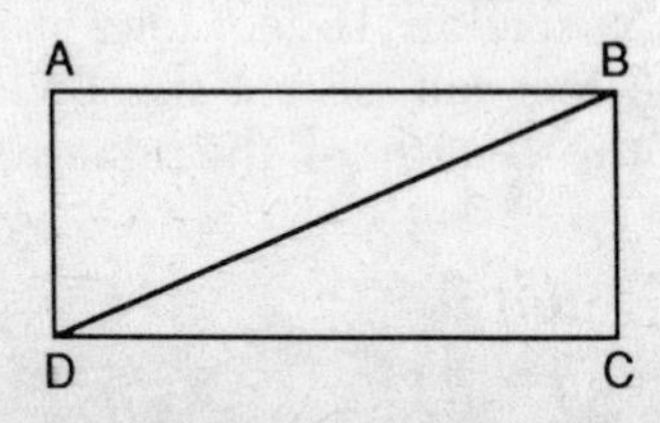

13_____

14 Solve for the positive value of x:

$$x^2 - 5x - 24 = 0$$

14____

15 Express $2x^2 - x - 3$ as the product of two binomials. 15____

16 From $5x^2 + 3x - 6$ subtract $4x^2 - 5x + 6$. 16____

17 Evaluate: ${}_7P_3$ 17____

18 The cumulative frequency table below shows the distribution of scores on a math test. How many scores were greater than 90?

Interval	Cumulative Frequency
61–70	4
61–80	10
61–90	12
61–100	16

18____

DIRECTIONS (**19–35**): *For* each *question chosen, write the* numeral *preceding the word or expression that best completes the statement or answers the question.*

19 A rotation of a figure can be considered
(1) a turning of the figure about some fixed point
(2) a slide of the figure
(3) an enlargement or a reduction of the figure
(4) a mirror image of the figure
19____

20 A tree 24 feet tall casts a shadow 16 feet long at the same time a man 6 feet tall casts a shadow x feet long. What is the length of the man's shadow?

(1) 6 (3) 3
(2) 5 (4) 4
20____

21 If a letter is chosen at random from the word "BASEBALL," what is the probability that the letter chosen is *not* an "L"?

(1) $\frac{1}{8}$ (3) $\frac{6}{8}$

(2) $\frac{2}{8}$ (4) $\frac{7}{8}$ 21_____

22 Which inequality is represented by the graph below?

(1) $-4 \leq x \leq 6$ (3) $-4 \leq x < 6$
(2) $-4 < x < 6$ (4) $-4 < x \leq 6$ 22_____

23 The quotient of $\frac{14x^6y}{2x^2y}$, $x \neq 0$, $y \neq 0$, is

(1) $7x^3$ (3) $7x^3y$
(2) $7x^4$ (4) $7x^4y$ 23_____

24 The expression $\frac{5}{2x - 10}$ is undefined when x is equal to
(1) 0 (3) 5
(2) –5 (4) 10 24_____

25 A quadrilateral with exactly one pair of parallel sides is a
(1) rhombus (3) square
(2) rectangle (4) trapezoid 25_____

26 When drawn on the same set of axes, the graph of the equations $y = x + 1$ and $y + x = 3$ intersect at the point whose coordinates are

(1) (2,1) (3) (2,3)
(2) (1,2) (4) (−1,4) 26_____

27 If the radius of a circle is doubled, then the circumference of the circle is multiplied by

(1) $\frac{1}{2}$ (3) 16

(2) 2 (4) 4 27_____

28 The number of feet in c inches is

(1) $\frac{c}{12}$ (3) $\frac{12}{c}$

(2) $\frac{c}{36}$ (4) $12c$ 28_____

29 Triangle $A'B'C'$ is the image of $\triangle ABC$ under a dilation such that $A'B' = \frac{1}{2}AB$. Triangles ABC and $A'B'C'$ are

(1) congruent but not similar
(2) similar but not congruent
(3) both congruent and similar
(4) neither congruent nor similar 29_____

30 The perimeter of a square is $4a$. What is the area of the square?

(1) a^2 (3) 16
(2) $4a^2$ (4) 4 30_____

31 Let p represent "$x > 10$" and let q represent "x is a multiple of 5." Which is true if $x = 26$?

(1) $p \vee q$
(2) $p \rightarrow q$
(3) $p \wedge q$
(4) $p \leftrightarrow q$

31_____

32 The sum of $\sqrt{50}$ and $\sqrt{2}$ is

(1) $\sqrt{52}$ (3) $6\sqrt{2}$
(2) 10 (4) 12

32_____

33 Which ordered pair is in the solution set of the system of inequalities shown in the graph below?

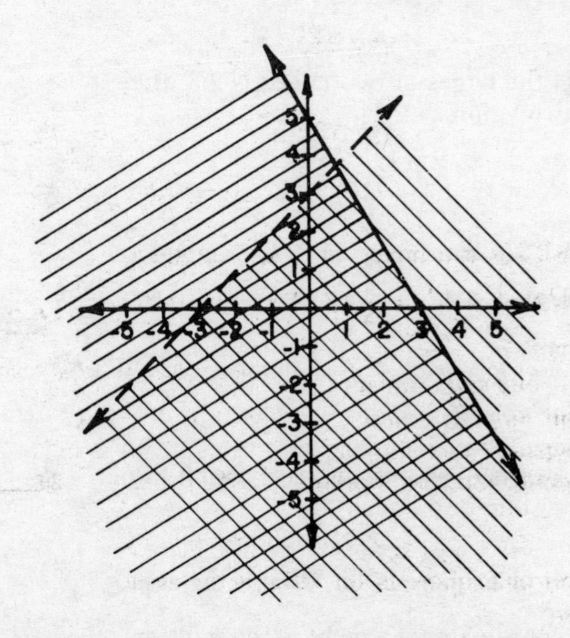

(1) (0,0) (3) (−3,3)
(2) (1,5) (4) (3,3)

33_____

34 Which statement would be a correct heading for column 3 in the table below?

Column 1	Column 2	Column 3
p	q	?
T	T	F
T	F	T
F	T	T
F	F	T

(1) $p \leftrightarrow q$ (3) $\sim p \rightarrow q$
(2) $\sim(p \vee q)$ (4) $\sim(p \wedge q)$ 34_____

35 If the ratio of the edges of two cubes is 2:3, the ratio of the two volumes is
(1) 2:3 (3) 8:27
(2) 4:9 (4) 2:5 35_____

PART TWO

DIRECTIONS: *Answer* four *questions from this part. Show all work unless otherwise directed.*

36 *a* On the same set of coordinate axes, graph the following system of inequalities:

$$y \geq -3$$
$$2y - x < 6$$ [8]

b Write the coordinates of a point in the solution set of the system of inequalities graphed in part *a*. [2]

37 If 3 is added to twice the square of an integer, the result is equal to seven times the integer. Find the integer. [*Only an algebraic solution will be accepted.*] [4,6]

38 The table shows the results of a math test given to a number of students.

Interval	Frequency
96–100	9
91–95	7
86–90	9
81–85	8
76–80	6
71–75	5

a Draw a frequency histogram based on the data. [4]

b In which interval is the median score? [2]

c How many students scored at or below the 25th percentile? [2]

d To get an A on this test, a student had to have a score greater than 90. What is the probability that a student selected at random from this distribution got an A on the test? [2]

39 One black marble and two red marbles are in a bag. Erika picks a marble from the bag at random. She looks at it, returns it, and makes a second random selection.

a Draw a tree diagram or list the sample space showing all possible outcomes. [2]

b What is the probability that two red marbles were selected? [2]

c What is the probability that two black marbles were selected? [2]

d What is the probability that one black and one red marble were selected? [2]

e What is the probability that *at most* one black marble was selected? [2]

40 In the accompanying diagram, *ABCD* is an isosceles trapezoid with altitude $\overline{BE}$, $AB = 10$, $AD = 15$, and $BE = 12$.

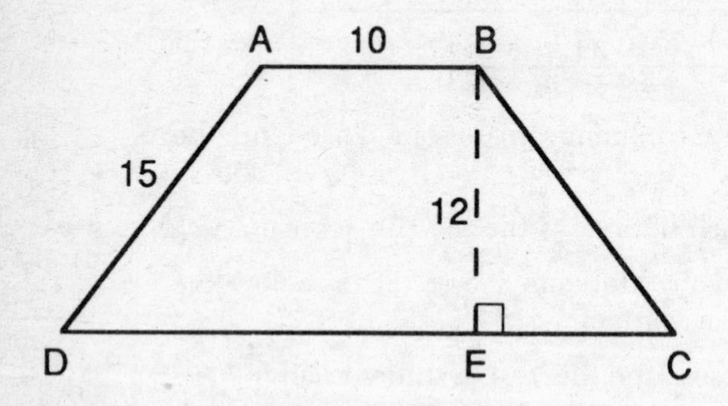

a Find *EC*. [2]

b Find the area of

(1) triangle *BEC* [2]

(2) trapezoid *ABCD* [3]

(3) trapezoid *ABED* [1]

c If diagonal $\overline{DB}$ is drawn, find the area of $\triangle ABD$. [2]

41 Use any method [algebraic, trial and error, making a table, etc.] to solve this problem. A written explanation of how you arrived at your answer is also acceptable. Show all work.

There are two pairs of integers that satisfy both of these conditions:

The larger integer is 9 more than the smaller integer.
The sum of the squares of the integers is 41.

a Find the two pairs of integers. [8]

b Show that one pair of integers found in part *a* satisfies both given conditions. [2]

42 Solve the following system of equations algebraically and check:

$$\begin{aligned} 4x + 3y &= 25 \\ 5x + 2y &= 33 \end{aligned}$$ [8,2]

Answers June 1991

Three-Year Sequence for High School Mathematics—Course I

ANSWER KEY

PART ONE

1. 20	**13.** 10	**25.** (4)
2. 1	**14.** 8	**26.** (2)
3. 13	**15.** $(2x - 3)(x + 1)$	**27.** (2)
4. $\frac{b-a}{2}$	**16.** $x^2 + 8x - 12$	**28.** (1)
5. 50	**17.** 210	**29.** (2)
6. 30	**18.** 4	**30.** (1)
7. 3	**19.** (1)	**31.** (1)
8. 47	**20.** (4)	**32.** (3)
9. $\sim p \rightarrow q$	**21.** (3)	**33.** (1)
10. 40	**22.** (4)	**34.** (4)
11. 18	**23.** (2)	**35.** (3)
12. $\frac{7a}{18}$	**24.** (3)	

PART TWO *See answers explained section.*

ANSWERS EXPLAINED

PART ONE

1. Let a, b, c, d, and e represent the five numbers.

The mean is the sum of all the numbers divided by the number of them; the mean is 10:

$$10 = \frac{a + b + c + d + e}{5}$$

If all the numbers are doubled, they become $2a$, $2b$, $2c$, $2d$, and $2e$ respectively:

$$\text{New mean} = \frac{2a + 2b + 2c + 2d + 2e}{5}$$

Factor out the common monomial factor 2 in the numerator:

$$\text{New mean} = \frac{2(a + b + c + d + e)}{5}$$

But $\frac{a + b + c + d + e}{5} = 10$:

$$\text{New mean} = 2(10)$$
$$\text{New mean} = 20$$

The mean of the new set is **20.**

2. The given equation contains decimals:

$$0.3x + 1.7 = 2$$

Clear decimals by multiplying each term on both sides of the equation by 10:

$$10(0.3x) + 10(1.7) = 10(2)$$
$$3x + 17 = 20$$

Add -17 (the additive inverse of 17) to both sides of the equation:

$$\begin{aligned} -17 &= -17 \\ \hline 3x \quad &= 3 \end{aligned}$$

Divide both sides of the equation by 3:

$$\frac{3x}{3} = \frac{3}{3}$$
$$x = 1$$

$x =$ **1.**

3. The given expression is:

$$a^2 - b$$

Substitute 3 for a and -4 for b:

$$3^2 - (-4)$$

$3^2 = 9$ and $-(-4) = 4$:

$$9 + 4$$
$$13$$

The value is **13.**

4. The given equation is a *literal equation*:

$$2x + a = b$$

Add $-a$ (the additive inverse of a) to both sides of the equation:

$$\begin{array}{r} -a = \quad -a \\ \hline 2x \quad = b - a \end{array}$$

Divide both sides of the equation by 2:

$$\frac{2x}{2} = \frac{b-a}{2}$$

$$x = \frac{b-a}{2}$$

$x = \dfrac{b-a}{2}$.

5.

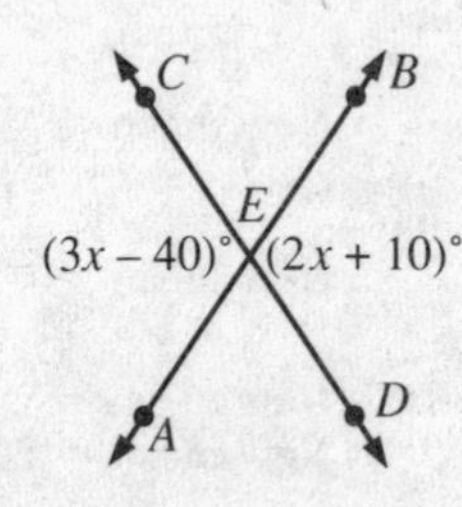

Vertical angles are congruent, that is, equal in measure:

$$m\angle AEC = m\angle BED$$

$$3x - 40 = 2x + 10$$

Add $-2x$ (the additive inverse of $2x$) and also add 40 (the additive inverse of -40) to both sides of the equation:

$$\begin{array}{rcl} -2x + 40 & = & -2x + 40 \\ \hline x & = & 50 \end{array}$$

$x = $ **50.**

6.

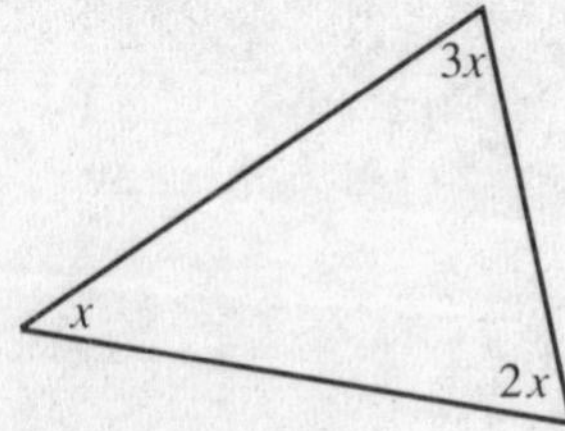

Let x, $2x$, and $3x$ represent the respective measures of the three angles of the triangle.

The sum of the measures of the three angles of a triangle is 180°:

$$x + 2x + 3x = 180$$

Combine like terms:

$$6x = 180$$

Divide both sides of the equation by 6:

$$\frac{6x}{6} = \frac{180}{6}$$

$$x = 30$$

The measure of the *smallest* angle is **30.**

7. If k is a constant, "x varies directly as y" may be represented as: $x = ky$

Since $x = 60$ when $y = 5$: $60 = 5k$

Divide both sides of the equation by 5: $\frac{60}{5} = \frac{5k}{5}$

$12 = k$

Since k is a constant, k will always equal 12 and thus 12 may be substituted for k: $x = 12y$

To find the value of y when $x = 36$, substitute 36 for x: $36 = 12y$

Divide both sides of the equation by 12: $\frac{36}{12} = \frac{12y}{12}$

$3 = y$

The value of y is **3** when $x = 36$.

8.

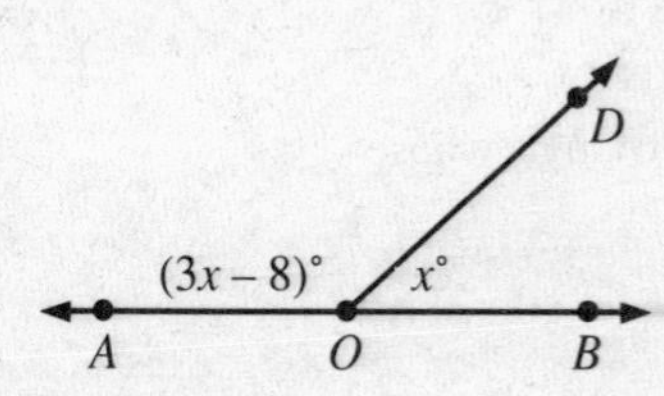

Since $\overleftrightarrow{AOB}$ is a straight line, $\angle AOB$ is a straight angle and $m\angle AOB = 180°$: $m\angle AOD + m\angle DOB = 180$

$3x - 8 + x = 180$

Combine like terms: $4x - 8 = 180$

Add 8 (the additive inverse of -8) to both sides of the equation: $+8 = +8$

$4x = 188$

Divide both sides of the equation by 4: $\frac{4x}{4} = \frac{188}{4}$

$x = 47$

$x = 47$.

9. The statement $p \to \sim q$ is the *implication* or *conditional* that "If p is true, then the *negation* of q is true." The *hypothesis* or *antecedent* of this conditional is p, and its *conclusion* or *consequent* is $\sim q$.

The *inverse* of a conditional is formed by negating both its hypothesis and conclusion. Thus, the inverse of the given conditional is: $\sim p \to \sim(\sim q)$

But $\sim(\sim q)$ is q: $\sim p \to q$

The inverse is $\boldsymbol{\sim p \to q}$.

10.

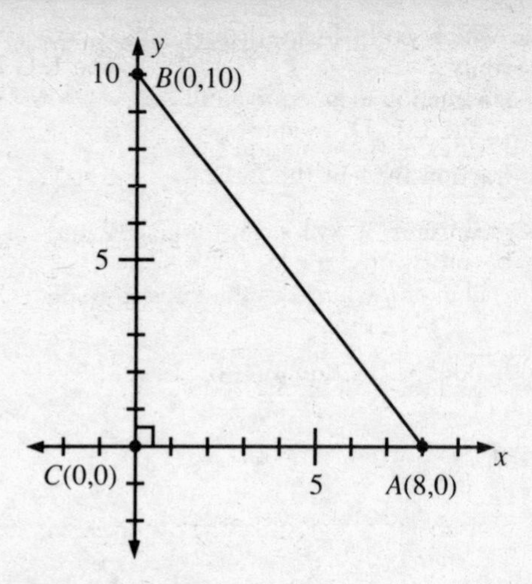

Triangle ABC is a right triangle whose legs are $\overline{AC}$ and $\overline{BC}$.

The area of a right triangle is equal to one-half the product of the lengths of its legs:

$$\text{Area of } \Delta ABC = \frac{1}{2}(AC)(BC)$$

$AC = 8$ and $BC = 10$:

$$\text{Area of } \Delta ABC = \frac{1}{2}(8)(10)$$

$$\text{Area of } \Delta ABC = 40$$

The area is **40.**

11. The given equation contains a fraction:

$$\frac{2}{3}x - 2 = 10$$

Clear fractions by multiplying each term on both sides of the equation by 3:

$$3\left(\frac{2}{3}x\right) - 3(2) = 3(10)$$

$$2x - 6 = 30$$

Add 6 (the additive inverse of −6) to both sides of the equation:

$$\begin{array}{r} +6 = +6 \\ \hline 2x \quad = 36 \end{array}$$

Divide both sides of the equation by 2:

$$\frac{2x}{2} = \frac{36}{2}$$

$$x = 18$$

x = **18.**

12. The given fractions have different denominators:

$$\frac{5a}{6} - \frac{4a}{9}$$

Find the least common denominator (L.C.D.). The L.C.D. is the smallest number into which both denominators will divide evenly: The L.C.D. for 6 and 9 is 18.

Convert each fraction to an equivalent fraction having the L.C.D. by multiplying the first fraction by 1 in the form $\frac{3}{3}$ and the second fraction by 1 in the form $\frac{2}{2}$:

$$\frac{3(5a)}{3(6)} - \frac{2(4a)}{2(9)}$$

$$\frac{15a}{18} - \frac{8a}{18}$$

Fractions with the same denominator may be combined by combining their numerators:

$$\frac{15a - 8a}{18}$$

Combine like terms in the numerator:

$$\frac{7a}{18}$$

The single fraction in simplest form is $\mathbf{\frac{7a}{18}}$.

13.

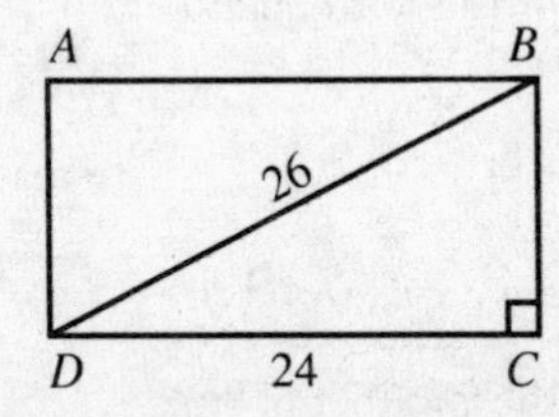

Since the angles of a rectangle are right angles, $\angle C$ is a right angle, and therefore ΔDBC is a right triangle.

Triangle DBC is a 5-12-13 right triangle with $DC = 2(12)$ and $DB = 2(13)$; therefore $BC = 2(5) = 10$.

$BC = \mathbf{10}$.

ALTERNATIVE SOLUTION: The Pythagorean Theorem may be used to find BC:

$$\begin{aligned} (BC)^2 + (24)^2 &= (26)^2 \\ (BC)^2 + 576 &= 676 \\ -576 &= -576 \\ \hline (BC)^2 &= 100 \\ BC &= 10 \end{aligned}$$

14. The given equation is a *quadratic equation:* $x^2 - 5x - 24 = 0$

The left side is a *quadratic trinomial* that can be factored into the product of two binomials. The factors of the first term are x and x, and they become the first terms of the binomials: $(x \quad)(x \quad) = 0$

The factors of the last term, -24, become the second terms of the binomials, but they must be chosen in such a way that the product of the inner terms and the product of the outer terms add up to the middle term, $-5x$, of the original trinomial.

Try -8 and $+3$ as the factors of -24:

$-8x$ = inner product

$(x - 8)(x + 3) = 0$

$+3x$ = outer product

Since $(-8x) + (+3x) = -5x$, these are the correct factors: $(x - 8)(x + 3) = 0$

If the product of two factors is zero, either factor may equal zero: $x - 8 = 0 \vee x + 3 = 0$

Add the appropriate additive inverse to both sides of each equation, 8 for the left equation and -3 for the right equation:

$$\begin{array}{rl} +8 = +8 & \quad -3 = -3 \\ \hline x \quad = 8 & \quad x \quad = -3 \end{array}$$

The positive value of x is **8**.

15. The given expression is a *quadratic trinomial:* $2x^2 - x - 3$

It may be factored into the product of two binomials. The factors of the first term are $2x$ and x, and they become the first terms of the binomials: $(2x \quad)(x \quad)$

The factors of the last term become the second terms of the binomials, but they must be chosen in such a way that the sum of the inner and outer cross-products equals the middle term, $-x$, of the original trinomial.

Try -3 and $+1$ as the factors of -3:

$-3x$ = inner product

$(2x - 3)(x + 1)$

$+2x$ = outer product

Since $(-3x) + (+2x) = -x$, these are the correct factors: $(2x - 3)(x + 1)$

The product of two binomials is **$(2x - 3)(x + 1)$**.

16. Write the trinomial to be subtracted under the one it is to be subtracted from, with like terms in the same column:

$$\begin{array}{r} 5x^2 + 3x - 6 \\ 4x^2 - 5x + 6 \\ \hline \end{array}$$

Rewrite the two trinomials, changing the sign of each term in the subtrahend (the lower trinomial):

$$\begin{array}{r} 5x^2 + 3x - 6 \\ -4x^2 + 5x - 6 \\ \hline \end{array}$$

Combine the numerical coefficients in each column:

$$x^2 + 8x - 12$$

The remainder is $\mathbf{x^2 + 8x - 12}$.

17. The symbol ${}_nP_r$ represents the number of permutations of n things taken r at a time:

$${}_nP_r = n(n-1)(n-2)\ldots \text{ to } r \text{ factors}$$

For ${}_7P_3$, $n = 7$ and $r = 3$:

$${}_7P_3 = 7(6)(5)$$
$${}_7P_3 = 42(5)$$
$${}_7P_3 = 210$$

${}_7P_3 = \mathbf{210}$.

18.

Interval	Cumulative Frequency
61–70	4
61–80	10
61–90	12
61–100	16

The 61–100 interval contains all scores of 61 and above; there are 16 such scores.

The 61–90 interval contains all scores of 61 and above that are not greater than 90; there are 12 such scores.

The number of scores above 90 is therefore the difference between these two; $16 - 12 = 4$.

There were **4** scores greater than 90.

19. Consider each choice in turn:

(1) A turning of a figure about some fixed point is a *rotation*. This is the correct choice.

(2) A slide of a figure is a *translation*.

(3) An enlargement or a reduction of a figure is a *dilation*.

(4) A mirror image of a figure is a *reflection*.

The correct choice is **(1)**.

20.

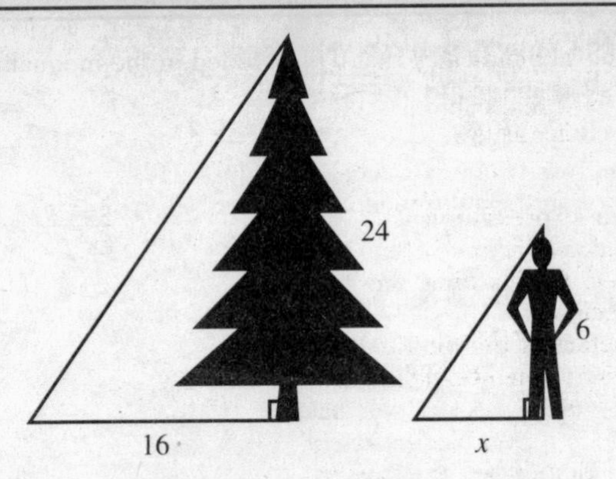

The tree and its shadow form the legs of a right triangle that is similar to the right triangle whose legs are formed by the man and his shadow.

Corresponding sides of similar triangles are in proportion:

$$\frac{x}{16} = \frac{6}{24}$$

In a proportion, the product of the means is equal to the product of the extremes (cross-multiply):

$$24x = 6(16)$$
$$24x = 96$$

Divide both sides of the equation by 24:

$$\frac{24x}{24} = \frac{96}{24}$$
$$x = 4$$

The length of the man's shadow is 4 feet.
The correct choice is (4).

21. $\text{The probability of an event occurring} = \dfrac{\text{number of successful outcomes}}{\text{total possible number of outcomes}}.$

"BASEBALL" contains 8 letters, 2 of which are "L"'s and 6 of which are *not* "L"'s.

For choosing a letter that is *not* an "L," the number of successful outcomes is 6. The total possible number of choices is 8.

The probability of choosing a letter that is *not* an "L" is $\frac{6}{8}$.
The correct choice is (3).

22.

The heavy shaded line represents all numbers greater than -4 and less than 6. The open circle at -4 indicates that -4 is not included in the inequality.

The shaded circle at 6 indicates that 6 is included in the inequality. The graph thus represents the inequality $-4 < x \leq 6$.

The correct choice is (**4**).

23. The given expression is: $\frac{14x^6y}{2x^2y}$, $x \neq 0$, $y \neq 0$

The numerical coefficient of the quotient is obtained by dividing the two numerical coefficients: $14 \div 2 = 7$

The literal factors of the quotient are obtained by dividing the literal factors of numerator and denominator. Note that division by x and y is possible because $x \neq 0$ and $y \neq 0$. Remember that powers of the same base are divided by subtracting their exponents:

$$x^6 \div x^2 = x^4$$
$$y \div y = 1$$

Combine the results above: $\frac{14x^6y}{2x^2y} = 7x^4$

The correct choice is (**2**).

24. The given expression is: $\frac{5}{2x - 10}$

Since division by zero is undefined, the expression will be undefined if its denominator equals zero: $2x - 10 = 0$

Add 10 (the additive inverse of -10) to both sides of the equation:

$$\begin{aligned} 2x - 10 &= 0 \\ +10 &= +10 \\ \hline 2x &= 10 \end{aligned}$$

Divide both sides of the equation by 2:

$$\frac{2x}{2} = \frac{10}{2}$$
$$x = 5$$

The expression is undefined when $x = 5$.

The correct choice is (**3**).

25.

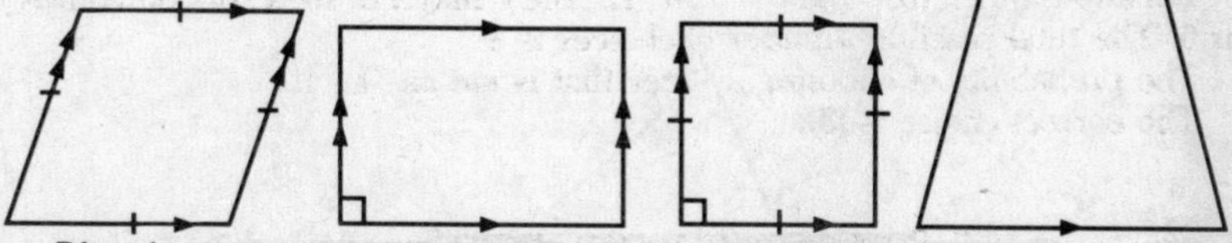

Rhombus Rectangle Square Trapezoid

Consider each choice in turn:

(1) A rhombus is an equilateral parallelogram.

(2) A rectangle is a parallelogram whose angles are right angles.

(3) A square is an equilateral rectangle and hence a square is also a parallelogram.

(4) A trapezoid is a quadrilateral with two and only two parallel sides.

Choices (1), (2) and (3) are all parallelograms, and thus each has two pairs of parallel sides. A trapezoid is the only choice with exactly one pair of parallel sides.

The correct choice is **(4)**.

26. The coordinates of the intersection point will be the common solution of the given equations, $y = x + 1$ and $y + x = 3$.

Substitute $x + 1$ for y in the second equation: $x + 1 + x = 3$

Combine like terms: $2x + 1 = 3$

Add -1 (the additive inverse of 1) to both sides of the equation:

$$\begin{aligned} -1 &= -1 \\ \hline 2x \quad &= 2 \end{aligned}$$

Divide both sides of the equation by 2:

$$\frac{2x}{2} = \frac{2}{2}$$

$$x = 1$$

Substitute 1 for x in the first equation: $y = 1 + 1$

$$y = 2$$

The coordinates of the intersection point are (1, 2).

The correct choice is **(2)**.

27. The circumference, C, of a circle whose radius is r is given by the formula: $C = 2\pi r$

If the radius is doubled, r will become $2r$, and the circumference C will be replaced by a new circumference, C':

$$C' = 2\pi(2r)$$

$$C' = 4\pi r$$

Since $4\pi r = 2(2\pi r)$: $C' = 2C$

The new circumference, C', is twice the old one, C.

The correct choice is **(2)**.

28. There are 12 inches in 1 foot, so c must be divided by 12 to obtain the number of feet in c inches.

The correct choice is **(1)**.

29.

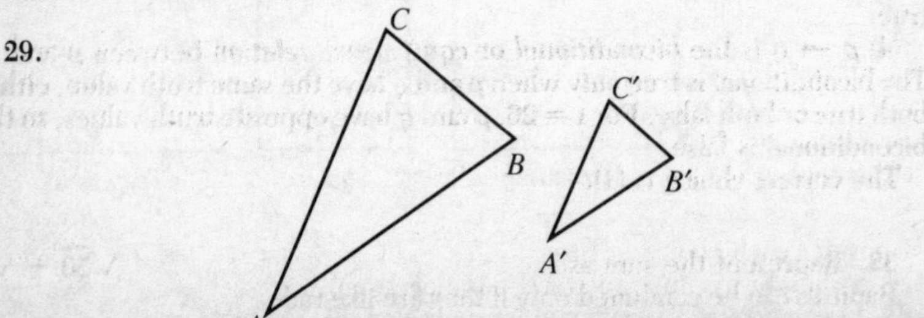

Since $\Delta A'B'C'$ is the image of ΔABC under a dilation such that $A'B' = \frac{1}{2}AB$, then $B'C' = \frac{1}{2}BC$ and $A'C' = \frac{1}{2}AC$. Thus, the corresponding sides of the two triangles are in proportion and hence the triangles are similar. They are not congruent since their corresponding sides are not congruent.

The correct choice is **(2)**.

30.

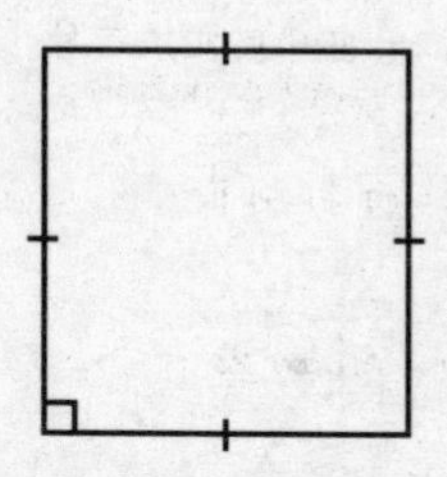

The perimeter is the sum of the lengths of all four sides of the square. Since all sides of a square are equal in length, the length of each side must be one fourth of the perimeter; $\frac{1}{4}$ of $4a$ is a. Thus, the length of one side is a.

The area, A, of a square is the square of the length of one side: $A = a^2$.

The correct choice is **(1)**.

31. Given: p represents "$x > 10$";
q represents "x is a multiple of 5."

If $x = 26$, p is true since $26 > 10$, but q is false since 26 is not a multiple of 5.

Consider each choice:

(1) $p \vee q$ is the *disjunction* of p and q. The disjunction is true if either p or q, or both, are true. This is the correct choice.

(2) $p \rightarrow q$ is the *implication* or *conditional* that if p is true then q is true. But we have p true and q false, so the implication is not true.

(3) $p \wedge q$ is the *conjunction* of p and q. The conjunction is true only when p and q are both true. Since q is false for $x = 26$, the conjunction is not true.

(4) $p \leftrightarrow q$ is the *biconditional* or *equivalence relation* between p and q. The biconditional is true only when p and q have the same truth value, either both true or both false. For $x = 26$, p and q have opposite truth values, so the biconditional is false.

The correct choice is **(1)**.

32. Represent the sum as: $\sqrt{50} + \sqrt{2}$

Radicals can be combined only if they are like radicals, that is, they must have the same index and the same radicand.

Simplify $\sqrt{50}$ by factoring out the largest perfect square in its radicand: $\sqrt{25(2)} + \sqrt{2}$

Remove the perfect square from under the radical sign by taking its square root and writing it as a coefficient of the radical: $5\sqrt{2} + \sqrt{2}$

The radicals are now like radicals; they have the same index, 2, since both are square roots, and they have the same radicand, 2. Like radicals are combined by combining their coefficients (remember that $\sqrt{2}$ has the coefficient "1" understood): $6\sqrt{2}$

The correct choice is **(3)**.

33.

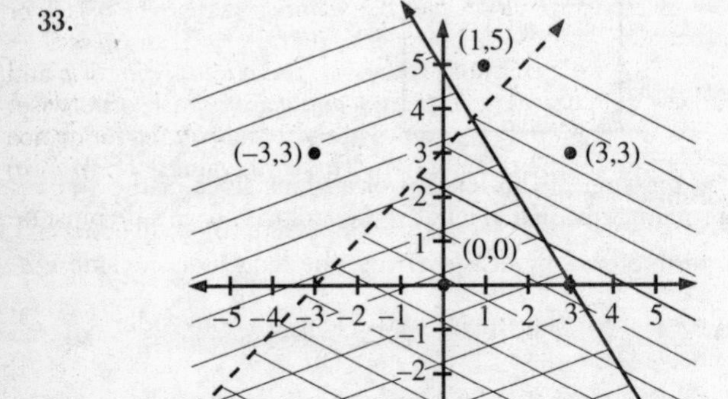

The solution set of the system of inequalities is represented by all points in the region covered by both types of cross-hatching or lying on the portion of the solid line that forms a boundary of this region.

Locate each of the pairs given as choices by plotting the points on the graph. Only (0,0) lies in the required region.

The correct choice is **(1)**.

34.

Col. 1	Col. 2	Col. 3	(1)		(2)		(3)		(4)
p	q	?	$p \leftrightarrow q$	$p \vee q$	$\sim(p \vee q)$	$\sim p$	$\sim p \rightarrow q$	$p \wedge q$	$\sim(p \wedge q)$
T	T	F	T	T	F	F	T	T	F
T	F	T	F	T	F	F	T	F	T
F	T	T	F	T	F	T	T	F	T
F	F	T	T	F	T	T	F	F	T

As shown above, prepare additional columns in the table for the truth values of each choice, determined according to the values of p and q in Columns 1 and 2.

Choice (1), $p \leftrightarrow q$, is the *biconditional or equivalence relation* between p and q. It has the value T whenever p and q are both T or both F; it has the value F when their values differ.

For choice (2), first prepare a column for the *disjunction*, $p \vee q$. It has the value T when either p or q, or both, are T, but has the value F when both are F. The column for choice (2), $\sim(p \vee q)$, represents the *negation* of $p \vee q$; its truth values are the opposite of those for $p \vee q$.

For choice (3), first prepare a column for $\sim p$, the *negation* of p. Use the values of $\sim p$ and q to prepare the column for choice (3), the *implication* or *conditional*, $\sim p \rightarrow q$. The implication has the value T whenever q is T or when $\sim p$ and q are both F; it has the value F when $\sim p$ is T and q is F.

For choice (4), first prepare a column for $p \wedge q$, the *conjunction* of p and q. The conjunction has the truth value T when p and q are both T; otherwise, it has the value F. Use the column for $p \wedge q$ to prepare the column for choice (4), $\sim(p \wedge q)$, the *negation* of the conjunction. The truth values of $\sim(p \wedge q)$ are the opposite of those for $p \wedge q$.

The entries in the column for choice (4) are identical to those in column 3. Hence, column 3 represents $\sim(p \wedge q)$.

The correct choice is **(4)**.

35. Let $2x$ = the length of the edge of one cube, and let V_1 = its volume.

Then $3x$ = the length of the other cube; let V_2 = its volume.

The volume of a cube is the cube of the length of one edge: $V_1 = (2x)^3$

$V_2 = (3x)^3$

Cube $2x$ and $3x$: $V_1 = 8x^3$

$V_2 = 27x^3$

The ratio of the volumes is: $\dfrac{V_1}{V_2} = \dfrac{8x^3}{27x^3} = \dfrac{8}{27}$

The correct choice is **(3)**.

PART TWO

36. a. To draw the graph of $y \geq -3$, first draw the graph of the equation $y = -3$. Its graph is a line parallel to the x-axis and 3 units below it. Draw this as a *solid* line to indicate that the points on it are part of the solution set of $y \geq -3$.

The solution set of the inequality $y > -3$ lies above the line $y = -3$ since the inequality states that y is greater than -3. Shade this region with cross-hatching extending up and to the right.

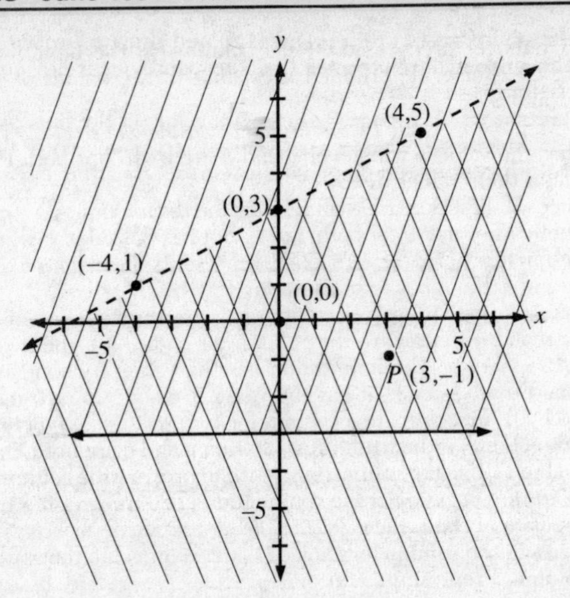

To draw the graph of the inequality $2y - x < 6$, first draw the graph of the equation $2y - x = 6$. Rearrange the equation so that it is solved for y. Add x (the additive inverse of $-x$) to both sides of the equation:

$$\begin{array}{r} 2y - x = 6 \\ +x = +x \\ \hline 2y = 6 + x \end{array}$$

Divide both sides of the equation by 2:

$$\frac{2y}{2} = \frac{6 + x}{2}$$

$$y = \frac{6 + x}{2}$$

Prepare a table of corresponding values of x and y by choosing three convenient values for x and substituting them in the equation to determine the corresponding values of y:

x	$\frac{6 + x}{2}$	$= y$
-4	$\frac{6 - 4}{2} = \frac{2}{2}$	$= 1$
0	$\frac{6 + 0}{2} = \frac{6}{2}$	$= 3$
4	$\frac{6 + 4}{2} = \frac{10}{2}$	$= 5$

Plot the points (−4,1), (0,3), and (4,5) and draw a *broken* line through them. The broken line indicates that the points on it are *not* part of the solution set of $2y - x < 6$.

The solution set of $2y - x < 6$ lies on one side of the line $2y - x = 6$. To find out on which side, choose a convenient test point, say (0,0), and substitute its coordinates in the inequality to see whether they satisfy it:

$$2y - x < 6$$

$$2(0) - 0 \overset{?}{<} 6$$

$$0 - 0 \overset{?}{<} 6$$

$$0 < 6 \checkmark$$

Since (0,0) satisfies the inequality, it lies on the side of the line $2y - x = 6$ that contains the points that represent $2y - x < 6$. Shade this side with cross-hatching extending down and to the right.

b. The points in the solution set of the system of inequalities are those that lie in the region covered by *both* types of cross-hatching, and also those that are on the portion of the solid line that forms a boundary of this region. An example of such a point is $P(3, -1)$.

(3, −1) are the coordinates of a point in the solution set.

37. Let x = the integer.

3 added to twice the square of the integer equals seven times the integer.

$$3 \quad + \quad 2x^2 \quad = \quad 7x$$

The equation to use is: $3 + 2x^2 = 7x$

This is a *quadratic equation*. Rearrange it so that all terms are on one side equal to 0 by adding $-7x$ (the additive inverse of $7x$) to both sides:

$$\begin{array}{r} -\ 7x = -7x \\ \hline 3 + 2x^2 - 7x = 0 \end{array}$$

Rearrange the terms in descending order of exponents: $2x^2 - 7x + 3 = 0$

The left side is a *quadratic trinomial* that can be factored into the product of two binomials. The factors of the first term are $2x$ and x, and they become the first terms of the binomials: $(2x \quad)(x \quad) = 0$

The factors of the last term, +3, become the second terms of the binomials, but they must be chosen in such a way that the sum of the inner and outer cross-products of the binomials equals the middle term, $-7x$, of the original trinomial. Try −1 and −3 as the factors of +3:

$$-x = \text{inner product}$$
$$(2x - 1)(x - 3) = 0$$
$$-6x = \text{outer product}$$

Since $(-x) + (-6x) = -7x$, these are the correct factors:

$$(2x - 1)(x - 3) = 0$$

If the product of two factors is zero, either factor may equal zero:

$$2x - 1 = 0 \lor x - 3 = 0$$

Add the appropriate additive inverse to both sides of each equation, 1 for the left equation and 3 for the right equation:

$$\begin{array}{rr} 1 = 1 & 3 = 3 \\ \hline 2x \quad = 1 & x \quad = 3 \end{array}$$

In the left equation, divide both sides by 2:

$$\frac{2x}{2} = \frac{1}{2}$$

Reject $\frac{1}{2}$ since x must be an integer:

$$x = \frac{1}{2}$$

The integer is 3.

$$x = 3$$

38.

a.

Interval	Frequency
96–100	9
91–95	7
86–90	9
81–85	8
76–80	6
71–75	5
Total = 44	

b. Add the frequencies. There is a total of 44 scores. The median is the midscore when all scores are arranged in numerical order of size.

Since $\frac{44}{2} = 22$, there must be 22 scores below the median and 22 above it. The median will be midway between the 22nd and 23rd scores.

The lowest three intervals contain 5 + 6 + 8 = 19 scores. Three more are needed to reach the 22nd score, and four more are needed to reach the 23rd.

Since the next higher interval, 86–90, contains 9 scores, both the 22nd and 23rd scores will lie in this interval. The median, midway between them, therefore lies in this interval.

The median lies in the **86–90** interval.

c. The 25th percentile is the score at or below which 25%, or $\frac{1}{4}$, of all the scores lie. Since $\frac{44}{4} = 11$, 11 scores lie at or below the 25th percentile.

11 students scored at or below the 25th percentile.

d. Probability of an event occurring $= \frac{\text{number of successful outcomes}}{\text{total possible number of outcomes}}$.

Scores greater than 90 are those in the 91–95 interval and those in the 96–100 interval. There are 7 + 9, or 16, such scores. The number of successful outcomes for a random selection of a score greater than 90 is thus 16.

The total possible number of outcomes is 44 since any one of the 44 scores could be selected at random.

The probability of selecting at random a student with an A $= \frac{16}{44}$.

The probability is $\mathbf{\frac{16}{44}}$.

39. a. TREE DIAGRAM

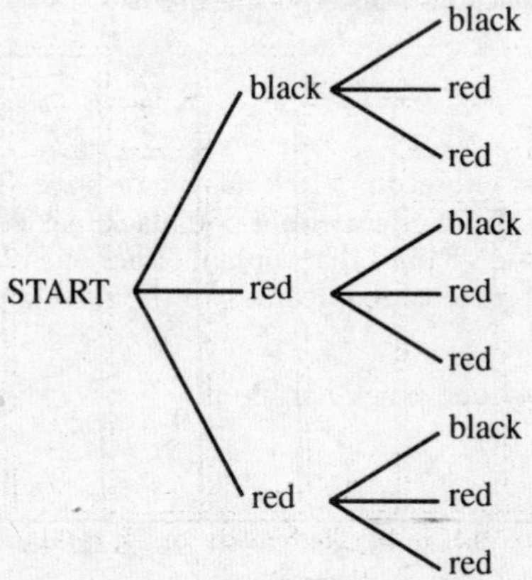

START
- black
  - black
  - red
  - red
- red
  - black
  - red
  - red
- red
  - black
  - red
  - red

SAMPLE SPACE

First Selection	Second Selection
black	black
black	red
black	red
red	black
red	red
red	red
red	black
red	red
red	red

Probability of an event occurring $= \frac{\text{number of successful outcomes}}{\text{total possible number of outcomes}}$.

b. The tree diagram shows 4 paths that lead to red marbles on both the first and second selections. Similarly, the sample space shows 4 lines that contain a red marble in both the first and second columns. Hence, there are 4 successful outcomes for the selection of two red marbles.

For the selection of two marbles, the tree diagram contains 9 complete paths that lead from the start through the first choice to a second choice. Similarly, the sample space contains a total of 9 lines. Thus, the total possible number of outcomes is 9.

The probability of selecting two red marbles is $\frac{4}{9}$.

The probability is $\mathbf{\frac{4}{9}}$.

c. There is only one path on the tree diagram that contains two black marbles and only one line in the sample space that involves two black marbles. Thus, the number of successful outcomes for the choice of two black marbles is 1.

The probability of choosing two black marbles is $\frac{1}{9}$.

The probability is $\mathbf{\frac{1}{9}}$.

d. The tree diagram contains 4 paths that include one black and one red marble. Similarly, the sample space has 4 lines that contain one black and one red marble. Thus, the number of successful outcomes for the selection of one black and one red marble is 4.

The probability of selecting one black and one red marble is $\frac{4}{9}$.

The probability is $\mathbf{\frac{4}{9}}$.

e. *At most* one black marble means either one black marble or none. The number of paths on the tree diagram that contain either one black marble or none is 8. Similarly, the sample space has 8 lines that contain either one black marble or none. Thus, the number of successful outcomes for the selection of *at most* one black marble is 8.

The probability of selecting *at most* one black marble is $\frac{8}{9}$.

The probability is $\mathbf{\frac{8}{9}}$.

40.

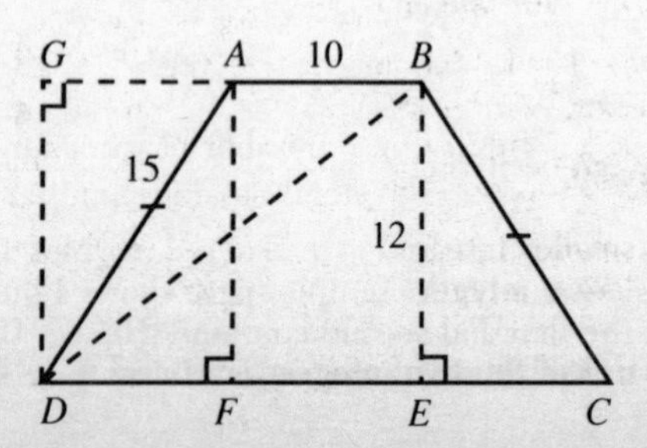

a. Since $ABCD$ is an isosceles trapezoid, $BC = AD = 15$.

Triangle BEC is a 3-4-5 right triangle with $BE = 3 \times 4 = 12$, and $BC = 3 \times 5 = 15$. Therefore, $EC = 3 \times 3 = 9$.

$EC =$ **9**.

b. (1) The area, A, of a right triangle equals one-half the product of the lengths of its legs:

$$\text{Area of } \Delta BEC = \frac{1}{2}(BE)(EC)$$

$$\text{Area of } \Delta BEC = \frac{1}{2}(12)(9)$$

$$\text{Area of } \Delta BEC = 6(9)$$

$$\text{Area of } \Delta BEC = 54$$

The area of ΔBEC is **54**.

(2) Draw $\overline{AF} \perp \overline{DC}$.

$AF = BE = 12$; $AB = FE = 10$

ΔAFD is congruent to ΔBEC, so $DF = EC = 9$.

$DC = DF + FE + EC = 9 + 10 + 9 = 28$.

The area, A, of a trapezoid whose bases are b_1 and b_2 and whose altitude is h is given by the formula:

$$A = \frac{1}{2}h(b_1 + b_2)$$

For trapezoid $ABCD$, $h = BE = 12$, $b_1 = AB = 10$, and $b_2 = DC = 28$:

$$A = \frac{1}{2}(12)(10 + 28)$$

$$A = 6(38)$$

$$A = 228$$

The area of trapezoid $ABCD$ is **228**.

(3) Area of trapezoid $ABED$ = area of trapezoid $ABCD$ − area of ΔBEC

Area of trapezoid $ABED = 228 - 54$

Area of trapezoid $ABED = 174$

The area of trapezoid $ABED =$ **174**.

c. Consider $\overline{AB}$ as the base of ΔABD. The altitude to $\overline{AB}$ will be a line segment $\overline{DG}$ dropped from D perpendicular to $\overline{BA}$ extended.

The area, A, of a triangle with base b and altitude h is given by the formula:

$$A = \frac{1}{2}bh$$

For ΔABD, $b = AB = 10$, and $h = DG = BE = 12$:

$$A = \frac{1}{2}(10)(12)$$

$$A = 60$$

The area of ΔABD is **60**.

41. a. Let x = the smaller integer.

Then $x + 9$ = the larger integer.

The sum of the squares of the two integers is 41: $x^2 + (x + 9)^2 = 41$.

Expand $(x + 9)^2$:

$$\begin{array}{r} x + 9 \\ \underline{x + 9} \\ x^2 + 9x \qquad \\ \underline{9x + 81} \\ x^2 + 18x + 81 \end{array}$$

$$x^2 + x^2 + 18x + 81 = 41$$

Combine like terms: $2x^2 + 18x + 81 = 41$

This is a *quadratic equation*. Rearrange it so that all terms are on one side equal to zero by adding -41 (the additive inverse of 41) to both sides:

$$\begin{array}{r} -41 = -41 \\ \hline 2x^2 + 18x + 40 = 0 \end{array}$$

To simplify, divide each term on both sides of the equation by 2: $x^2 + 9x + 20 = 0$

The left side is a *quadratic trinomial* that can be factored into the product of two binomials. The factors of the first term are x and x, and they become the first terms of the binomials: $(x \qquad)(x \qquad) = 0$

The factors of the last term, $+20$, become the second terms of the binomials, but they must be chosen in such a way that the sum of the inner and outer cross-products of the binomials equals the middle term, $+9x$, of the original trinomial. Try $+4$ and $+5$ as the factors of $+20$:

$+4x$ = inner product

$$(x + 4)(x + 5) = 0$$

$+5x$ = outer product

Since $(+4x) + (+5x) = +9x$, these are the correct factors: $(x + 4)(x + 5) = 0$

If the product of two factors is zero, either factor may equal zero: $x + 4 = 0 \quad \vee \quad x + 5 = 0$

Add the appropriate additive inverse to both sides of each equation, -4 for the left equation and -5 for the right equation:

$$\begin{array}{rr} -4 = -4 & -5 = -5 \\ \hline x \quad = -4 & x \quad = -5 \end{array}$$

The larger integer is $x + 9$: $x + 9 = 5 \qquad x + 9 = 4$

The pairs of integers are **(−4,5)** and **(−5,4)**.

b. The pair $(-4,5)$ satisfies *both* conditions:

The larger number, 5, is 9 more than the smaller number, -4, since $-4 + 9 = 5$. ✓

The squares of the integers are $(-4)^2$, or 16, and 5^2, or 25. Their sum, $16 + 25$, equals 41.✔

42. The given system of equations is:

$$\begin{cases} 4x + 3y = 25 \\ 5x + 2y = 33 \end{cases}$$

Multiply each term of the first equation by -2, and each term of the second equation by 3:

$$\begin{cases} -2(4x) - 2(3y) = -2(25) \\ 3(5x) + 3(2y) = 3(33) \end{cases}$$

$$\begin{cases} -8x - 6y = -50 \\ 15x + 6y = 99 \end{cases}$$

Add the two equations to eliminate y:

$$7x = 49$$

Divide both sides of the equation by 7:

$$\frac{7x}{7} = \frac{49}{7}$$

$$x = 7$$

Substitute 7 for x in the first equation:

$$4(7) + 3y = 25$$

$$28 + 3y = 25$$

Add -28 (the additive inverse of 28) to both sides of the equation:

$$\underline{-28 \qquad = -28}$$

$$3y = -3$$

Divide both sides of the equation by 3:

$$\frac{3y}{3} = \frac{-3}{3}$$

$$y = -1$$

The solution is $\mathbf{x = 7, y = -1}$.

CHECK: The solution must satisfy both original equations. Substitute 7 for x and -1 for y in both of the original equations to see whether they are satisfied:

$$\underline{4x + 3y = 25} \qquad\qquad \underline{5x + 2y = 33}$$

$$4(7) + 3(-1) \stackrel{?}{=} 25 \qquad\qquad 5(7) + 2(-1) \stackrel{?}{=} 33$$

$$28 - 3 \stackrel{?}{=} 25 \qquad\qquad 35 - 2 \stackrel{?}{=} 33$$

$$25 = 25 ✔ \qquad\qquad 33 = 33 ✔$$

Topic	Question Numbers	Number of Points	Your Points	Your Percentage
1. Numbers (rat'l, irrat'l); Percent	—	0		
2. Properties of No. Systems	24	2		
3. Operations on Rat'l Nos. and Monomials	12, 23	2 + 2 = 4		
4. Operations on Multinomials	16	2		
5. Square Root; Operations involving Radicals	32	2		
6. Evaluating Formulas and Expressions	3	2		
7. Linear Equations (simple cases incl. parentheses)	—	0		
8. Linear Equations Containing Decimals or Fractions	2, 11	2 + 2 = 4		
9. Graphs of Linear Functions (slope)	—	0		
10. Inequalities	22	2		
11. Systems of Eqs. & Inequalities (alg. & graphic solutions)	26, 33, 36a, b, 42	2 + 2 + 8 + 2 + 10 = 24		
12. Factoring	15	2		
13. Quadratic Equations	14	2		
14. Verbal Problems	37, 41a, b	10 + 8 + 2 = 20		
15. Variation	7	2		
16. Literal Eqs.; Expressing Relations Algebraically	4	2		
17. Factorial n	—	0		
18. Areas, Perims., Circums., Volumes of Common Figures	10, 27, 30, 35, 40b, c	2 + 2 + 2 + 2 + 6 + 2 = 16		

Topic	Question Numbers	Number of Points	Your Points	Your Percentage
19. Geometry (congruence, parallel lines, compls., suppls.)	5, 8	2 + 2 = 4		
20. Ratio & Proportion (incl. similar triangles & polygons)	20	2		
21. Pythagorean Theorem	13	2		
22. Logic (symbolic rep., logical forms, truth tables)	9, 31, 34	2 + 2 + 2 = 6		
23. Probability (incl. tree diagrams & sample spaces)	21, 38d, 39a, b, c, d, e	2 + 2 + 2 + 2 + 2 + 2 + 2 = 14		
24. Combinations, Arrangements, Permutations	17	2		
25. Statistics (central tend., freq. dist., histogr., quartiles, percentiles)	1, 18, 38a, b, c	2 + 2 + 4 + 2 + 2 = 12		
26. Properties of Triangles and Quadrilaterals	6, 25, 40a	2 + 2 + 2 = 6		
27. Transformations (reflect., translations, rotations, dilations)	19, 29	2 + 2 = 4		
28. Symmetry	—	0		
29. Area from Coordinate Geometry	—	0		
30. Dimensional Analysis	28	2		
31. Scientific Notation; Negative & Zero Exponents	—	0		

Examination January 1992

Three-Year Sequence for High School Mathematics—Course I

PART ONE

DIRECTIONS: *Answer 30 questions from this part. Each correct answer will receive 2 credits. No partial credit will be allowed. Write your answers in the spaces provided. Where applicable, answers may be left in terms of π or in radical form.* [60]

1 Thirteen students took a math test. The number of errors was 3, 7, 4, 0, 4, 1, 5, 4, 7, 3, 4, 5, and 7. What is the mode of this distribution? 1_____

2 Express the sum of $4x^2 - 7x + 6$ and $-3x^2 + 9x - 11$ as a trinomial. 2_____

3 The circumference of a circle is 12π. What is the radius of the circle? 3_____

4 In the diagram below, $m\angle BCD = 140$ and $m\angle BAC = 80$. Find $m\angle ABC$.

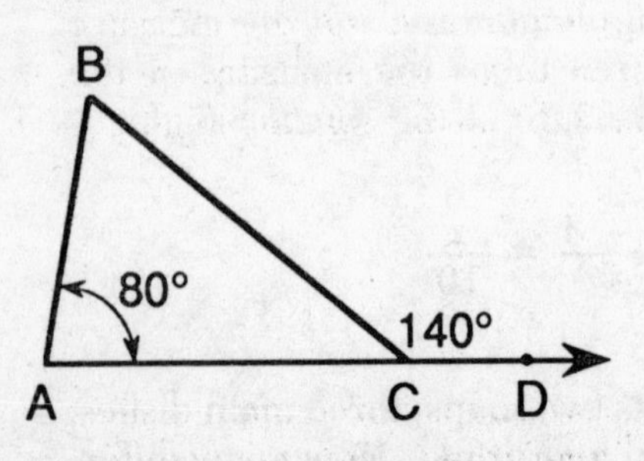

4_____

5 In the accompanying diagram, transversal $\overleftrightarrow{EF}$ intersects parallel lines $\overleftrightarrow{AB}$ and $\overleftrightarrow{CD}$ at G and H, respectively. If $m\angle AGH = 3x + 40$ and $m\angle GHD = 6x - 17$, what is the value of x?

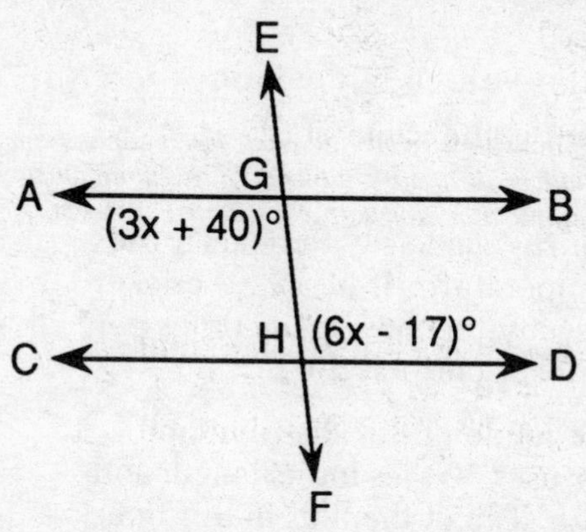

5_____

6 Solve for x: $8x - 5(x - 1) = 20$ 6_____

7 If 340,000 is expressed in the form 3.4×10^n, what is the value of n? 7_____

8 If $m = 3$ and $v = -4$, find the value of $\frac{1}{2}mv^2$. 8_____

9 Two angles are supplementary, and the measure of one angle is three times the measure of the other. Find the measure of the *smaller* angle. 9_____

10 Solve for x: $\frac{x - 3}{4} = \frac{x}{10}$ 10_____

11 A school menu lists two soups, three main dishes, three desserts, and four drinks. How many different meals consisting of one soup, one main dish, one dessert, and one drink are possible? 11_____

12 If a number is picked at random from the set $\{-4,-3,-2,-1,0,1,2,3,4\}$, what is the probability that the number satisfies the equation $x^2 - 9 = 0$? 12_____

13 If $(2x + 3)(x - 2)$ is written in the form $ax^2 + bx + c$, what is the value of c? 13_____

14 The number of chirps made by a cricket varies directly as the temperature. If at 12° a cricket chirps 30 times per minute, how many times per minute will the cricket chirp at 20°? 14_____

15 A rectangular floor uses 50 tiles for its length and 20 tiles for its width. Ten of the tiles in this floor are cracked. In an inspection, one tile is selected at random. What is the probability that this tile is cracked? 15_____

16 The perimeter of a square is $4a + 12$. Express the length of a side of the square in terms of a. 16_____

17 Express as a single fraction in simplest form:

$$\frac{5x + 2}{6} + \frac{2x - 3}{3}$$

17_____

DIRECTIONS (**18–35**): *For each question chosen, write the* numeral *preceding the word or expression that best completes the statement or answers the question.*

18 If p represents "It will rain" and q represents "We go to the movies," the statement "If we do not go to the movies, then it will rain" can be expressed by

(1) $p \to q$
(2) $q \to \sim p$
(3) $\sim q \to p$
(4) $\sim p \to \sim q$

18_____

19 Let p represent "x is an odd integer," and let q represent "x is a multiple of 3." For which value of x will $p \wedge q$ be true?

(1) 1 (3) 9
(2) 6 (4) 12

19_____

20 The product of $-4a^2b^3$ and $5ab^4$ is

(1) a^2b^{12} (3) $-20a^2b^{12}$
(2) $-20a^2b^7$ (4) $-20a^3b^7$

20_____

21 Maria is twice as old as Sue. If x represents Sue's age, which expression represents how old Maria will be in three years?

(1) $2x$ (3) $\frac{1}{2}x - 3$
(2) $x + 3$ (4) $2x + 3$

21_____

22 Which measure is *always* the same as the 50th percentile?

(1) mean (3) mode
(2) median (4) lower quartile

22_____

23 Which statement represents the inverse of the statement "If I do not study, then I will fail"?

(1) If I study, then I will not fail.
(2) If I fail, then I did not study.
(3) If I study, then I will fail.
(4) If I do not fail, then I did study.

23_____

24 The width and length of a rectangle are represented by x and $3x + 5$, respectively. If the area of the rectangle is 24, which equation can be used to find the dimensions of the rectangle?

(1) $x(3x + 5) = 24$
(2) $2x(3x + 5) = 24$
(3) $x + (3x + 5) = 24$
(4) $2x + 2(3x + 5) = 24$

24_____

25 In which graph is the slope of line ℓ negative?

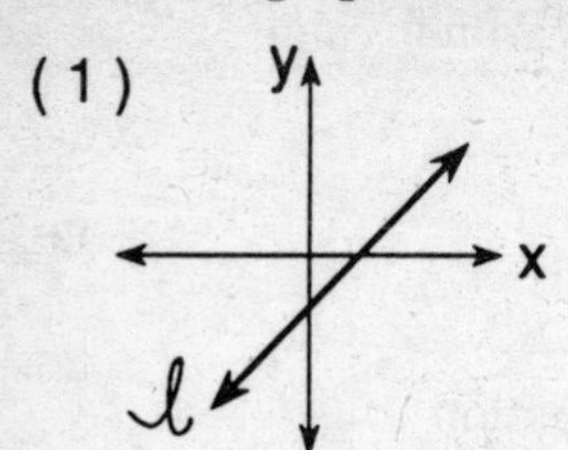

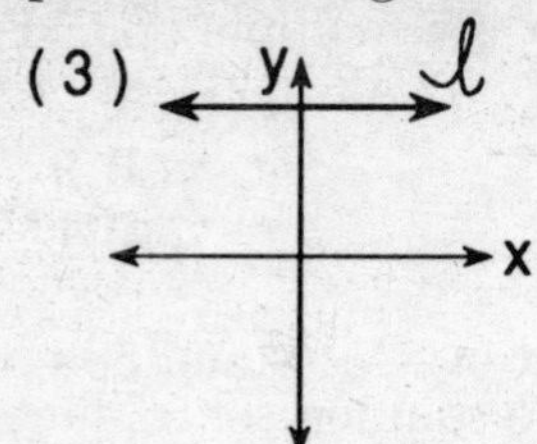

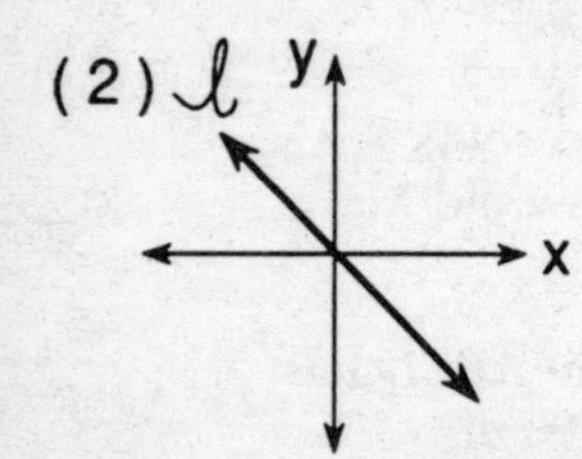

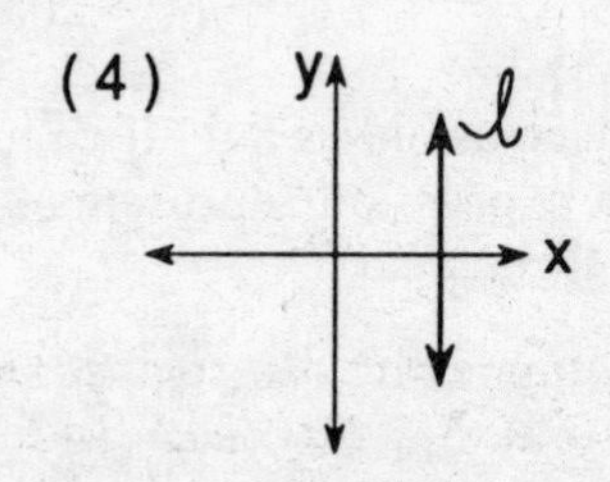

25_____

26 Which figure has one and only one line of symmetry?

(1) rhombus
(2) circle
(3) square
(4) isosceles triangle

26_____

27 What are the coordinates of the point where the graph of the equation $x + 2y = 8$ crosses the y-axis?

(1) (0,8)
(2) (8,0)
(3) (0,4)
(4) (4,0)

27_____

28 Which inequality is equivalent to $2x + 6 > 2$?

(1) $x > -2$
(2) $x < -2$
(3) $x > 2$
(4) $x < 2$

28_____

29 If the legs of a right triangle are 4 and 7, the length of the hypotenuse is

(1) $\sqrt{3}$
(2) $\sqrt{11}$
(3) $\sqrt{33}$
(4) $\sqrt{65}$

29_____

30 Which is the additive inverse of $-\frac{a}{3}$?

(1) $\frac{a}{3}$ (3) $-\frac{3}{a}$

(2) $\frac{3}{a}$ (4) 0

30_____

31 Which is the solution set of the equation $2x^2 + 3x - 2 = 0$?

(1) $\left\{-\frac{1}{2}, 2\right\}$ (3) $\left\{\frac{1}{2}, 2\right\}$

(2) $\left\{\frac{1}{2}, -2\right\}$ (4) $\left\{-\frac{1}{2}, -2\right\}$

31_____

32 In the accompanying diagram, square *ABCD* has vertices *A*(0,0), *B*(*a*,0), *C*(*a*,*a*), and *D*(0,*a*).

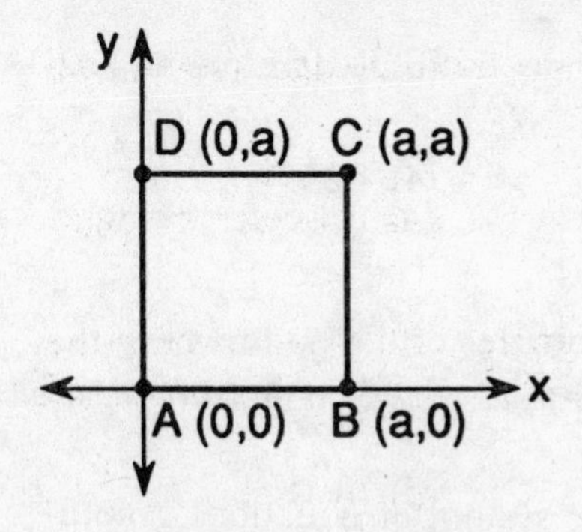

32_____

What is the area of square *ABCD*?

(1) $4a$ (3) a^2

(2) $2a$ (4) $a\sqrt{2}$

33 Which property is *not* common to all parallelograms?

(1) Opposite sides are parallel.
(2) Opposite angles are congruent.
(3) Consecutive angles are supplementary.
(4) Diagonals are congruent.

33_____

34 Which expression could be used to change 8 kilometers per hour to meters per minute?

(1) $\frac{8 \text{ km}}{\text{hr}} \cdot \frac{\text{km}}{1000 \text{ m}} \cdot \frac{\text{hr}}{60 \text{ min}}$

(2) $\frac{8 \text{ km}}{\text{hr}} \cdot \frac{1000 \text{ m}}{\text{km}} \cdot \frac{60 \text{ min}}{\text{hr}}$

(3) $\frac{8 \text{ km}}{\text{hr}} \cdot \frac{1000 \text{ m}}{\text{km}} \cdot \frac{\text{hr}}{60 \text{ min}}$

(4) $\frac{8 \text{ km}}{\text{hr}} \cdot \frac{\text{km}}{1000 \text{ m}} \cdot \frac{60 \text{ min}}{\text{hr}}$ 34_____

35 The value of 5^{-2} is

(1) $-\frac{1}{25}$ (3) -10

(2) $\frac{1}{25}$ (4) -25 35_____

PART TWO

DIRECTIONS: *Answer* four *questions from this part. Show all work unless otherwise directed.* [40]

36 Solve the following system of equations graphically and check:

$$y = -x + 2$$
$$3y - 2x = -9$$
[8,2]

37 A jar contains one dime, two quarters, and three nickels. Without looking, Andrew picks one coin from the jar. Without replacing this coin, he picks another coin.

a Draw a tree diagram or list the sample space of all possible outcomes. [3]

b What is the probability Andrew picked a dime first and then a nickel? [2]

c What is the probability he picked two dimes from the jar? [2]

d What is the probability he picked two coins such that the sum of their values is greater than or equal to 35 cents? [3]

38 In the accompanying diagram, right triangle *DEF* is similar to right triangle *ABC*. The measure of $\overline{AC}$ is 2 more than the measure of $\overline{BC}$, the measure of $\overline{EF}$ is 3 less than the measure of $\overline{BC}$, and $DF = 4$.

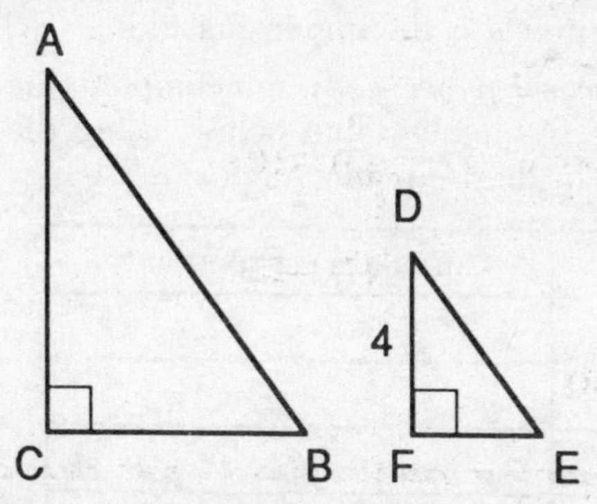

a Find the measure of $\overline{BC}$. [6]

b Find the measure of $\overline{AB}$. [2]

c What is the ratio of the area of $\triangle ABC$ to the area of $\triangle DEF$? [2]

39 Brian has $78 and wants to purchase tapes through a music club. Each tape costs $7.50. The music club will add a total postage and handling charge of $3.50 to his order. What is the greatest number of tapes he can purchase? [*Only an algebraic solution will be accepted.*] [5,5]

40 The table below shows the distribution of scores that 20 math students received on a classroom test.

Interval	Frequency
90–99	3
80–89	8
70–79	6
60–69	2
50–59	1

a In which interval does the median lie? [2]

b In which interval is the upper quartile? [2]

c *On your answer paper*, copy and complete the cumulative frequency table below, using the data given in the frequency table. [2]

Interval	Cumulative Frequency
50–99	
50–89	
50–79	
50–69	
50–59	1

d Construct a cumulative frequency histogram using the table completed in part *c*. [4]

41 Each part below consists of a set of three statements. The truth values for two statements in each set are given. Based on this information, determine the truth value of the remaining statement. *On your answer paper*, write the letters *a* through *e*, and next to each letter, write the miss-

ing truth value (TRUE or FALSE). If the truth value cannot be determined from the information given, write "CANNOT BE DETERMINED."

	Statements	Truth Value	
a	(1) $p \vee q$	TRUE	
	(2) q	?	[2]
	(3) p	FALSE	
b	(1) p	?	
	(2) q	FALSE	[2]
	(3) $p \leftrightarrow q$	TRUE	
c	(1) $p \wedge q$	FALSE	
	(2) p	FALSE	[2]
	(3) q	?	
d	(1) $q \rightarrow p$	TRUE	
	(2) p	FALSE	[2]
	(3) $\sim q$	?	
e	(1) $p \vee q$	TRUE	
	(2) $p \rightarrow q$	TRUE	[2]
	(3) q	?	

42 Answer *both a* and *b*.

a The measure of the vertex angle of an isosceles triangle exceeds 3 times the measure of a base angle by 20. Write an equation or system of equations that could be used to find the measure of *each* angle of the triangle. State what the variable(s) represents. [*Solution of the equation(s) is not required.*] [5]

b Write an equation that could be used to find three consecutive positive even integers such that the product of the first and third is six more than nine times the second. State what the variable represents. [*Solution of the equation is not required.*] [5]

Answers January 1992

Three-Year Sequence for High School Mathematics—Course I

ANSWER KEY

PART ONE

1. 4

2. $x^2 + 2x - 5$

3. 6

4. 60

5. 19

6. 5

7. 5

8. 24

9. 45

10. 5

11. 72

12. $\frac{2}{9}$

13. –6

14. 50

15. $\frac{1}{100}$

16. $a + 3$

17. $\frac{9x - 4}{6}$

18. (3)

19. (3)

20. (4)

21. (4)

22. (2)

23. (1)

24. (1)

25. (2)

26. (4)

27. (3)

28. (1)

29. (4)

30. (1)

31. (2)

32. (3)

33. (4)

34. (3)

35. (2)

PART TWO *See answers explained section.*

ANSWERS EXPLAINED

PART ONE

1. The mode of a distribution is the item that occurs most often. In the distribution 3, 7, 4, 0, 4, 1, 5, 4, 7, 3, 4, 5, and 7, the item 4 appears four times; 7 appears three times; 3 and 5 each appear twice; and 0 and 1 each appear once.

The mode is **4**.

2. Write one trinomial beneath the other with like terms in the same column:

$$\begin{array}{r} 4x^2 - 7x + 6 \\ -3x^2 + 9x - 11 \\ \hline \end{array}$$

In each column, combine the numerical coefficients algebraically and bring down the literal factor:

$$x^2 + 2x - 5$$

The sum is $\mathbf{x^2 + 2x - 5}$.

3. The circumference, C, of a circle whose radius is r is given by this formula:

$$C = 2\pi r$$

C is given to be 12π:

$$12\pi = 2\pi r$$

Divide both sides of the equation by 2π:

$$\frac{12\pi}{2\pi} = \frac{2\pi r}{2\pi}$$

$$6 = r$$

The radius is **6**.

4.

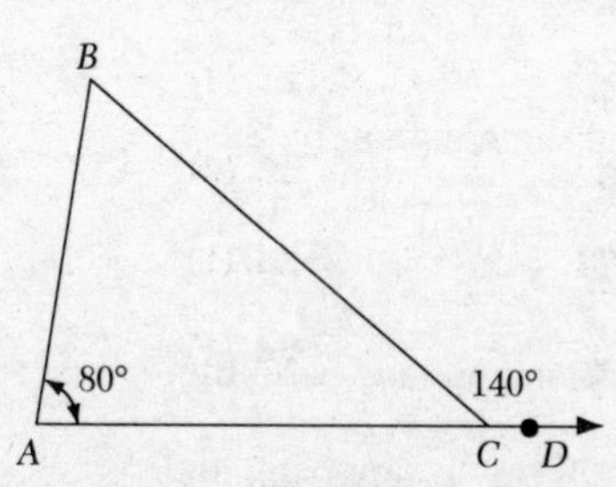

$\angle BCD$ is an exterior angle of ΔABC. The measure of an exterior angle of a triangle equals the sum of the measures of the two remote interior angles:

$$m\angle BCD = m\angle A + m\angle ABC$$

$$140 = 80 + m\angle ABC$$

Add -80 (the additive inverse of 80) to both sides of the equation:

$$\begin{array}{rl} -80 = & -80 \\ \hline 60 = & m\angle ABC \end{array}$$

$m\angle ABC = \mathbf{60}$.

ALTERNATIVE SOLUTION: $\angle BCA$ is the supplement of exterior angle BCD. Thus m$\angle BCA$ = 40. The measure of $\angle ABC$ can then be found by applying the fact that the sum of the measures of the three angles of a triangle is 180.

5.

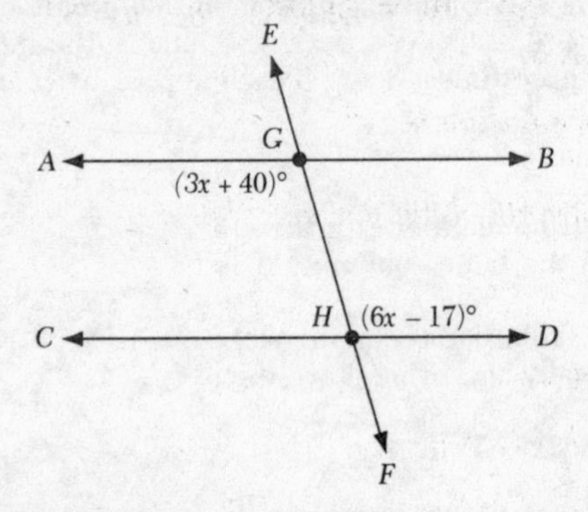

If two lines are parallel, a transversal to them makes a pair of alternate interior angles congruent (hence equal in measure):

$$m\angle GHD = m\angle AGH$$
$$6x - 17 = 3x + 40$$

Add $-3x$ (the additive inverse of $3x$) and also add 17 (the additive inverse of -17) to both sides of the equation:

$$\frac{\begin{array}{rcl} -3x + 17 &=& -3x + 17 \end{array}}{\begin{array}{rcl} 3x &=& 57 \end{array}}$$

Divide both sides of the equation by 3:

$$\frac{3x}{3} = \frac{57}{3}$$

$$x = 19$$

The value of x is **19.**

6. The given equation contains parentheses:

$$8x - 5(x - 1) = 20$$

Remove the parentheses by applying the distributive law, multiplying each term within the parentheses by -5:

$$8x - 5x + 5 = 20$$

Combine like terms:

$$3x + 5 = 20$$

Add -5 (the additive inverse of 5) to both sides of the equation:

$$\frac{\begin{array}{rcl} -5 &=& -5 \end{array}}{\begin{array}{rcl} 3x &=& 15 \end{array}}$$

Divide both sides of the equation by 3:

$$\frac{3x}{3} = \frac{15}{3}$$

$$x = 5$$

x = **5.**

7. To have 3.4 become 340,000, the decimal point in 3.4 must be moved five places to the right. Each time the decimal point is moved one place to the right, the result is equivalent to a multiplication by 10. Therefore, 3.4 must be multiplied by 10^5 to equal 340,000:

$$3.4 \times 10^5 = 340{,}000, \quad \text{or} \quad n = 5.$$

The value of n is **5.**

8. The given expression is: $\frac{1}{2}mv^2$

To evaluate, substitute 3 for m and -4 for v: $\frac{1}{2}(3)(-4)^2$

Square -4 first: $\frac{1}{2}(3)(16)$

Multiply 3 by 16: $\frac{1}{2}(48)$

Multiply 48 by $\frac{1}{2}$: 24

The value of $\frac{1}{2}mv^2$ is **24.**

9. Let x = the measure of the *smaller* angle.
Then $3x$ = the measure of the larger angle.

The sum of the measures of two supplementary angles is 180°: $x + 3x = 180$

Combine like terms: $4x = 180$

Divide both sides of the equation by 4: $\frac{4x}{4} = \frac{180}{4}$

$x = 45$

The measure of the *smaller* angle is **45.**

10. The given equation is in the form of a proportion: $\frac{x-3}{4} = \frac{x}{10}$

In a proportion, the product of the means equals the product of the extremes (cross-multiply): $4x = 10(x - 3)$

Remove the parentheses by applying the distributive law: $4x = 10x - 30$

Add $-4x$ (the additive inverse of $4x$) and also add 30 (the additive inverse of -30) to both sides of the equation:

$$\begin{array}{r} +30 - 4x = -4x + 30 \\ \hline 30 \quad = \quad 6x \end{array}$$

Divide both sides of the equation by 6: $\frac{30}{6} = \frac{6x}{6}$

$5 = x$

$x =$ **5.**

11. A meal may be put together by choosing either of the two soups, any one of the three main dishes, any one of the three desserts, and any one of the four drinks. The total number of different meals is the product of the number of choices for each of the four components of the menu: $2 \times 3 \times 3 \times 4$

Multiply out: 6×12

72

72 different meals are possible.

12. First solve the equation to find the numbers that satisfy it: $x^2 - 9 = 0$

Add 9 (the additive inverse of -9) to both sides of the equation:

$$\begin{array}{r} x^2 - 9 = 0 \\ 9 = 9 \\ \hline x^2 \quad = 9 \end{array}$$

Take the square root of each side of the equation: $x = \pm\sqrt{9}$

Two numbers, 3 and -3, satisfy the equation: $x = \pm 3$

$$\text{Probability of an event occurring} = \frac{\text{number of successful outcomes}}{\text{total possible number of outcomes}}.$$

There are 2 successful outcomes, 3 and -3, for picking a number that satisfies the equation. The total possible number of outcomes is 9, the total number of items in the set $\{-4,-3,-2,-1,0,1,2,3,4\}$. Therefore, the probability of picking a number that satisfies the equation is: $\frac{2}{9}$

The probability is $\mathbf{\frac{2}{9}}$.

13. The indicated product is: $(2x + 3)(x - 2)$

Multiply out by applying the distributive law, multiplying each term of $2x + 3$ by each term of $x - 2$:

$$\begin{array}{r} 2x + 3 \\ x - 2 \\ \hline 2x^2 + 3x \quad\quad \\ -4x - 6 \\ \hline 2x^2 - \ \ x - 6 \end{array}$$

The product is in the form $ax^2 + bx + c$ with $a = 2$, $b = -1$, and $c = -6$.

The value of c is $\mathbf{-6}$.

14. Let n = the number of chirps made per minute.
Let t = the temperature in degrees.

If k is a constant, the relationship that n varies directly with t is expressed by the equation: $n = kt$

Since $n = 30$ when $t = 12$: $30 = 12k$

Divide both sides of the equation by 12: $\frac{30}{12} = \frac{12k}{12}$

Reduce $\frac{30}{12}$ on the left side by dividing the numerator and the denominator by 6: $\frac{5}{2} = k$

Since k is a constant, we know that k is *always* $\frac{5}{2}$: $n = \frac{5}{2}t$

To find n when $t = 20$, substitute 20 for t: $n = \frac{5}{2}(20)$

Divide 2 into 20: $n = 5\,(10)$

$n = 50$

The cricket will chirp **50** times per minute.

15.

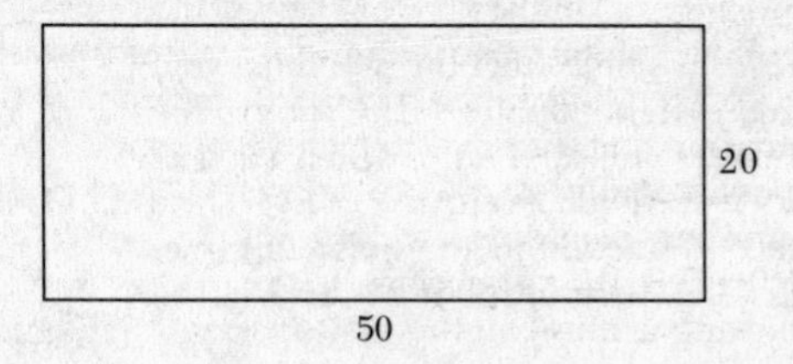

Since 50 tiles are used along the length and 20 tiles are used along the width, the entire floor is covered by 50×20, or 1,000, tiles.

$$\text{Probability of an event occurring} = \frac{\text{number of successful outcomes}}{\text{total possible number of outcomes}}.$$

For the selection of a cracked tile, there are 10 successful outcomes. The total possible number of outcomes is the total number of tiles on the floor, that is, 1,000. Therefore, the probability of picking a cracked tile is: $\frac{10}{1{,}000}$

Reduce the fraction by dividing the numerator and the denominator by 10: $\frac{1}{100}$

The probability is $\mathbf{\frac{1}{100}}$.

16.

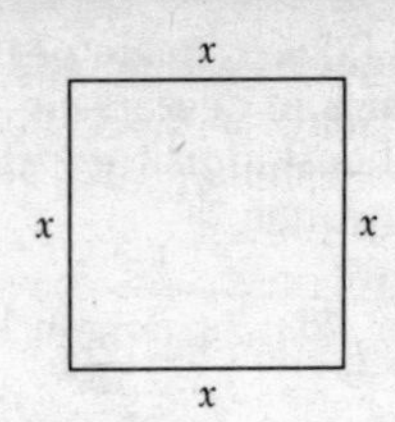

Let x = the length of one side of the square.

Since a square is equilateral, each side's length is x.

The perimeter of a square is the sum of the lengths of all four sides:

$$4x = 4a + 12$$

Divide each term on both sides of the equation by 4:

$$\frac{4x}{4} = \frac{4a}{4} + \frac{12}{4}$$

$$x = a + 3$$

The length of a side is $\boldsymbol{a + 3}$.

17. The given expression contains fractions with different denominators:

$$\frac{5x + 2}{6} + \frac{2x - 3}{3}$$

To express this as a single fraction, we must find the least common denominator (L.C.D.). The L.C.D. is the smallest number into which each of the denominators will divide evenly: The L.C.D. for 6 and 3 is 6.

Convert the second fraction to an equivalent fraction having the L.C.D. by multiplying it by 1 in the form $\frac{2}{2}$:

$$\frac{5x + 2}{6} + \frac{2(2x - 3)}{2(3)}$$

$$\frac{5x + 2}{6} + \frac{4x - 6}{6}$$

Since the fractions now have the same denominator, they may be combined by combining their numerators and writing the sum over the common denominator:

$$\frac{5x + 2 + 4x - 6}{6}$$

Combine like terms in the numerator:

$$\frac{9x - 4}{6}$$

The single fraction in simplest form is $\boldsymbol{\frac{9x - 4}{6}}$.

18. The statement p represents "It will rain."

The statement q represents "We go to the movies."

The statement "If we do not go to the movies, then it will rain" is the *implication* or *conditional* that the *negation* of q implies p. The negation of q is expressed by $\sim q$. The statement is expressed as $\sim q \rightarrow p$.

The correct choice is **(3)**.

19. The statement p represents "x is an odd integer."
The statement q represents "x is a multiple of 3."

The statement $p \wedge q$ is the *conjunction* of p and q. The conjunction is true if and only if both p and q are true.

Consider each choice in turn:

(1) $x = 1$: 1 is an odd integer, so p is true; but 1 is not a multiple of 3, so q is false.

(2) $x = 6$: 6 is not an odd integer, so p is false, although q is true since $6 = 2 \times 3$.

(3) $x = 9$: 9 is an odd integer, so p is true; and 9 is also a multiple of 3 since $9 = 3 \times 3$, so q is true. Since both p and q are true, $p \wedge q$ is true. This is the correct choice.

(4) $x = 12$: 12 is not an odd integer, so p is false, although q is true since $12 = 4 \times 3$.

The correct choice is **(3)**.

20. Indicate the product as: $(-4a^2b^3)(5ab^4)$

The numerical coefficient of the product is obtained by multiplying the numerical coefficients together: $-4 \times 5 = -20$

The literal factor of the product is obtained by multiplying the literal factors together. Remember that, to multiply powers of the same base, the exponents are added. Also remember that the exponent of a is understood to be 1: $a^2b^3 \times ab^4 = a^3b^7$

Combine the results from the steps above: $(-4a^2b^3)(5ab^4) = -20a^3b^7$

The correct choice is **(4)**.

21. Sue's age $= x$.

Maria is twice as old as Sue, so Maria's age now is $2x$.

In 3 years, Maria will add 3 to her present age, so her age then will be $2x + 3$.

The correct choice is **(4)**.

22. The *50th percentile* is the score below which 50% (or half) of all the scores fall.

The *median* is the middle score when all scores are arranged in order of size. Therefore, half of all scores fall below the median, and half are above it. The median is therefore *always* the same as the 50th percentile. This is choice (2), which is correct.

The *mean* [choice (1)] is the sum of all the scores divided by the number of scores. It may be greater than, equal to, or less than the 50th percentile, depending on how the scores are distributed over the entire range.

The *mode* [choice (3)] is the score that occurs most often and therefore has no particular relation to the 50th percentile.

The *lower quartile* [choice (4)] is the score below which 25% of all scores fall.

The correct choice is **(2)**.

23. The statement "If I do not study, then I will fail" is an *implication* or *conditional* whose *hypothesis* or *antecedent* (the "if" clause) is "I do not study." Its *conclusion* or *consequent* (the "then" clause) is "I will fail."

The *inverse* of an implication is formed by negating both its hypothesis and its conclusion. Therefore, the inverse of "If I do not study, then I will fail" is "If I study, then I will not fail." This is choice (1), which is correct.

Choice (2), "If I fail, then I did not study" is the *converse* of the original implication since it is formed by interchanging the hypothesis and the conclusion of the original.

Choice (3), "If I study, then I will fail" is an implication that has a hypothesis that negates the original, but the conclusion is the same as that of the original.

Choice (4), "If I do not fail, then I did study" is the *contrapositive* of the original implication. It is formed by negating both the hypothesis and the conclusion of the original implication and then interchanging them.

The correct choice is **(1)**.

24.

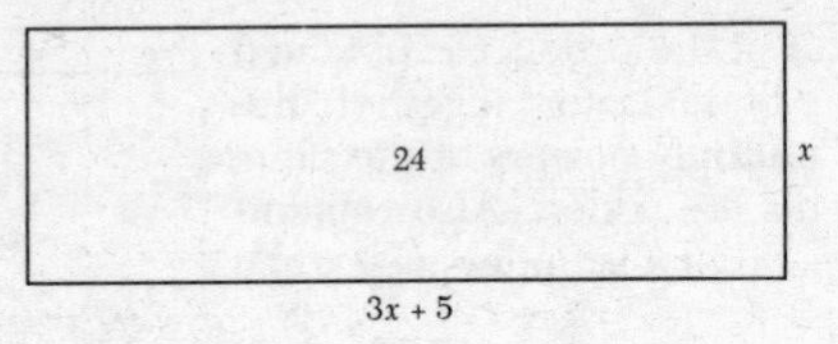

The area of a rectangle equals the product of its length and width: $x(3x + 5) = 24$

The correct choice is **(1)**.

25.

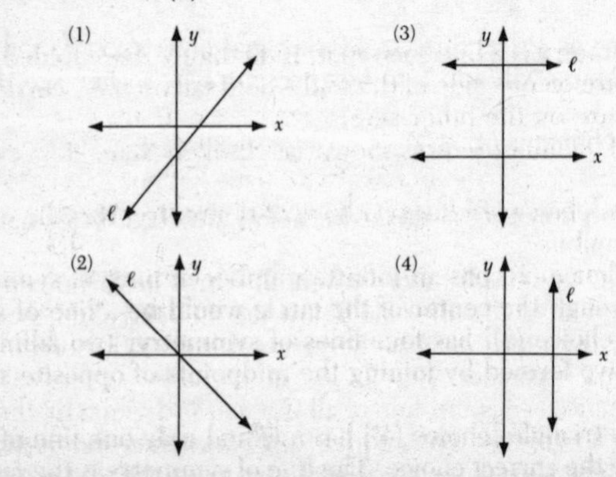

The slope of a line $= \dfrac{\text{change in } y}{\text{change in } x} = \dfrac{\Delta y}{\Delta x}$. Examine each choice in turn:

(1) As a point moves to the right along line ℓ, its y-coordinate increases and so does its x-coordinate. Thus, Δy and Δx are both positive, and the slope is positive.

(2) As a point moves to the right along line ℓ, its y-coordinate decreases, so Δy is negative. Moving to the right increases x, so Δx is positive. The slope, $\frac{\Delta y}{\Delta x}$, is therefore negative. This is the correct choice.

(3) Moving to the right along ℓ causes x to increase, so Δx is positive. But y does not change, so $\Delta y = 0$. The slope, $\frac{\Delta y}{\Delta x}$, is $\frac{0}{\Delta x}$, or 0.

(4) If we let a point move up the line, the y-coordinate increases, so Δy is positive. But the x-coordinate cannot change since the line is vertical, so $\Delta x = 0$. However, we cannot divide a positive number by 0, so $\frac{\Delta y}{\Delta x}$ cannot be determined.

The correct choice is **(2)**.

26.

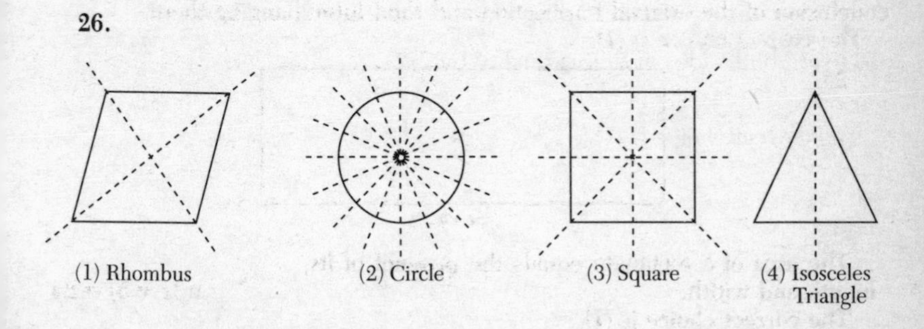

A line of symmetry is a line such that, if the figure were folded on it, each point of the figure on one side of the fold would fall on (i.e., correspond to) a point of the figure on the other side.

The lines of symmetry are shown as broken lines for each of the choices.

The rhombus [choice (1)] has two lines of symmetry; they lie on the diagonals of the rhombus.

The circle [choice (2)] has an infinite number of lines of symmetry. Any line passing through the center of the circle would be a line of symmetry.

The square [choice (3)] has four lines of symmetry, two falling along its diagonals and two formed by joining the midpoints of opposite sides of the square.

The isosceles triangle [choice (4)] has one and only one line of symmetry and is therefore the correct choice. The line of symmetry is the median from the vertex angle to the base of the triangle.

The correct choice is **(4)**.

27. The given equation is: $x + 2y = 8$

All points on the y-axis have x-coordinates of 0, so at the point where a graph crosses the y-axis, the x-coordinate must be 0.

Substitute 0 for x in the equation: $0 + 2y = 8$

Divide both sides of the equation by 2: $\frac{2y}{2} = \frac{8}{2}$

$$y = 4$$

The y-coordinate of the point where the graph of the equation crosses the y-axis is 4, so the point is (0,4).

The correct choice is **(3)**.

28. The given inequality is: $2x + 6 > 2$

Add -6 (the additive inverse of 6) to both sides of the inequality:

$$\begin{array}{r} 2x + 6 > 2 \\ -6 = -6 \\ \hline 2x > -4 \end{array}$$

Divide both sides of the inequality by 2: $\frac{2x}{2} > \frac{-4}{2}$

$$x > -2$$

The correct choice is **(1)**.

29.

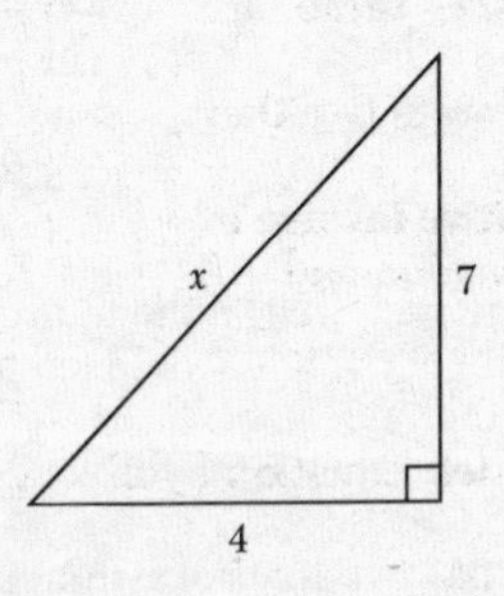

Let x = the length of the hypotenuse.

By the Pythagorean Theorem, in a right triangle the square of the length of the hypotenuse is equal to the sum of the squares of the lengths of the legs: $x^2 = 4^2 + 7^2$

Since $4^2 = 16$, and $7^2 = 49$: $x^2 = 16 + 49$

Combine like terms: $x^2 = 65$

Take the square root of each side of the equation: $x = \pm\sqrt{65}$

Reject the negative value as meaningless for a length: $x = \sqrt{65}$

The correct choice is **(4)**.

30. Let x = the additive inverse of $-\frac{a}{3}$.

The sum of a number and its additive inverse is 0: $-\frac{a}{3} + x = 0$

Add $\frac{a}{3}$ to both sides of the equation to isolate x on one side: $\frac{a}{3} \quad = \frac{a}{3}$

$$x = \frac{a}{3}$$

The additive inverse of $-\frac{a}{3}$ is $\frac{a}{3}$.

The correct choice is **(1)**.

31. The given equation is a *quadratic equation:* $2x^2 + 3x - 2 = 0$

The left side is a *quadratic trinomial* that can be factored into the product of two binomials. The factors of the first term, $2x^2$, are $2x$ and x, and they become the first terms of the binomials: $(2x \quad)(x \quad) = 0$

The factors of the last term become the second terms of the binomials, but they must be chosen and placed within the binomials in such a way that the sum of the inner and outer cross-products of the binomials equals the middle term, $+3x$, of the trinomial. Try -1 and 2 as the factors of -2:

$-x$ = inner product
$$(2x - 1)(x + 2) = 0$$
$+4x$ = outer product

Since $(-x) + (+4x) = +3x$, these are the correct factors: $(2x - 1)(x + 2) = 0$

If the product of two factors is 0, either factor may equal 0: $2x - 1 = 0 \vee x + 2 = 0$

Add the appropriate additive inverse to each side, 1 for the left equation and -2 for the right equation:

$$\begin{array}{rr} 1 = 1 & -2 = -2 \\ \hline 2x \quad = 1 & x \quad = -2 \end{array}$$

Divide both sides of the left equation by 2: $\frac{2x}{2} = \frac{1}{2}$ $\qquad x = -2$

$$x = \frac{1}{2}$$

The solution set is $\left\{\frac{1}{2}, -2\right\}$.

The correct choice is **(2)**.

ALTERNATIVE SOLUTION: The question can also be solved by substituting each of the two numbers of the set given in each choice to see whether these numbers satisfy the equation. *Both* members of the solution set must satisfy the equation. This method has the disadvantage that it may require as many as eight trial substitutions to find the correct pair of roots.

,32.

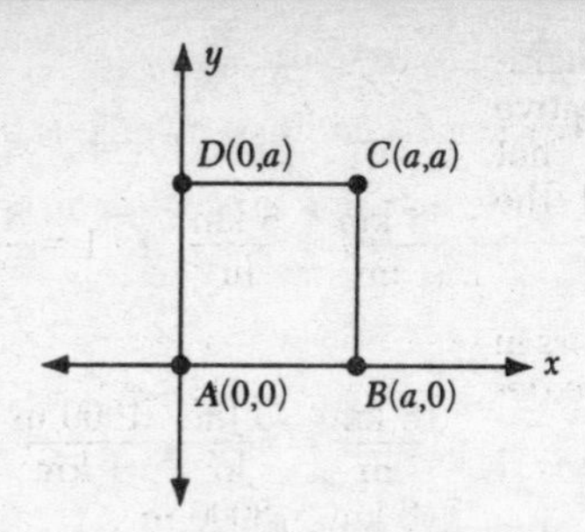

The length of a side of the square is equal to AB. Since $A(0,0)$ is at the origin, the x-coordinate of B (which is given as a) is the same as the length AB. Thus the side of the square is of length a.

The area of a square is equal to the square of the length of one side:

$$\text{Area} = a^2$$

The correct choice is (3).

33.

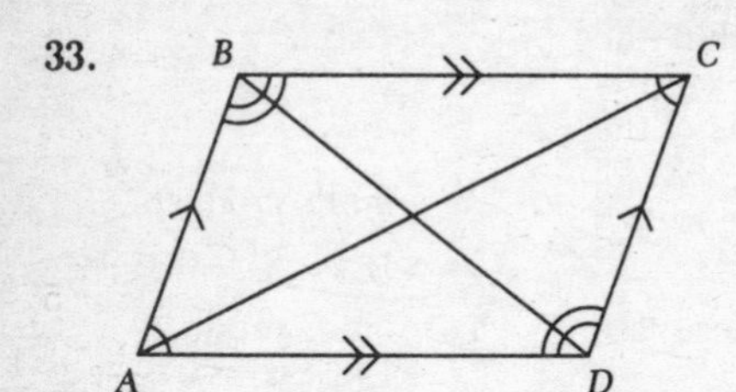

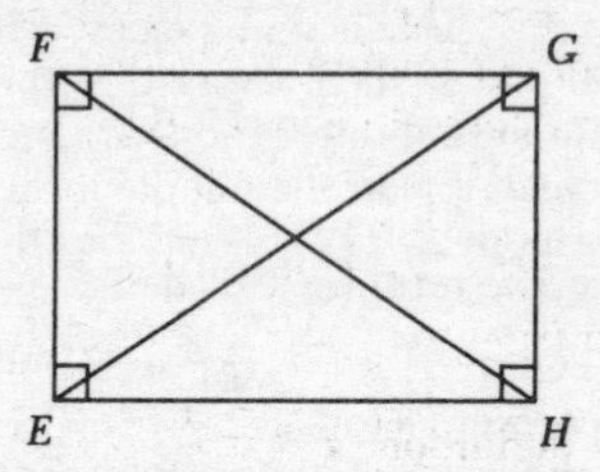

For a parallelogram such as $ABCD$, consider each choice in turn:

(1) The opposite sides are parallel by definition.

(2) It can be proved that the opposite angles are congruent.

(3) Consecutive angles are supplementary since in each case they are the interior angles on the same side of the transversal cutting two parallel lines.

(4) In certain parallelograms, such as rectangle $EFGH$ illustrated above, the diagonals are congruent. This is *not* true of all parallelograms, however, as is evident with diagonals $\overline{AC}$ and $\overline{BD}$ of $ABCD$, where $AC > BD$.

The correct choice is (4).

34. Since 1 kilometer = 1000 meters, $\frac{1000 \text{ m}}{1 \text{ km}} = 1$.

Since 1 hour = 60 minutes, $\frac{1 \text{ hr}}{60 \text{ min}} = 1$.

Multiplying any quantity by the multiplicative identity, 1, does not change the value of the quantity:

$$\frac{8 \text{ km}}{\text{hr}} = \frac{8 \text{ km}}{\text{hr}} \cdot 1 \cdot 1 = \frac{8 \text{ km}}{\text{hr}} \cdot \frac{1000 \text{ m}}{1 \text{ km}} \cdot \frac{1 \text{ hr}}{60 \text{ min}}$$

Divide out like factors in the numerator and the denominator:

$$\frac{8 \text{ km}}{\text{hr}} = \frac{8 \cancel{\text{km}}}{\cancel{\text{hr}}} \cdot \frac{1000 \text{ m}}{1 \cancel{\text{km}}} \cdot \frac{1 \cancel{\text{hr}}}{60 \text{ min}}$$

$$\frac{8 \text{ km}}{\text{hr}} = \frac{8000 \text{ m}}{60 \text{ min}}$$

Divide the numerator and the denominator on the right by 60:

$$\frac{8 \text{ km}}{\text{hr}} = \frac{133.3 \text{ m}}{\text{min}}$$

The last equation, obtained by using choice (3), expresses $\frac{8 \text{ km}}{\text{hr}}$ as 133.3 meters per minute.

The correct choice is **(3)**.

35. The term to be evaluated is:

$$5^{-2}$$

By definition, $x^{-n} = \frac{1}{x^n}$:

$$5^{-2} = \frac{1}{5^2}$$

$5^2 = 25$:

$$5^{-2} = \frac{1}{25}$$

The correct choice is **(2)**.

PART TWO

36. STEP 1: First draw the graph of $y = -x + 2$. Prepare a table of values for x and y by choosing any three convenient values for x and substituting them in the equation to find the corresponding values of y:

x	$-x + 2$	$= y$
-4	$-(-4) + 2 = 4 + 2$	$= 6$
0	$-0 + 2$	$= 2$
5	$-5 + 2$	$= -3$

Plot points $(-4,6)$, $(0,2)$, and $(5,-3)$, and draw a straight line through them. This line is the graph of $y = -x + 2$.

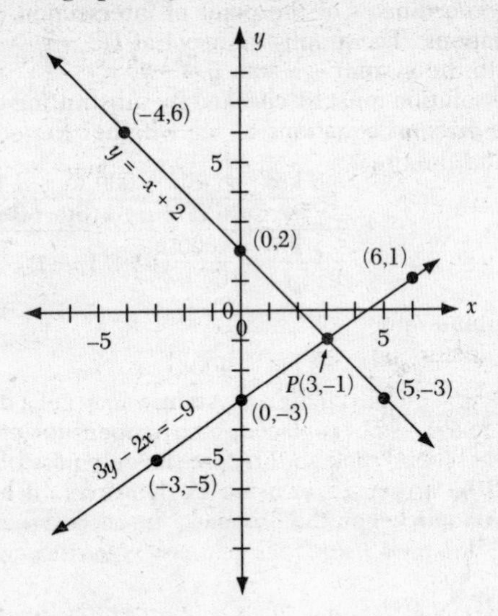

STEP 2: To draw the graph of $3y - 2x = -9$, it is convenient to first solve the equation for y:

$$3y - 2x = -9$$

Add $2x$ (the additive inverse of $-2x$) to both sides of the equation:

$$\frac{\quad +2x = \quad +2x}{3y \quad = -9 + 2x}$$

Divide both sides of the equation by 3:

$$\frac{3y}{3} = \frac{-9 + 2x}{3}$$

$$y = \frac{-9 + 2x}{3}$$

Prepare a table of values for x and y by selecting any three convenient values for x and substituting them in the equation to find the corresponding values of y:

x	$\frac{-9 + 2x}{3}$	$= y$
-3	$\frac{-9 + 2(-3)}{3} = \frac{-9 - 6}{3} = \frac{-15}{3}$	$= -5$
0	$\frac{-9 + 2(0)}{3} = \frac{-9 + 0}{3} = \frac{-9}{3}$	$= -3$
6	$\frac{-9 + 2(6)}{3} = \frac{-9 + 12}{3} = \frac{3}{3}$	$= 1$

Plot points $(-3,-5)$, $(0,-3)$, and $(6,1)$, and draw a straight line through them. This line is the graph of $3y - 2x = -9$.

STEP 3: The coordinates of the point of intersection of the two graphs satisfy *both* equations. The graphs intersect at $P(3,-1)$.

The solution to the system is $\mathbf{x = 3,\ y = -1}$.

CHECK: The solution must be checked by substituting the values of x and y in *both* of the *original* equations to see whether the equations are satisfied:

	$y = -x + 2$	$3y - 2x = -9$
$x = 3,\ y = -1$:	$-1 \overset{?}{=} -3 + 2$	$3(-1) - 2(3) \overset{?}{=} -9$
	$-1 = -1\ \checkmark$	$-3 - 6 \overset{?}{=} -9$
		$-9 = -9\ \checkmark$

37. a. On the first pick from the jar, Andrew may get a dime, a quarter, or a nickel. Since the coin is not replaced, if he happens to get a dime, no dimes will be left for the second pick; in this case the only possible results from the second pick will be a quarter or a nickel. However, if he happens to get either a quarter or a nickel on the first pick, the second pick could result in a dime, a quarter, or a nickel since there are two quarters and three nickels in the jar.

Since any one of the six coins may be selected on the first pick, the *tree diagram* requires six branches from "Start" to "First Pick." After each first pick, any one of the five remaining coins may be selected on the second pick, so each first-pick branch requires five secondary branches to represent the possibilities under "Second Pick" that can follow it.

The *sample space* requires one line to show each possible combination of first pick followed by second pick. Since there are six possible first picks, each followed by five possible second picks, 6×5, or 30, lines are needed in the sample space.

b. Probability of an event occurring $= \dfrac{\text{number of successful outcomes}}{\text{total possible number of outcomes}}$.

There is only one dime, so the number of successful outcomes for the first pick being a dime is 1. The total possible number of outcomes is the total number of coins in the jar; one dime, two quarters, and three nickels represent a total of six coins, so the total possible number of outcomes is 6. The probability of getting a dime on the first pick is $\dfrac{1}{6}$.

For getting a nickel on the second pick, the number of successful outcomes is 3 (the three nickels), but the total possible number of outcomes is now 5 since the first coin picked is not replaced. The probability of getting a nickel on the second pick is $\dfrac{3}{5}$.

TREE DIAGRAM

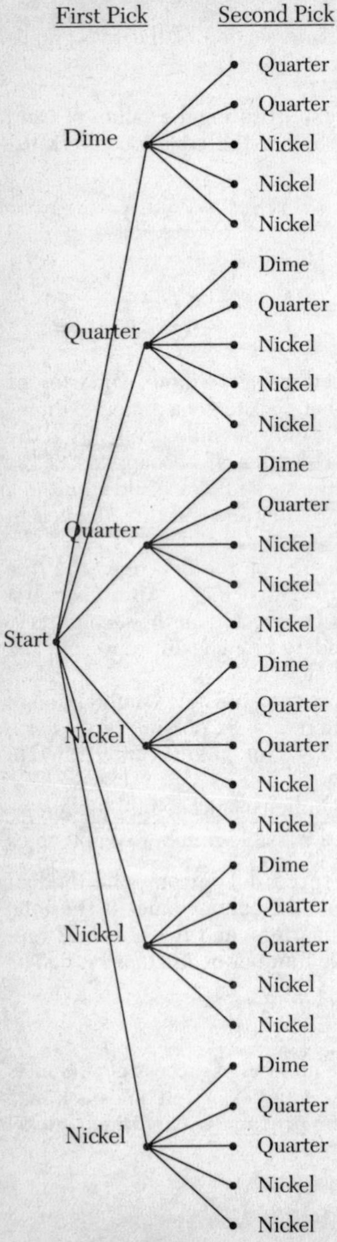

First Pick | Second Pick

Start

Dime: Quarter, Quarter, Nickel, Nickel, Nickel

Quarter: Dime, Quarter, Nickel, Nickel, Nickel

Quarter: Dime, Quarter, Nickel, Nickel, Nickel

Nickel: Dime, Quarter, Quarter, Nickel, Nickel

Nickel: Dime, Quarter, Quarter, Nickel, Nickel

Nickel: Dime, Quarter, Quarter, Nickel, Nickel

SAMPLE SPACE

First Pick	Second Pick
Dime	Quarter
Dime	Quarter
Dime	Nickel
Dime	Nickel
Dime	Nickel
Quarter	Dime
Quarter	Quarter
Quarter	Nickel
Quarter	Nickel
Quarter	Nickel
Quarter	Dime
Quarter	Quarter
Quarter	Nickel
Quarter	Nickel
Quarter	Nickel
Nickel	Dime
Nickel	Quarter
Nickel	Quarter
Nickel	Nickel
Nickel	Nickel
Nickel	Dime
Nickel	Quarter
Nickel	Quarter
Nickel	Nickel
Nickel	Nickel
Nickel	Dime
Nickel	Quarter
Nickel	Quarter
Nickel	Nickel
Nickel	Nickel

The probability of getting a dime on the first pick followed by a nickel on the second pick is the product of the separate probabilities of these events:

$$\frac{1}{6} \times \frac{3}{5} = \frac{3}{30}$$

The probability of a dime followed by a nickel is $\mathbf{\frac{3}{30}}$ or $\mathbf{\frac{1}{10}}$.

c. Since there is only one dime in the jar, the probability of picking two dimes must be 0. Note that, although the probability of getting a dime on the first pick is $\frac{1}{6}$, the probability of getting a dime on the second pick is $\frac{0}{5}$; the product of $\frac{1}{6}$ and $\frac{0}{5}$ is 0.

The probability of two dimes is **0.**

d. A dime and quarter combination has a value of 35¢. A two-quarter combination has a value of 50¢. All other possible combinations (two nickels or a quarter and a nickel) have values less than 35¢.

The tree diagram has six branches from "Start" through "Second Pick" that yield values of 35¢ or more: dime-quarter, dime-quarter, quarter-dime, quarter-quarter, quarter-dime, and quarter-quarter. Thus there are 6 successful outcomes for a combination whose value is 35¢ or more. There is a total of six branches from "Start" to "First Pick," and each has five branches to "Second Pick," making a total of 6 × 5, or 30, possible distinct paths. The probability that the sum of the values of the two coins picked is greater than or equal to 35¢ is thus $\frac{6}{30}$.

If the sample space is used to solve this question, it will be noted that there are six lines containing dime-quarter (or quarter-dime) and quarter-quarter combinations out of a total of 30 lines in the sample space.

The probability of two coins whose sum is a value ≥ 35¢ is $\mathbf{\frac{6}{30}}$ or $\mathbf{\frac{1}{5}}$.

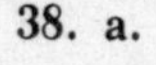

38. a.

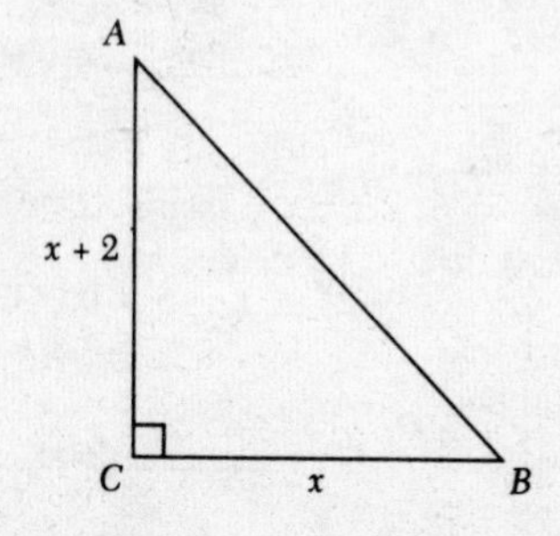

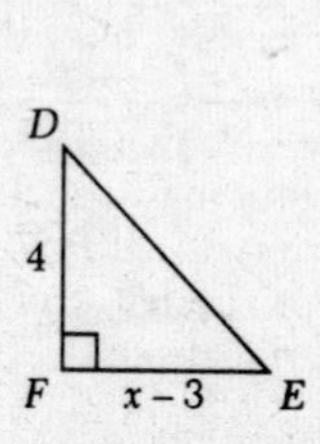

Let x = the measure of $\overline{BC}$.
Then $x + 2$ = the measure of $\overline{AC}$.
And $x - 3$ = the measure of $\overline{EF}$.

If two triangles are similar, the measures of their corresponding sides are in proportion:

$$\frac{x}{x-3} = \frac{x+2}{4}$$

In a proportion, the product of the means equals the product of the extremes (cross-multiply):

$$(x + 2)(x - 3) = 4x$$

Multiply out the indicated product on the left:

$$\begin{array}{r} x - 3 \\ x + 2 \\ \hline x^2 - 3x \qquad \\ + 2x - 6 \\ \hline x^2 - \ \ x - 6 \end{array}$$

$$x^2 - x - 6 = 4x$$

This is a *quadratic equation*. Rearrange it so that all terms are on one side equal to 0 by adding $-4x$ (the additive inverse of $4x$) to both sides:

$$\begin{array}{r} -4x \quad = -4x \\ \hline x^2 - 5x - 6 = 0 \end{array}$$

The left side is a *quadratic trinomial* that can be factored into the product of two binomials. The factors of the first term, x^2, are x and x, and they become the first terms of the binomials:

$$(x \quad)(x \quad) = 0$$

The factors of the last term, -6, become the second terms of the binomials, but they must be chosen in such a way that the sum of the inner and outer cross-products of the binomials equals the middle term, $-5x$, of the quadratic trinomial. Try -6 and 1 as the factors of -6:

$-6x$ = inner product

$$(x - 6)(x + 1) = 0$$

$+x$ = outer product

Since $(-6x) + (+x) = -5x$, these are the correct factors:

$$(x - 6)(x + 1) = 0$$

If the product of two factors is 0, either of the factors may equal 0:

$$x - 6 = 0 \lor x + 1 = 0$$

Add the appropriate additive inverse to each side, 6 for the left equation and -1 for the right equation:

$$\begin{array}{rr} 6 = 6 & -1 = -1 \\ \hline x \quad = 6 & x \quad = -1 \end{array}$$

Reject the negative value as meaningless for a length: $x = 6$

The measure of $\overline{BC}$ is **6.**

b. $AC = x + 2 = 6 + 2 = 8$. ΔABC is a 3-4-5 right triangle with $BC = 2 \times 3$ and $AC = 2 \times 4$; therefore, $AB = 2 \times 5 = 10$.
The measure of $\overline{AB}$ is **10.**
ALTERNATIVE SOLUTION: AB can be found by employing the Pythagorean Theorem:

$$(AB)^2 = 6^2 + 8^2$$

c. The ratio of the areas of two similar triangles is equal to the ratio of the squares of the lengths of any two corresponding sides:

$$\frac{\text{Area } \Delta ABC}{\text{Area } \Delta DEF} = \frac{(AC)^2}{(DF)^2}$$

$$\frac{\text{Area } \Delta ABC}{\text{Area } \Delta DEF} = \frac{8^2}{4^2} = \frac{64}{16} = \frac{4}{1}$$

The ratio of the areas of ΔABC and ΔDEF is **4:1.**

39. Let x = the number of tapes purchased.
Then $7.50x$ = the price of the tapes.
And $7.50x + 3.50$ = the total cost of the order, including postage and handling.

The total cost must be less than or equal to \$78: $7.50x + 3.50 \leq 78$

Clear the decimals by multiplying all terms on both sides of the inequality by 10: $75x + 35 \leq 780$

Add -35 (the additive inverse of 35) to both sides of the inequality:

$$\begin{array}{r} 75x + 35 \leq 780 \\ -35 = -35 \\ \hline 75x \quad \leq 745 \end{array}$$

Divide both sides of the inequality by 75: $\frac{75x}{75} \leq \frac{745}{75}$

$$\begin{array}{r} 9.9 \\ 75\overline{)745.0} \\ 675 \\ \hline 70\ 0 \\ 67\ 5 \\ \hline \end{array}$$

$x \leq 9.9^+$

But x must be an integer: $x \leq 9$

The greatest integer value of x is 9.

The greatest number of tapes Brian can purchase is **9.**

40. a.

Interval	Frequency
90–99	3
80–89	8
70–79	6
60–69	2
50–59	1
Total	20

Total the frequencies. There is a total of 20 scores.

The median is the middle score when all scores are arranged in order of size. Half the scores are less than the median, and half are greater than the median. Since one-half of 20 is 10, 10 scores will be below the median and 10 will be above it. The median will lie between the 10th and 11th scores. Counting up from the bottom of the frequency distribution, we find that there are 1 + 2 + 6, or 9, scores in the lowest three intervals. We need one more to reach the 10th score and two more to reach the 11th. Since the next higher interval, 80–89, contains eight scores, both the 10th and 11th scores lie in that interval. Thus the median lies in the 80–89 interval.

The median lies in the **80–89** interval.

b. The upper quartile is the score that is higher than three-fourths of all scores and lower than one fourth. Since $\frac{3}{4}$ of 20 is 15 and $\frac{1}{4}$ of 20 is 5, the upper quartile lies between the 15th and 16th scores from the bottom. Counting up from the bottom, we find that the lowest three intervals contain a total of nine scores. We need six more to reach the 15th score and seven more to reach the 16th. Since the 80–89 interval contains eight scores, both the 15th and 16th scores lie in this interval, and so does the upper quartile, which is between them.

The upper quartile is in the **80–89** interval.

c. The cumulative frequencies are obtained by adding the frequency of each interval to the cumulative frequency immediately below it. Notice that the top cumulative frequency must equal the total of all frequencies, 20.

Interval	Cumulative Frequency
50–99	17 + 3 = 20
50–89	9 + 8 = 17
50–79	3 + 6 = 9
50–69	1 + 2 = 3
50–59	1

d.

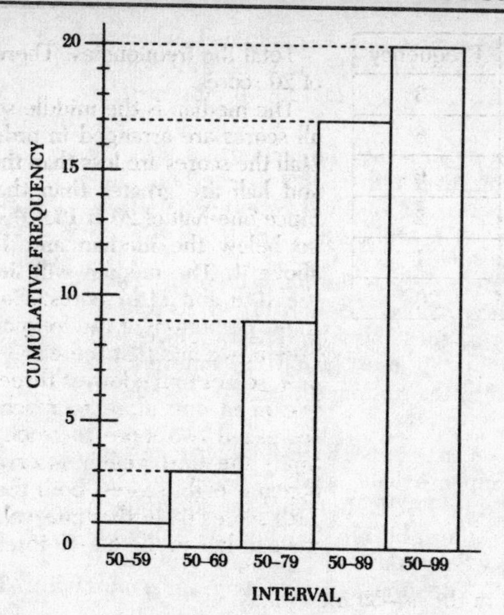

41.

	Statement	Truth Value
a	(1) $p \vee q$	TRUE
	(2) q	?
	(3) p	FALSE
b	(1) p	?
	(2) q	FALSE
	(3) $p \leftrightarrow q$	TRUE
c	(1) $p \wedge q$	FALSE
	(2) p	FALSE
	(3) q	?
d	(1) $q \rightarrow p$	TRUE
	(2) p	FALSE
	(3) $\sim q$	?
e	(1) $p \vee q$	TRUE
	(2) $p \rightarrow q$	TRUE
	(3) q	?

a. The statement $p \vee q$ is the *disjunction* of p and q. It is TRUE if p or q, or both, are TRUE. Since p is FALSE, q must be TRUE to make $p \vee q$ TRUE.

a. TRUE

b. The statement $p \leftrightarrow q$ is a *biconditional*, which is TRUE if and only if p and q have the same truth value. Since $p \leftrightarrow q$ is TRUE and q is FALSE, p must also be FALSE.

b. FALSE

c. The statement $p \wedge q$ is the *conjunction* of p and q. A conjunction is FALSE if either p or q, or both, are FALSE. Since q is FALSE, p could be either TRUE or FALSE and $p \wedge q$ would still be FALSE.

c. CANNOT BE DETERMINED

d. The statement $q \rightarrow p$ is the *implication* or *conditional* that p implies q. Since $q \rightarrow p$ is TRUE, its *contrapositive*, $\sim p \rightarrow \sim q$, must also be TRUE. Since p is FALSE, its *negation*, $\sim p$, must be TRUE, and from the conditional formed by the contrapositive, $\sim p \rightarrow \sim q$, $\sim q$ must be TRUE.

d. TRUE

e. The *disjunction*, $p \vee q$, is TRUE if either p or q, or both, are TRUE. Suppose that q is FALSE. Since $p \vee q$ is TRUE, then p must be TRUE. But since the *conditional*, $p \rightarrow q$, is TRUE, q must be TRUE. Thus, by supposing that q is FALSE, we are led to the contradiction that it must also be TRUE. The supposition is wrong, and q must be TRUE.

e. TRUE

42.

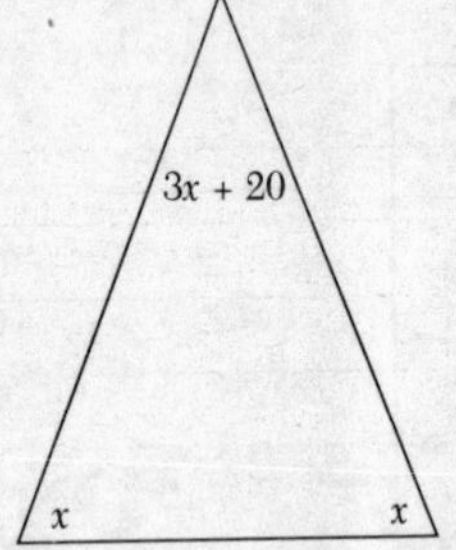

a. Let x = the measure of a base angle.

Then x is also the measure of the other base angle since base angles of an isosceles triangle are congruent (equal in measure).

Then $3x + 20$ = the measure of the vertex angle.

The sum of the measures of the three angles of a triangle is 180°.

The equation is:

$$\mathbf{x + x + 3x + 20 = 180}$$

b. Let x = the smallest even integer.

Then $x + 2$ = the next consecutive even integer.

And $x + 4$ = the third consecutive even integer.

The product of the first and third integers is 9 times the second and 6 more.

$x(x + 4) \quad = 9(x + 2) \quad + 6$

The equation is: $\mathbf{x(x + 4) = 9(x + 2) + 6}$

Topic	Question Numbers	Number of Points	Your Points	Your Percentage
1. Numbers (rat'l, irrat'l); Percent	—	0		
2. Properties of No. Systems	30	2		
3. Operations on Rat'l Nos. and Monomials	20	2		
4. Operations on Multinomials	2, 13, 17	2 + 2 + 2 = 6		
5. Square Root; Operations Involving Radicals	—	0		
6. Evaluating Formulas and Expressions	8, 35	2 + 2 = 4		
7. Linear Equations (simple cases incl. parentheses)	6	2		
8. Linear Equations Containing Decimals or Fractions	10	2		
9. Graphs of Linear Functions (slope)	25, 27	2 + 2 = 4		
10. Inequalities	28, 39	2 + 10 = 12		
11. Systems of Eqs. & Inequalities (alg. & graphic solutions)	36	10		
12. Factoring	—	0		
13. Quadratic Equations	31	2		
14. Verbal Problems	42a, b	5 + 5 = 10		
15. Variation	14	2		
16. Literal Eqs.; Expressing Relations Algebraically	21	2		
17. Factorial *n*	—	0		
18. Areas, Perims., Circums., Vols. of Common Figures	3, 16, 24	2 + 2 + 2 = 6		
19. Geometry (congruence, parallel lines, compls., suppls.)	5, 9	2 + 2 = 4		
20. Ratio & Proportion (incl. similar triangles & polygons)	38a, b, c	6 + 2 + 2 = 10		
21. Pythagorean Theorem	29	2		
22. Logic (symbolic rep., logical forms, truth tables)	18, 19, 23, 41a, b, c, d, e	2 + 2 + 2 + 10 = 16		

Topic	Question Numbers	Number of Points	Your Points	Your Percentage
23. Probability (incl. tree diagrams & sample spaces)	12, 15, 37a, b, c, d	2 + 2 + 3 + 2 + 2 + 3 = 14		
24. Combinations, Arrangements, Permutations	11	2		
25. Statistics (central tend., freq. dist., histogr., quartiles, percentiles)	1, 22, 40a, b, c, d	2 + 2 + 2 + 2 + 2 + 4 = 14		
26. Properties of Triangles and Quadrilaterals	4, 33	2 + 2 = 4		
27. Transformations (reflect., translations, rotations, dilations)	—	0		
28. Symmetry	26	2		
29. Area from Coordinate Geometry	32	2		
30. Dimensional Analysis	34	2		
31. Scientific Notation; Negative & Zero Exponents	7	2		

Examination June 1992

Three-Year Sequence for High School Mathematics—Course I

PART ONE

DIRECTIONS: *Answer 30 questions from this part. Each correct answer will receive 2 credits. No partial credit will be allowed. Write your answers in the spaces provided. Where applicable, answers may be left in terms of* π *or in radical form.* [60]

1 A letter is chosen at random from the word "REGENTS." Find the probability that the letter chosen is an E. 1_____

2 Let p represent the statement "The triangle is isosceles," and let q represent the statement "The triangle is scalene." Write in symbolic form: "If the triangle is isosceles, then the triangle is *not* scalene." 2_____

3 Solve for x: $3(2x - 1) = x + 2$ 3_____

4 In a basketball game, the number of points scored by five members of the team were 28, 20, 16, 15, and 8. How many players scored fewer than the mean number of points? 4_____

5 Solve for x: $0.5x - 12 = 3.5$ 5_____

6 If $x = 5$ and $y = -2$, what is the value of $\frac{2x - y}{3}$? 6_____

7 The histogram below shows the distribution of temperatures for ten days. Which temperature is the mode?

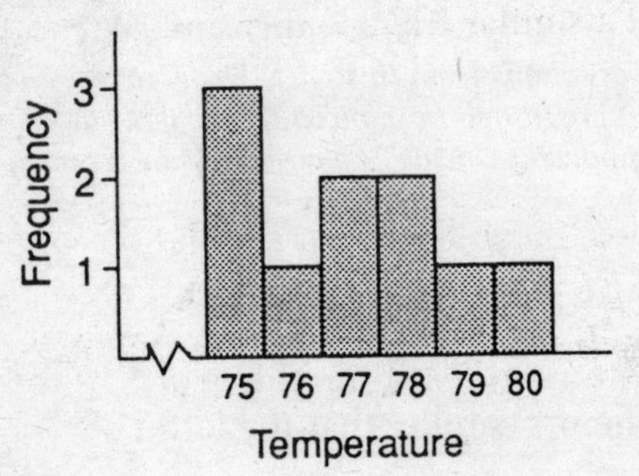

7_____

8 A rectangle has an area of 20. If the length of the rectangle is doubled and the width remains the same, what is the area of the new rectangle? 8_____

9 Solve for y: $\frac{3}{4}y - 8 = 1$ 9_____

10 In the accompanying diagram, $m\angle A = 2x - 30$, $m\angle B = x$, and $m\angle C = x + 10$. Find the number of degrees in $\angle B$.

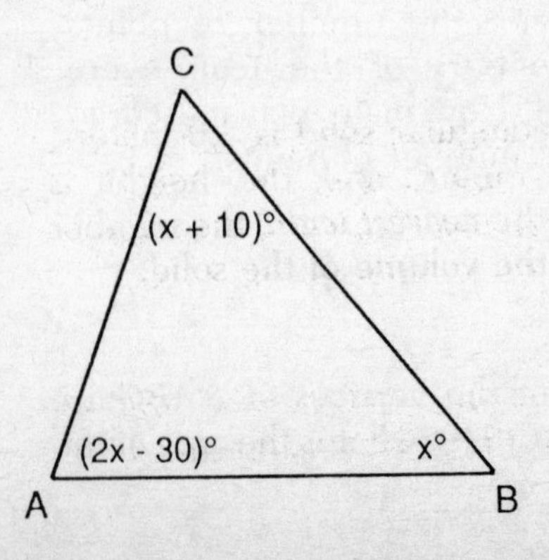

10_____

11 If $x = 4y$, what is the value of $\frac{x}{y}$, $y \neq 0$? 11_____

12 What is the inverse of $\sim s \rightarrow t$? 12_____

13 The sides of a triangle measure 6, 11, and 15. If the smallest side of a similar triangle measures 4, find the length of its longest side. 13_____

14 In the accompanying diagram, $\overleftrightarrow{AB}$ is parallel to $\overleftrightarrow{CD}$ and transversal $\overleftrightarrow{EF}$ intersects $\overleftrightarrow{AB}$ and $\overleftrightarrow{CD}$ at G and H, respectively. If $m\angle DHG$:$m\angle BGH$ = 1:2, find $m\angle DHG$.

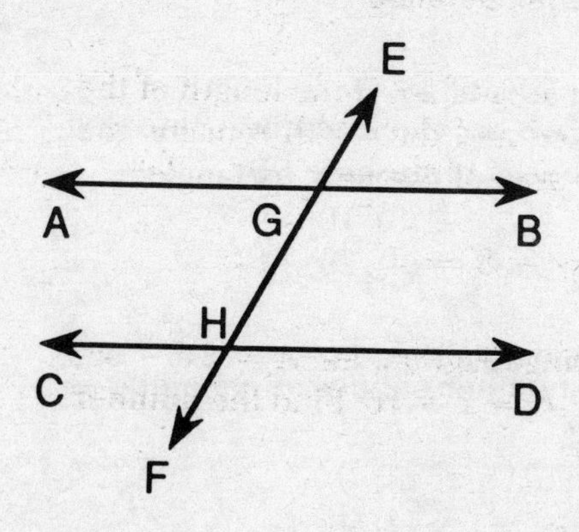

14_____

15 The length of a rectangular solid is 3.0 meters, the width is 0.6 meter, and the height is 0.4 meter. Find, to the *nearest tenth*, the number of cubic meters in the volume of the solid. 15_____

16 If the coordinates of the vertices of $\triangle ABC$ are $A(-5,0)$, $B(5,0)$, and $C(0,8)$, what is the area of the triangle? 16_____

17 Which value for n makes this sentence true?

$$0.00045 = 4.5 \times 10^n$$

17_____

18 The area of a circle is 49π. Find, in terms of π, the circumference of the circle.

18_____

19 If x varies directly as y and $x = 3$ when $y = 4$, find x when $y = 20$.

19_____

DIRECTIONS (**20–35**): *For each question chosen, write the* numeral *preceding the word or expression that best completes the statement or answers the question.*

20 If $n + 7$ represents an even number, the next larger even number is represented by

(1) $n + 8$
(2) $n + 9$
(3) $10n + 7$
(4) $2n + 7$

20_____

21 What is the total number of lines of symmetry in a square?

(1) 1
(2) 2
(3) 0
(4) 4

21_____

22 If $3x + c = 4$, then x equals

(1) $4 - c$
(2) $\frac{4 - c}{3}$
(3) $\frac{c - 4}{3}$
(4) $c - 4$

22_____

23 If $-21a^6b$ is divided by $-3a^2b$, the quotient is

(1) $7a^4$ (3) $7a^3b$
(2) $-7a^3$ (4) $7a^4b$ 23_____

24 Which is the greatest integer that makes the inequality $3 - 2x > 9$ a true statement?

(1) −2 (3) 5
(2) 2 (4) −4 24_____

25 Let p represent "x is prime," and let q represent "x is even." Which statement is true if $x = 2$?

(1) $\sim p \land q$ (3) $p \land q$
(2) $\sim q \land p$ (4) $\sim(p \land q)$ 25_____

26 A line is represented by the equation $y = 3x - 7$. Which statement about the line is true?

(1) The slope of the line is $\frac{1}{3}$.

(2) The y-intercept is −7.

(3) Point (1,4) lies on the line.

(4) This line is parallel to the line whose equation is $y = 2x - 7$. 26_____

27 Which graph shows the solution set of $-2 \leq x < 4$?

(1) 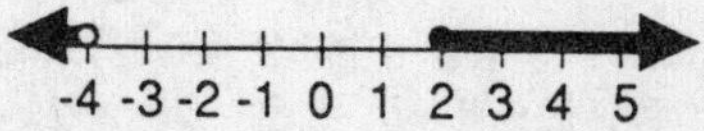 -4 -3 -2 -1 0 1 2 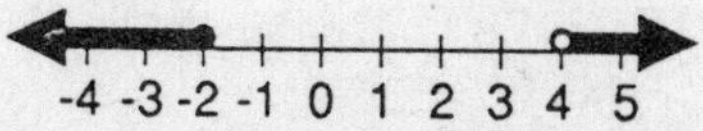 3  4  5

(2) -4 -3 -2 -1 0 1 2 3 4 5

(3) -4 -3 -2 -1 0 1 2 3 4 5

(4) -4 -3 -2 -1 0 1 2 3 4 5 27_____

28 The expression $\sqrt{500}$ is equivalent to

(1) $5\sqrt{10}$ (3) $500\sqrt{2}$
(2) $10\sqrt{5}$ (4) $5\sqrt{100}$ 28_____

29 If the length and width of a rectangle are 8 and 5, the length of a diagonal is

(1) 89 (3) $\sqrt{89}$
(2) $\sqrt{39}$ (4) $\sqrt{13}$ 29_____

30 Which inequality is illustrated in the accompanying graph?

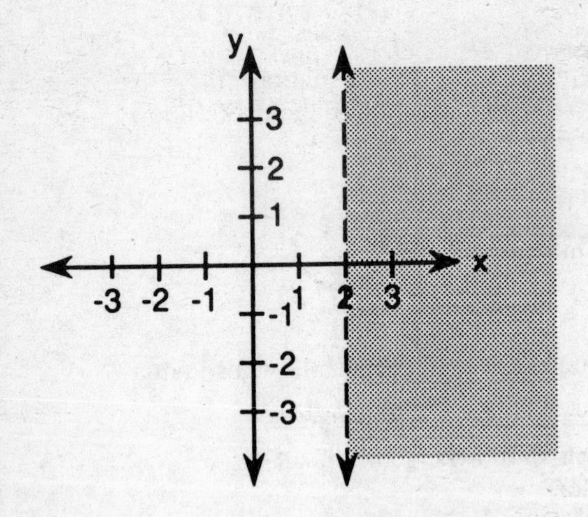

(1) $x > 2$ (3) $y > 2$
(2) $x < 2$ (4) $y < 2$ 30_____

31 The perimeter of a rectangle is $12x + 4$. If the width is $2x$, the length of the rectangle is

(1) $6x + 2$ (3) $4x + 2$
(2) $6x - 2$ (4) $4x - 2$ 31_____

32 The solution set for $2x^2 - 7x - 4 = 0$ is

(1) $\{2,-1\}$ (3) $\{-2,1\}$

(2) $\left\{-\frac{1}{2},4\right\}$ (4) $\left\{\frac{1}{2},-4\right\}$ 32_____

33 In parallelogram $ABCD$, $m\angle A = 2x + 50$ and $m\angle C = 3x + 40$. The measure of $\angle A$ is

(1) 18° (3) 70°
(2) 20° (4) 86° 33_____

34 In the accompanying diagram, the faces are congruent.

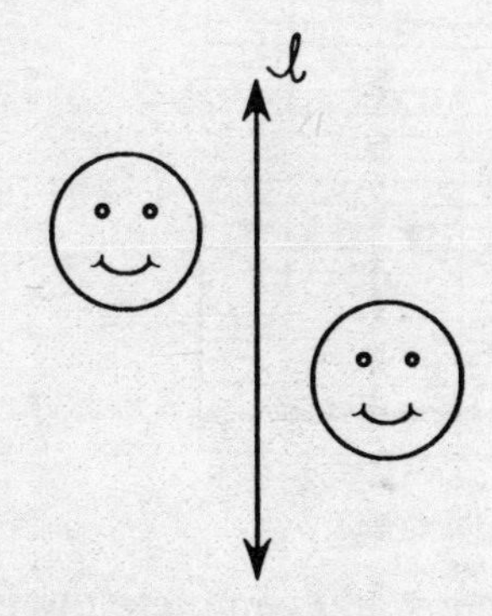

Which transformation is illustrated?
(1) a reflection in line ℓ
(2) a dilation
(3) a translation
(4) a rotation 34_____

35 If the probability that an event will occur is p, what is the probability that the event will *not* occur?

(1) $1 - p$ (3) p

(2) $p - 1$ (4) $\frac{1}{p}$ 35_____

PART TWO

DIRECTIONS: *Answer* three *questions from this part. Clearly indicate the necessary steps, including appropriate formula substitutions, diagrams, graphs, charts, etc. Calculations that may be obtained by mental arithmetic or the calculator do not need to be shown.* [30]

36 The cumulative frequency histogram below shows the number of mistakes 28 students in a French language class made on a test.

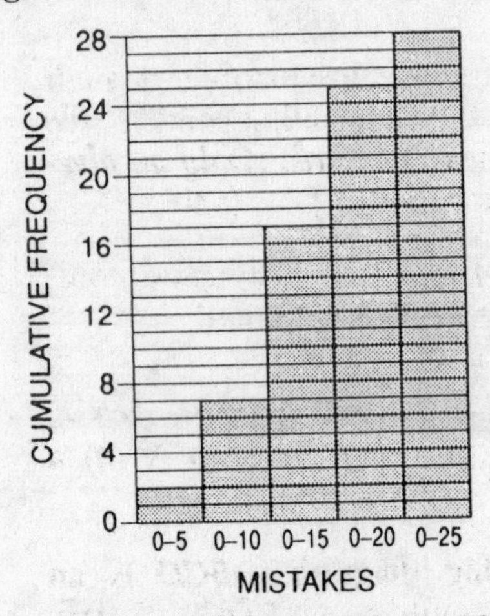

a *On your answer paper,* copy and complete the frequency table below using the data shown in the cumulative frequency histogram. [4]

Frequency Table

Number of Mistakes	Number of Students
0–5	
6–10	
11–15	
16–20	
21–25	

b If the number of mistakes John made is included in the interval that contains the median, what is the maximum number of mistakes that John could have made? [2]

c What percent of the French class made fewer than 11 mistakes? [2]

d What is the probability that a student selected at random made *at least* 16 mistakes? [2]

37 Find three positive consecutive integers such that the product of the first and second is two more than three times the third. [*Only an algebraic solution will be accepted.*] [5,5]

38 *a* *On your answer paper*, construct and complete a truth table for the statement $(p \leftrightarrow q) \rightarrow (\sim p \vee q)$. [9]

b From the truth table constructed in part *a*, is the statement $(p \leftrightarrow q) \rightarrow (\sim p \vee q)$ a tautology? [1]

39 In the accompanying diagram, *ABCD* is an isosceles trapezoid with bases $\overline{AB}$ and $\overline{DC}$, $DA = 13$, *CDEF* is a square, and circle *O* has a diameter of 12. The length of a side of the square is equal to the diameter of the circle.

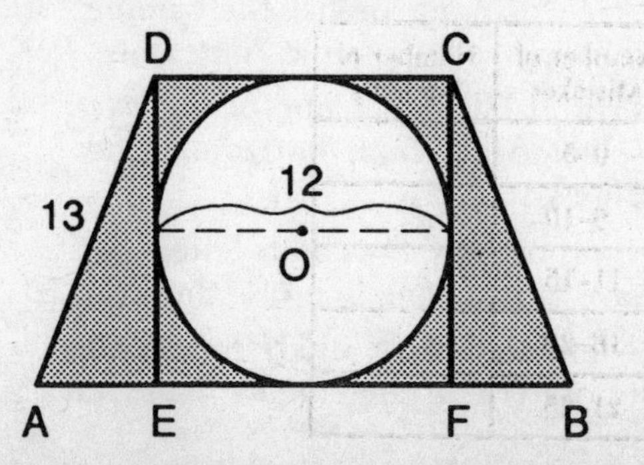

a Find the measure of

(1) $\overline{AE}$ [2] (2) $\overline{AB}$ [1]

b Find the area of trapezoid *ABCD*. [2]

c Find the area of the circle in terms of π. [2]

d Using $\pi = 3.14$, find, to the *nearest integer*, the area of the shaded region. [3]

40 The width of a rectangle is two more than a side of a square. The length of the rectangle is one less than twice the side of the square. If the area of the rectangle is 68 more than the area of the square, find the measure of a side of the square. [*Show or explain the procedure used to obtain your answer.*] [10]

41 *a* On the same set of coordinate axes, graph the following system of inequalities. Label the region that represents the solution set S.

$$y - 2x \geq 0$$
$$x + y < 6$$
[8]

b Write the coordinates of a point that does *not* satisfy either inequality graphed in part *a*. [2]

42 The letters **M**, **A**, **T**, and **H** are put in a jar.

a One letter is drawn at random from the jar, not replaced, and then a second letter is drawn.

(1) Draw a tree diagram or list the sample space showing all possible outcomes. [4]

(2) Find the probability that one of the two letters selected has both horizontal and vertical line symmetry. [2]

(3) Find the probability that both letters selected have at least one line of symmetry. [2]

b Using all the letters **M**, **A**, **T**, and **H**, how many different four-letter arrangements can be made? [2]

Answers June 1992

Three-Year Sequence for High School Mathematics—Course I

ANSWER KEY

PART ONE

1. $\frac{2}{7}$	**13.** 10	**25.** (3)
2. $p \rightarrow \sim q$	**14.** 60	**26.** (2)
3. 1	**15.** 0.7	**27.** (4)
4. 3	**16.** 40	**28.** (2)
5. 31	**17.** –4	**29.** (3)
6. 4	**18.** 14π	**30.** (1)
7. 75	**19.** 15	**31.** (3)
8. 40	**20.** (2)	**32.** (2)
9. 12	**21.** (4)	**33.** (3)
10. 50	**22.** (2)	**34.** (3)
11. 4	**23.** (1)	**35.** (1)
12. $s \rightarrow \sim t$	**24.** (4)	

PART TWO *See answers explained section.*

ANSWERS EXPLAINED

PART ONE

1. $\text{Probability of an event occurring} = \dfrac{\text{number of successful outcomes}}{\text{total possible number of outcomes}}$.

The word "REGENTS" contains two E's, so there are two possible successful outcomes for choosing an E. There is a total of seven letters in "REGENTS," so the total possible number of outcomes for the choice of one letter is seven. The probability of choosing an E is $\frac{2}{7}$.

The probability is $\frac{2}{7}$.

2. The symbol p represents "The triangle is isosceles."

The symbol q represents "The triangle is scalene."

"If the triangle is isosceles, then the triangle is *not* scalene" is the *conditional* that p implies the *negation* of q. It is represented symbolically by $p \rightarrow \sim q$, where $\sim q$ is the negation of q.

The symbolic representation is $\boldsymbol{p \rightarrow \sim q}$.

3. The given equation contains parentheses: $3(2x - 1) = x + 2$

Remove the parentheses by applying the distributive law; multiply each term within the parentheses by the factor 3: $6x - 3 = x + 2$

Add $-x$ (the additive inverse of x) and also add 3 (the additive inverse of -3) to both sides of the equation:

$$\begin{array}{rcr} -x + 3 & = & -x + 3 \\ \hline 5x & = & 5 \end{array}$$

Divide both sides of the equation by 5:

$$\frac{5x}{5} = \frac{5}{5}$$

$$x = 1$$

The solution is $x = 1$.

4. The number of points scored by the five players were:

$$28, 20, 16, 15, 8$$

The mean is the sum of all five scores divided by the number of scores:

$$\text{Mean} = \frac{28 + 20 + 16 + 15 + 8}{5}$$

$$\text{Mean} = \frac{87}{5}$$

$$\text{Mean} = 17\frac{2}{5}$$

The three players who scored 16, 15, and 8, respectively, scored fewer points than the mean.

3 players scored fewer points than the mean.

5. The given equation contains decimals:

$$0.5x - 12 = 3.5$$

Clear decimals by multiplying each term on both sides of the equation by 10:

$$10(0.5x) - 10(12) = 10(3.5)$$

$$5x - 120 = 35$$

Add 120 (the additive inverse of -120) to both sides of the equation:

$$\underline{\quad 120 = 120}$$

$$5x \quad = 155$$

Divide both sides of the equation by 5:

$$\frac{5x}{5} = \frac{155}{5}$$

$$x = 31$$

The solution is $x = $ **31**.

6. The given expression is:

$$\frac{2x - y}{3}$$

Substitute 5 for x and -2 for y:

$$\frac{2(5) - (-2)}{3}$$

Simplify each term in the numerator:

$$\frac{10 + 2}{3}$$

Combine like terms in the numerator:

$$\frac{12}{3}$$

Divide 12 by 3:

$$4$$

The value is **4**.

7. The *mode* of a distribution is the item that occurs most frequently. The temperature 75 occurs three times. The temperatures 77 and 78 each occur two times, and 76, 79, and 80 each occur once. Therefore, 75 is the mode.

The mode is **75**.

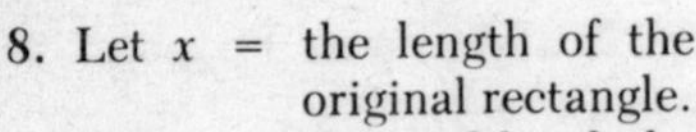

8. Let x = the length of the original rectangle.
Let y = the width of the original rectangle.
Then $2x$ = the length of the new rectangle.
And y = the width of the new rectangle.

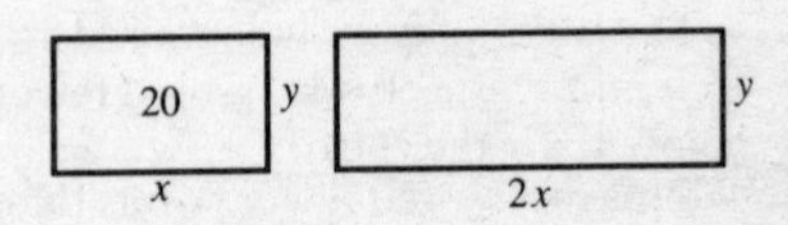

The area of a rectangle is equal to the product of its length and width; the area of the original rectangle is 20:

$$xy = 20$$

$$\text{Area of new rectangle} = (2x)y = 2xy$$

Substitute 20 for xy:

$$\text{Area of new rectangle} = 2(20)$$

$$\text{Area of new rectangle} = 40$$

The area is **40**.

9. The given equation contains a fraction:

$$\frac{3}{4}y - 8 = 1$$

Clear fractions by multiplying each term on both sides of the equation by 4:

$$4\left(\frac{3}{4}y\right) - 4(8) = 4(1)$$

$$3y - 32 = 4$$

Add 32 (the additive inverse of −32) to both sides of the equation:

$$\underline{\phantom{3y - {}} 32 = 32}$$

$$3y = 36$$

Divide both sides of the equation by 3:

$$\frac{3y}{3} = \frac{36}{3}$$

$$y = 12$$

The value of y is **12**.

10.

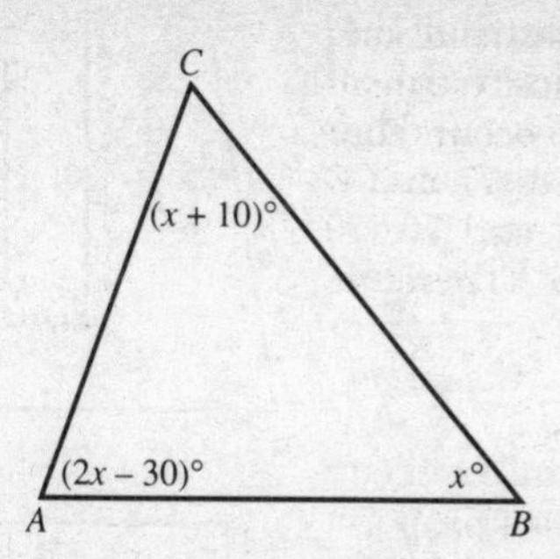

The sum of the measures of the three angles of a triangle is 180°:

$$m\angle A + m\angle B + m\angle C = 180$$
$$2x - 30 + x + x + 10 = 180$$

Combine like terms:

$$4x - 20 = 180$$

Add 20 (the additive inverse of –20) to both sides of the equation:

$$20 = 20$$
$$4x = 200$$

Divide both sides of the equation by 4:

$$\frac{4x}{4} = \frac{200}{4}$$
$$x = 50$$

The number of degrees in $\angle B$ is **50**.

11. It is given that:

$$x = 4y,\ y \neq 0$$

Divide both sides of the equation by y; this is possible since $y \neq 0$:

$$\frac{x}{y} = \frac{4y}{y}$$
$$\frac{x}{y} = 4$$

The value of $\frac{x}{y}$ is 4.

12. The given *conditional*, $\sim s \rightarrow t$, has $\sim s$ as its *hypothesis* or *antecedent*, and t as its *conclusion* or *consequent*.

The *inverse* of a conditional is another conditional formed by *negating* both the hypothesis and the conclusion of the original conditional:

The inverse of $\sim s \rightarrow t$ is $\sim(\sim s) \rightarrow \sim t$.

But $\sim(\sim s)$ is s: The inverse of $\sim s \rightarrow t$ is $s \rightarrow \sim t$.

The inverse is $s \rightarrow \sim t$.

13. Let x = the length of the longest side of the smaller triangle.

The smallest side, 4, of the smaller triangle will correspond to the smallest side, 6, of the larger triangle, and the longest side, x, of the smaller triangle will correspond to the longest side, 15, of the larger triangle.

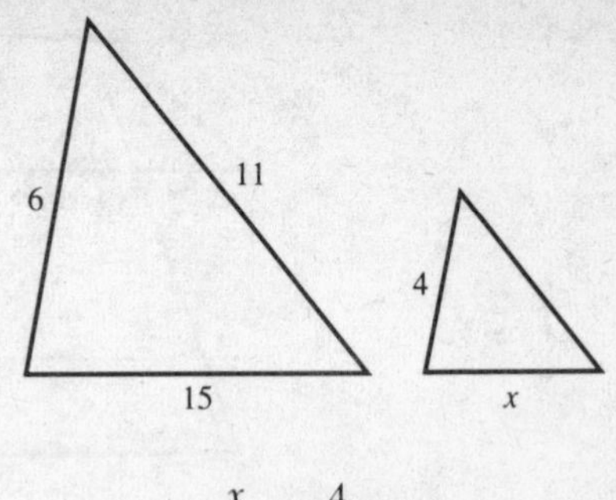

If two triangles are similar, their corresponding sides are in proportion:

$$\frac{x}{15} = \frac{4}{6}$$

In a proportion, the product of the means equals the product of the extremes (cross-multiply):

$$6x = 4(15)$$

$$6x = 60$$

Divide both sides of the equation by 6:

$$\frac{6x}{6} = \frac{60}{6}$$

$$x = 10$$

The length of the longest side is **10.**

14. Let $x = m\angle DHG$.
Then $2x = m\angle BGH$.

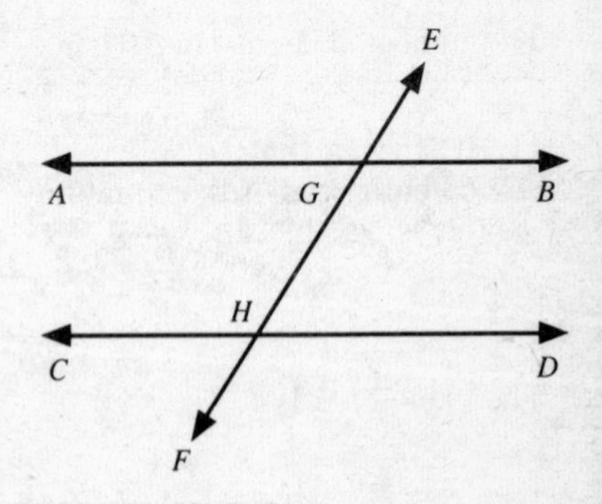

$\angle DHG$ and $\angle BGH$ are the interior angles on the same side of a transversal to two parallel lines. The interior angles on the same side of a transversal to two parallel lines are supplementary, that is, the sum of their measures is 180°:

$$m\angle DHG + m\angle BGH = 180$$

$$x + 2x = 180$$

Combine like terms:

$$3x = 180$$

Divide both sides of the equation by 3:

$$\frac{3x}{3} = \frac{180}{3}$$

$$x = 60$$

$m\angle DHG$ = **60.**

15.

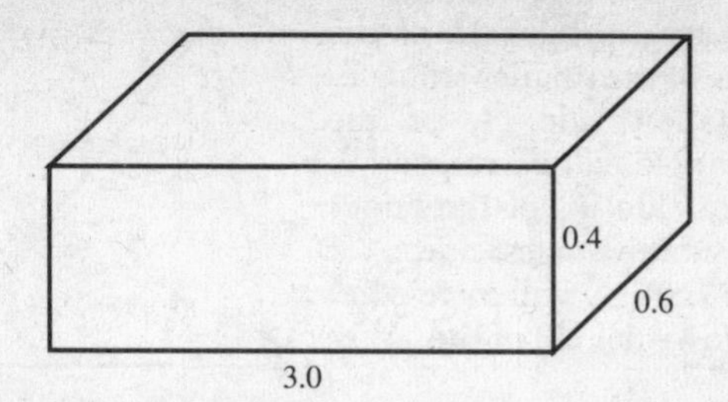

If ℓ = the length of a rectangular solid, w = its width, and h = its height, then its volume, V, is given by this formula: $V = \ell wh$

Here, $\ell = 3.0$, $w = 0.6$, and $h = 0.4$: $V = 3.0(0.6)(0.4)$

$(0.6)(0.4) = 0.24$: $V = 3.0(0.24)$

$V = 0.72$

Round off to the *nearest tenth*: $V = 0.7$

The volume is **0.7** cubic meters to the *nearest tenth.*

16.

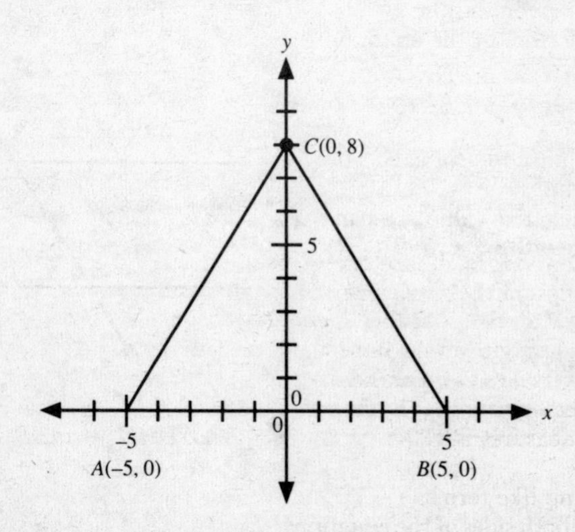

The area, A, of a triangle whose base is b and whose altitude is h is given by this formula: $A = \frac{1}{2}bh$

Here, $b = AB$, and $h = CO$: $A = \frac{1}{2}(AB)(CO)$

$AB = 10$; $CO = 8$: $A = \frac{1}{2}(10)(8)$

$A = 5(8)$

$A = 40$

The area of the triangle is **40**.

17. The given sentence is: $0.00045 = 4.5 \times 10^n$

The decimal point in 4.5 must be moved 4 places to the left to equal 0.00045. Each move of one place to the left is equivalent to dividing by 10, or multiplying by 10^{-1}. Therefore, to move the decimal point in 4.5 four places to the left requires multiplying by 10^{-4}: $0.00045 = 4.5 \times 10^{-4}$

$n = $ **−4.**

18. The area, A, of a circle whose radius is r is given by this formula: $A = \pi r^2$

The area is given as 49π: $49\pi = \pi r^2$

Divide both sides of the equation by π: $\frac{49\pi}{\pi} = \frac{\pi r^2}{\pi}$

$49 = r^2$

Take the square root of each side of the equation: $\pm 7 = r$

Reject the negative value as meaningless for a length: $7 = r$

The circumference, c, of a circle whose radius is r is given by this formula: $c = 2\pi r$

Substitute 7 for r: $c = 2\pi(7)$

$c = 14\pi$

The circumference is **14π**.

19. If k is a constant, x varies directly as y may be represented by: $x = ky$

It is given that $x = 3$ when $y = 4$: $3 = 4k$

Divide both sides of the equation by 4: $\frac{3}{4} = \frac{4k}{4}$

$\frac{3}{4} = \text{k}$

Since k is a constant, k always equals $\frac{3}{4}$:

$$x = \frac{3}{4}y$$

Let $y = 20$.

$$x = \frac{3}{4}(20)$$

$$x = \frac{60}{4}$$

$$x = 15$$

When $y = 20$, $x =$ **15.**

20. For any even number, the next larger even number is obtained by adding 2 to the first number. If $n + 7$ is an even number, the next larger even number is $n + 7 + 2$, or $n + 9$.

The correct choice is **(2)**.

21.

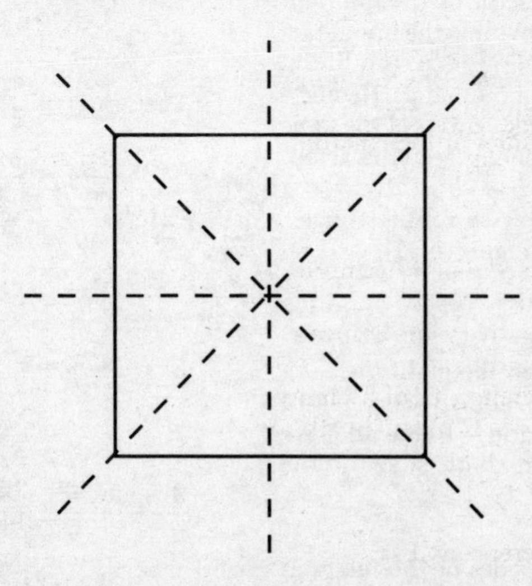

A line of symmetry is a line such that, if the square were folded on it, each point of the square on one side of the fold would correspond to a point of the square on the other side of the fold.

The square has 2 lines of symmetry that are parallel to two sides of the square and bisect the other two sides. The 2 diagonals of the square are also lines of symmetry. Thus, the square has 4 lines of symmetry.

The correct choice is **(4)**.

22. The given equation is a literal equation:

$$3x + c = 4$$

Add $-c$ (the additive inverse of c) to both sides of the equation:

$$\begin{array}{rcl} -c & = & -c \\ \hline 3x & = & 4 - c \end{array}$$

Divide both sides of the equation by 3:

$$\frac{3x}{3} = \frac{4 - c}{3}$$

The correct choice is (2).

23. The division may be indicated as:

$$\frac{-21a^6b}{-3a^2b}$$

The numerical coefficient of the quotient is obtained by dividing the numerical coefficient of the numerator by the numerical coefficient of the denominator:

$$(-21) \div (-3) = 7$$

The literal factor of the quotient is obtained by dividing the literal factor of the numerator by the literal factor of the denominator. Remember that, to divide powers of the same base, their exponents are subtracted. Also note that $b \div b = 1$:

$$(a^6b) \div (a^2b) = a^4$$

Combine the two results above:

$$(-21a^6b) \div (-3a^2b) = 7a^4$$

The correct choice is (1).

24. The given inequality is:

$$3 - 2x > 9$$

Add $2x$ (the additive inverse of $-2x$) and also add -9 (the additive inverse of 9) to both sides of the inequality:

$$\begin{array}{rcl} -9 + 2x & = & -9 + 2x \\ \hline -6 & > & 2x \end{array}$$

Divide both sides of the inequality by 2:

$$\frac{-6}{2} > \frac{2x}{2}$$

$$-3 > x$$

This inequality states that x must be less than -3. The greatest integer value of x is thus -4.

The correct choice is (4).

25. The symbol p represents "x is prime."
The symbol q represents "x is even."

If $x = 2$, p is true since 2 is a prime. Also, q is true since 2 is an even number.

Examine each choice in turn:

(1) $\sim p \land q$. This is the *conjunction* of the *negation* of p with q. For the conjunction to be true, both $\sim p$ and q must be true. Since p is true, the negation of p, that is, $\sim p$, is false. Thus the conjunction is false.

(2) $\sim q \land p$. This is the *conjunction* of the *negation* of q with p. Since q is true, $\sim q$ is false. Thus the conjunction is false since both $\sim q$ and p must be true to make the conjunction true.

(3) $p \land q$. This is the *conjunction* of p and q. The conjunction is true when p and q are both true. Since p and q are both true for $x = 2$, this conjunction is true.

(4) $\sim (p \land q)$. This is the *negation* of the *conjunction* of p and q. The discussion for choice (3) above shows that $(p \land q)$ is true for $x = 2$. Therefore, the negation of $(p \land q)$ is false.

The correct choice is **(3)**.

26. The equation given for the line is:

$$y = 3x - 7$$

This equation is in the form $y = mx + b$, where m represents the slope of the line and b represents its y-intercept. Here, $m = 3$, so the slope is 3, and $b = -7$, so the y-intercept is -7.

Examine each choice in turn:

(1) The slope of the line is 3, not $\frac{1}{3}$, so this choice is false.

(2) The y-intercept is -7. This is the correct choice.

(3) To determine whether point (1,4) is on the line, substitute 1 for x and 4 for y to see whether these coordinates satisfy the equation:

$$4 \stackrel{?}{=} 3(1) - 7$$
$$4 \stackrel{?}{=} 3 - 7$$
$$4 \neq -4$$

The point (1,4) does not lie on the line.

(4) The equation $y = 2x - 7$ is in the $y = mx + b$ form with $m = 2$. Therefore the slope of this line is 2. The slope of the original line is 3. The two lines are not parallel since they have different slopes.

The correct choice is **(2)**.

27. The given inequality is: $-2 \le x < 4$

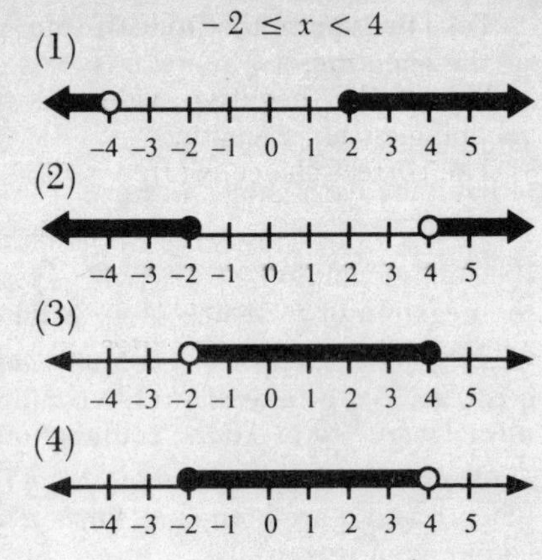

This inequality states that x must be greater than or equal to -2, but less than 4.

Examine each choice in turn:

(1) This graph shows values for x that are less than -4 or greater than or equal to 2. Note that the open circle at -4 indicates that -4 is not included in the set represented by the inequality, while the shaded circle at 2 indicates that 2 is included as a member of the set.

(2) This graph shows values for x that are less than or equal to -2 or greater than 4.

(3) This graph shows values for x that are greater than -2 but less than or equal to 4.

(4) This graph shows values for x that are greater than or equal to -2 but less than 4.

The correct choice is **(4)**.

28. The given expression is: $\sqrt{500}$

To simplify the radicand, 500, factor out the highest possible perfect square factor in it: $\sqrt{100\ (5)}$

Remove the perfect square factor from under the radical sign by taking its square root and writing it as a coefficient of the radical: $10\sqrt{5}$

The correct choice is **(2)**.

29. Let x = the length of the diagonal.

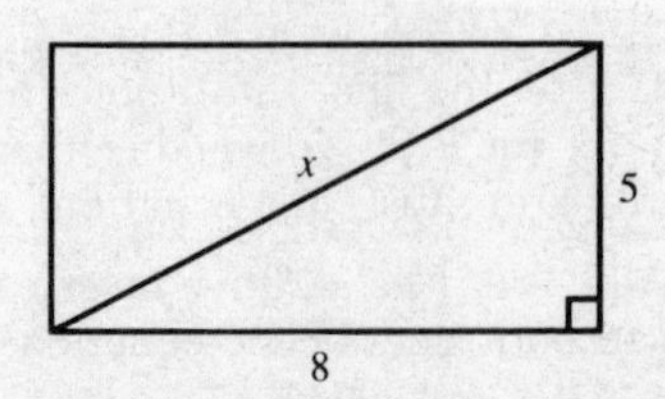

Since all angles of a rectangle are right angles, the diagonal of a rectangle divides it into two right triangles. By the Pythagorean Theorem, in a right triangle the square of the length of the hypotenuse equals the sum of the squares of the lengths of the legs. Here:

$$x^2 = 8^2 + 5^2$$
$$x^2 = 64 + 25$$
$$x^2 = 89$$

Take the square root of each side of the equation: $x = \pm\sqrt{89}$

Reject the negative value as meaningless for a length: $x = \sqrt{89}$

The correct choice is (3).

30. The shaded region shows all points where x-coordinates are greater than 2. Points where $x = 2$ are not included in the shaded region since the vertical line at $x = 2$ is a broken line. Therefore, for all points in the shaded region, $x > 2$.

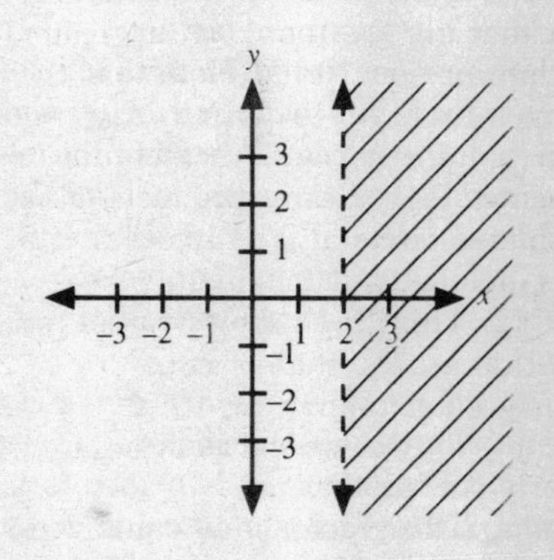

The correct choice is (1).

31. Let y = the length of the rectangle.

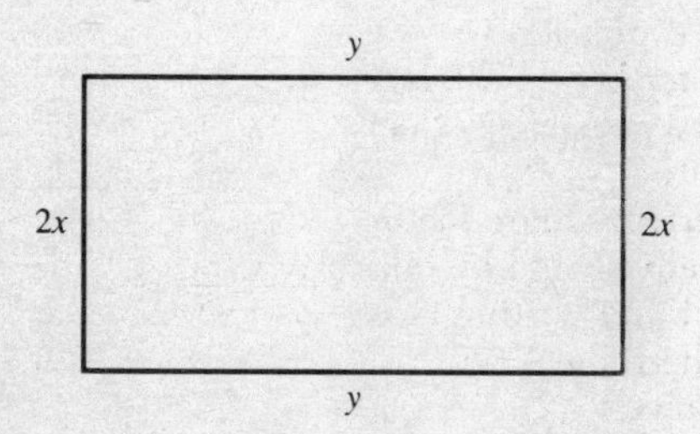

The sum of the lengths of all four sides of the rectangle is the perimeter of the rectangle: $y + 2x + y + 2x = 12x + 4$

Combine like terms: $2y + 4x = 12x + 4$

Add $-4x$ (the additive inverse of $4x$) to both sides of the equation:

$$\begin{aligned} -4x &= -4x \\ \hline 2y \quad &= 8x + 4 \end{aligned}$$

Divide each term on both sides of the equation by 2:

$$\frac{2y}{2} = \frac{8x}{2} + \frac{4}{2}$$

$$y = 4x + 2$$

The correct choice is (3).

32. The given equation is a *quadratic equation*:

$$2x^2 - 7x - 4 = 0$$

The left side is a *quadratic trinomial* that can be factored into the product of two binomials. The factors of the first term, $2x^2$, become the first terms of the binomials:

$$(2x\quad)(x\quad) = 0$$

The factors of the last term, -4, become the second terms of the binomials, but they must be chosen in such a way that the sum of the inner and outer cross-products of the binomials equals the middle term, $-7x$, of the original trinomial.

Try $+1$ and -4 as the factors of -4:

$$+x = \text{inner product}$$
$$(2x + 1)(x - 4) = 0$$
$$-8x = \text{outer product}$$

Since $(+x) + (-8x) = -7x$, these are the correct factors:

$$(2x + 1)(x - 4) = 0$$

If the product of two factors is zero, either factor may equal zero:

$$2x + 1 = 0 \quad \vee \quad x - 4 = 0$$

Add the appropriate additive inverse to both sides of the equations, -1 for the left equation and 4 for the right equation:

$$\begin{array}{rl} -1 = -1 & \qquad 4 = 4 \\ \hline 2x \quad = -1 & \qquad x \quad = 4 \end{array}$$

Divide both sides of the left equation by 2:

$$\frac{2x}{2} = \frac{-1}{2}$$

$$x = -\frac{1}{2}$$

The solution set is $\left\{-\frac{1}{2}, 4\right\}$.

The correct choice is (2).

33.

B, C, A, D; $3x + 40$; $2x + 50$

In a parallelogram, the opposite

angles are congruent:

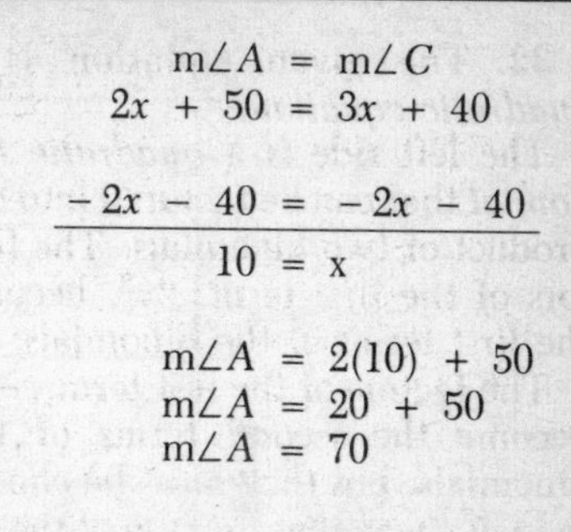

$$m\angle A = m\angle C$$
$$2x + 50 = 3x + 40$$

Add $-2x$ (the additive inverse of $2x$) and also add -40 (the additive inverse of 40) to both sides of the equation:

$$\frac{-2x - 40 = -2x - 40}{10 = x}$$

Substitute 10 for x in the expression for $m\angle A$:

$$m\angle A = 2(10) + 50$$
$$m\angle A = 20 + 50$$
$$m\angle A = 70$$

The correct choice is (3).

34. The diagram represents a translation, that is, a sliding of the figure from position A to position B, or vice versa.

A reflection of figure A in line ℓ would map it onto an image to the right of line ℓ and on the same horizontal level as A.

A dilation of figure A would change its size (either expanding or contracting it) while leaving its position unchanged.

A rotation would revolve figure A about some fixed point.

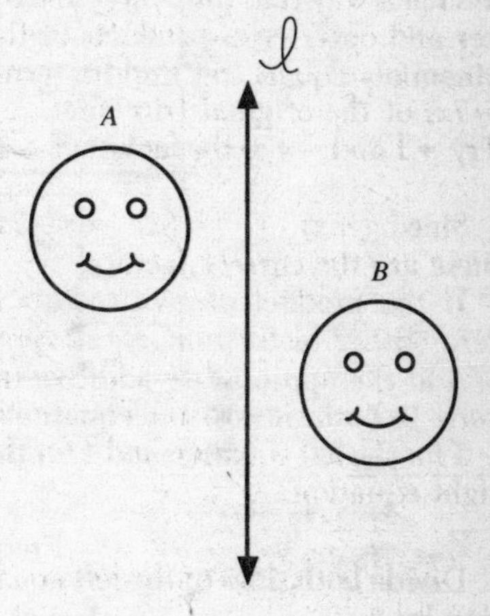

The correct choice is (3).

35. The sum of the probability that an event will occur and that it will not occur is always 1, which represents certainty. Therefore, if the probability that an event will occur is p, the probability that it will not occur is $1 - p$.

The correct choice is (1).

PART TWO

36. **a.** To obtain the number of students making the number of mistakes covered in each interval, subtract the cumulative frequency of the preceding interval from the cumulative frequency that covers the number of errors up to and including the interval in question:

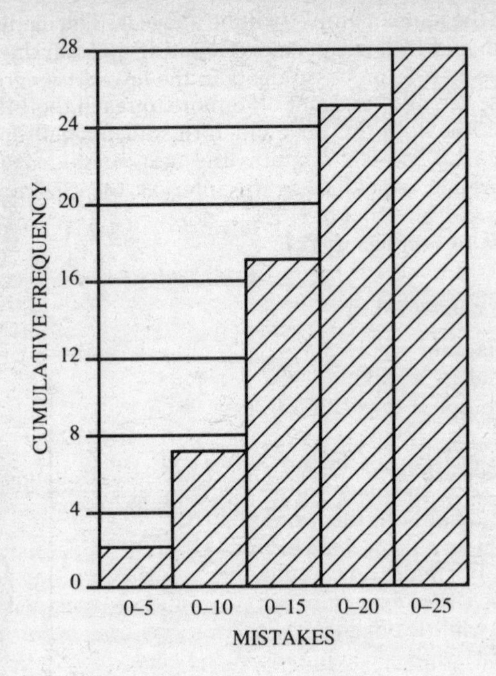

Frequency Table

Number of Mistakes	Number of Students
0–5	2
6–10	7 − 2 = 5
11–15	17 − 7 = 10
16–20	25 − 17 = 8
21–25	28 − 25 = 3
Total	28

b. We are told that the total number of students is 28. Total the column for the number of students to check that the sum is 28.

The *median* is the middle number of mistakes when the frequencies of the mistakes are arranged in order of size; half the frequencies are less than the median and half are greater. Since one-half of 28 is 14, 14 frequencies will be below the median and 14 will be above it. The median will lie between the 14th and 15th frequencies. Counting up from the bottom of the table, we have 3 + 8, or 11, students in the lowest two groups with the most mistakes. We need to count off 3 more to reach the 14th and 4 more to reach the 15th. Both the 14th and 15th students will lie in the 11–15 interval, since this interval contains the next 10 students. Since John's number of mistakes is included in this interval, the maximum number of mistakes he could have made is 15.

The maximum number is **15**.

c. The 0–5 interval and the 6–10 interval represent the students with fewer than 11 mistakes. There are 2 + 5, or 7, students in these two intervals. Since the total number of students in the class is 28, this group represents $\frac{7}{28}$, or $\frac{1}{4}$, or 25%, of the class.

The percent is **25**.

d. $$\text{Probability of an event occurring} = \frac{\text{number of successful outcomes}}{\text{total possible number of outcomes}}.$$

The students making *at least* 16 mistakes are included in the 16–20 and 21–25 intervals. There are 8 + 3, or 11, students in these two intervals. The number of successful outcomes for the random choice of a student making *at least* 16 mistakes is thus 11. The total possible number of outcomes is 28, the total number of students in the class.

The probability of choosing a student who made *at least* 16 mistakes is $\frac{11}{28}$.

The probability is $\mathbf{\frac{11}{28}}$.

37. Let x = the first positive integer.

Then $x + 1$ = the second consecutive positive integer.

And $x + 2$ = the third consecutive positive integer.

The first integer times the second is 3 times the third plus 2 more.

$$x \cdot (x + 1) = 3 \cdot (x + 2) + 2$$

The equation to use is: $x(x + 1) = 3(x + 2) + 2$

Remove the parentheses by applying the distributive law, multiplying each term within the parentheses by the factor in front: $x^2 + x = 3x + 6 + 2$

Combine like terms: $x^2 + x = 3x + 8$

This is a *quadratic equation*. Rearrange it so that all terms are on one side equal to zero by adding $-3x$ (the additive inverse of $3x$) and -8 (the additive inverse of 8) to both sides:

$$\begin{aligned} x^2 + x &= 3x + 8 \\ -3x - 8 &= -3x - 8 \\ \hline x^2 - 2x - 8 &= 0 \end{aligned}$$

The left side is a *quadratic trinomial* that can be factored into the product of two binomials. The factors of the first term, x^2, become the first terms of the binomials:

$$(x \qquad)(x \qquad) = 0$$

The factors of the last term, -8, become the second terms of the binomials, but they must be chosen in such a way that the sum of the inner and outer cross-products of the binomials equals the middle term, $-2x$, of the trinomial. Try -4 and $+2$ as the factors of -8:

$$\begin{aligned} -4x &= \text{inner product} \\ (x - 4)(x + 2) &= 0 \\ +2x &= \text{outer product} \end{aligned}$$

Since $(-4x) + (+2x) = -2x$, these are the correct factors:

$$(x - 4)(x + 2) = 0$$

If the product of two factors is zero, either factor may equal zero:

$$x - 4 = 0 \lor x + 2 = 0$$

Add the appropriate additive inverse to both sides of each equation, 4 for the left equation and -2 for the right equation:

$$\begin{aligned} 4 &= 4 \qquad & -2 &= -2 \\ \hline x &= 4 \qquad & x &= -2 \end{aligned}$$

Reject the negative value since the question calls for positive integers:

$$\begin{aligned} x &= 4 \\ x + 1 &= 5 \\ x + 2 &= 6 \end{aligned}$$

The numbers are **4, 5, 6.**

38. a. Begin with the columns for p and q. Fill them with truth values, T or F, in such a way that all possible combinations of values for p and q are included.

Prepare a column for $(p \leftrightarrow q)$, the *equivalence relation*, or *biconditional*, between p and q. The biconditional has the truth value T when the truth values of p and q are the same, that is, both are T or both are F. The biconditional has the truth value F when p and q have different truth values.

Prepare a column for $\sim p$, the *negation* of p. Each truth value in this column will be the opposite of the truth value for p on the same line.

Prepare a column for $(\sim p \lor q)$, the *disjunction* of the negation of p with q. The truth value of the disjunction is T when either $\sim p$ or q, or both, are T, but the value is F when $\sim p$ and q are both F.

The final column should be headed $(p \leftrightarrow q) \rightarrow (\sim p \lor q)$, the *conditional* that the biconditional, $(p \leftrightarrow q)$, implies the disjunction, $(\sim p \lor q)$.

The conditional has the truth value T whenever $(\sim p \vee q)$ has the value T, and also when both $(p \leftrightarrow q)$ and $(\sim p \vee q)$ have the value F. The conditional has the value F when $(p \leftrightarrow q)$ is T and $(\sim p \vee q)$ is F.

p	q	$(p \leftrightarrow q)$	$\sim p$	$(\sim p \vee q)$	$(p \leftrightarrow q) \rightarrow (\sim p \vee q)$
T	T	T	F	T	T
T	F	F	F	F	T
F	T	F	T	T	T
F	F	T	T	T	T

b. A *tautology* is a statement formed by combining other propositions $(p, q, r, \ldots)$ such that the statement is true regardless of the truth or falsity of $p, q, r, \ldots$. The truth values in the last column of the table in part a show that $(p \leftrightarrow q) \rightarrow (\sim p \vee q)$ has the truth value T for every possible combination of truth values of p and q.
Therefore, $(p \leftrightarrow q) \rightarrow (\sim p \vee q)$ is a tautology.

The answer is **yes**.

39. a. (1) Since $CDEF$ is a square, $CD = DE = EF = CF = 12$.

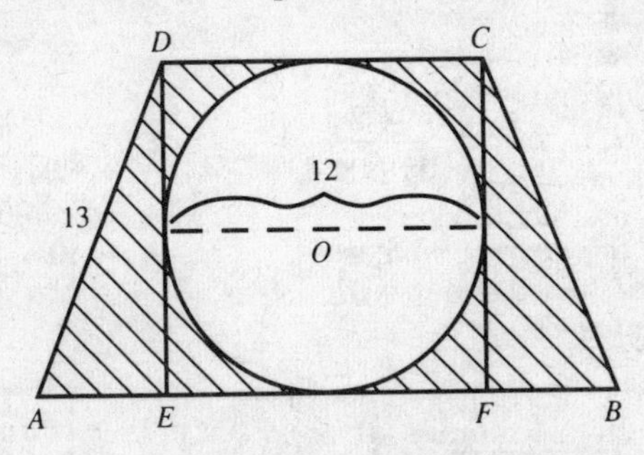

Since the angles of a square are right angles, $\overline{DE} \perp \overline{AB}$, making $\triangle AED$ a right triangle. Triangle AED is a 5-12-13 right triangle; $AD = 13$, $DE = 12$, so $AE = 5$

The measure of $\overline{AE}$ is **5**.

(2) Triangle $CFB \cong \triangle AED$, so $FB = AE = 5$.
$AB = AE + EF + FB = 5 + 12 + 5 = 22$.

The measure of $\overline{AB}$ is **22**.

b. The area, A, of a trapezoid whose altitude is h and the lengths of whose bases are b_1 and b_2 is given by this formula:

$$A = \frac{1}{2}h(b_1 + b_2)$$

For trapezoid $ABCD$, $h = DE = 12$, $b_1 = DC = 12$, and $b_2 = AB = 22$:

$$A = \frac{1}{2}(12)(12 + 22)$$
$$A = 6(34)$$
$$A = 204$$

The area of trapezoid $ABCD$ is **204**.

c. The area, A, of a circle whose radius is r is given by this formula:

$$A = \pi r^2$$

The diameter of circle O is 12. The radius is half the diameter, so $r = \frac{1}{2}(12) = 6$:

$$A = \pi(6)^2$$
$$A = 36\pi$$

The area of circle O is **36π**.

d. The area, A of the shaded region in the diagram is equal to the area of the trapezoid minus the area of the circle:

$$A = 204 - 36\pi$$

Let π equal 3.14:

$$A = 204 - 36(3.14)$$

$$\begin{array}{r} 3.14 \\ \times\ 36 \\ \hline 18\ 84 \\ 94\ 2 \\ \hline 113.04 \end{array}$$

$$A = 204 - 113.04$$
$$A = 90.96$$

Round off to the *nearest integer*:

$$A = 91$$

The area of the shaded region is **91** to the *nearest integer*.

40.

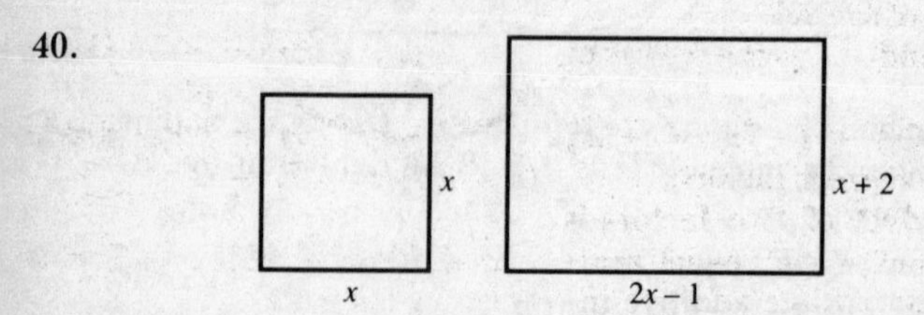

Let x = the length of a side of the square.
Then $x + 2$ = the width of the rectangle.
And $2x - 1$ = the length of the rectangle.
The area of the rectangle = $(2x - 1)(x + 2)$.
The area of the square = x^2.

The area of the rectangle equals the area of the square plus 68.

$$(2x - 1)(x + 2) = x^2 + 68$$

The equation to use is: $(2x - 1)(x + 2) = x^2 + 68$

Multiply:

$$\begin{array}{r} 2x - 1 \\ x + 2 \\ \hline 2x^2 - x \\ 4x - 2 \\ \hline 2x^2 + 3x - 2 \end{array}$$

$$2x^2 + 3x - 2 = x^2 + 68$$

This is a *quadratic equation*. Rearrange it so that all terms are on one side equal to zero by adding $-x^2$ (the additive inverse of x^2) and also -68 (the additive inverse of 68) to both sides.

$$\begin{array}{r} -x^2 \quad - 68 = -x^2 - 68 \\ \hline x^2 + 3x - 70 = 0 \end{array}$$

The left side is a *quadratic trinomial* that can be factored into the product of two binomials. The factors of the first term, x^2, are x and x, and they become the first terms of the binomials: $(x \quad)(x \quad) = 0$

The factors of the last term, -70 become the second terms of the binomials, but they must be chosen in such a way that the sum of the inner and outer cross-products of the binomials equals the middle term, $+3x$, of the trinomial.

Try $+10$ and -7 as the factors of -70:

$+10x$ = inner product

$(x + 10)(x - 7) = 0$

$-7x$ = outer product

Since $(+10x) + (-7x) = +3x$, these are the correct factors: $(x + 10)(x - 7) = 0$

If the product of two factors is zero, either factor may equal zero: $x + 10 = 0 \lor x - 7 = 0$

Add the appropriate additive inverse to both sides of each equation, -10 for the left equation and 7 for the right equation:

$$\begin{array}{rr} -10 = -10 & 7 = 7 \\ \hline x \quad = -10 & x \quad = 7 \end{array}$$

Reject the negative value as meaningless for a length: $x = 7$

The measure of a side of the square is 7.

41. a. To graph the inequality $y - 2x \geq 0$, first graph the equation $y - 2x = 0$. It is convenient to rewrite the equation as $y = 2x$, which is obtained by adding $2x$ to both sides. Prepare a table of values for x and y by choosing any three convenient values for x and substituting them in the equation to find the corresponding values for y:

x	$2x$	$= y$
-3	$2(-3)$	$= -6$
0	$2(0)$	$= 0$
3	$2(3)$	$= 6$

Plot the points $(-3, -6)$, $(0,0)$, and $(3,6)$, and draw a *solid line* through them. The *solid line* indicates that the points on it are part of the graph of the inequality $y - 2x \geq 0$.

The solution set of the inequality $y - 2x > 0$ lies on one side of the line $y - 2x = 0$. To find out on which side, choose a convenient point, say $(0,2)$ and substitute its coordinates in the inequality to see if they satisfy it:

$$2 - 2(0) \stackrel{?}{>} 0$$

$$2 - 0 \stackrel{?}{>} 0$$

$$2 > 0 \checkmark$$

Since $(0,2)$ satisfies the inequality, it lies in the solution set of $y - 2x > 0$. Shade this side of the line $y - 2x = 0$ with cross-hatching extending up and to the left.

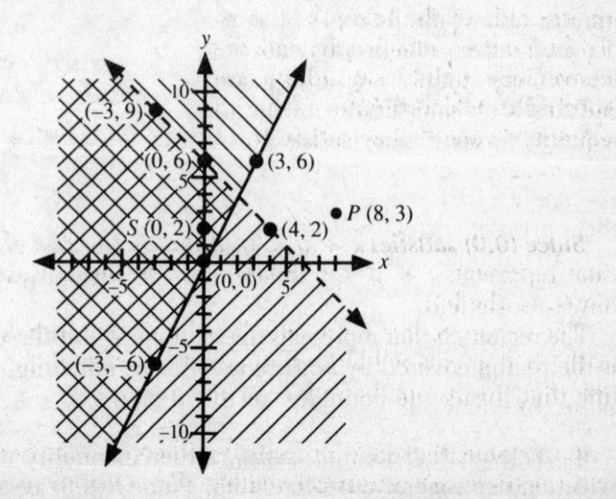

To graph the second inequality, $x + y < 6$, first graph the equation $x + y = 6$. It is convenient to rearrange the equation so that it is solved for y by adding $-x$ to each side:

$$\begin{array}{r} x + y = 6 \\ -x \quad = -x \\ \hline y = 6 - x \end{array}$$

Prepare a table of values for x and y by choosing any three convenient values for x and substituting them in the equation to find the corresponding values of y:

x	$6 - x$	$= y$
-3	$6 - (-3) = 6 + 3$	$= 9$
0	$6 - 0$	$= 6$
4	$6 - 4$	$= 2$

Plot the points $(-3,9)$, $(0,6)$, and $(4,2)$, and draw a *broken line* through them. The *broken line* indicates that the points on it are *not* part of the solution set of $x + y < 6$.

The solution set for $x + y < 6$ lies on one side of the line $x + y = 6$. To find out on which side, choose a convenient point, say $(0,0)$, and substitute its coordinates in the inequality to see if they satisfy it:

$$0 + 0 \overset{?}{<} 6$$

$$0 < 6 \checkmark$$

Since $(0,0)$ satisfies $x + y < 6$, it lies on the side of the line $x + y = 6$ that represents $x + y < 6$. Shade this side with cross-hatching extending down to the left.

The region, S, that represents the solution set for the system of inequalities is the region covered by *both* types of cross-hatching, including the solid line that forms one boundary of this region.

b. A point that does *not* satisfy either inequality must lie in a region with neither type of cross-hatching. Point $P(8,3)$ is such a point.

$P(8,3)$ does *not* satisfy either inequality.

42. a. (1) Tree Diagram

START → M → A, T, H
START → A → M, T, H
START → T → M, A, H
START → H → M, A, T

Sample Space

First Draw	Second Draw
M	A
M	T
M	H
A	M
A	T
A	H
T	M
T	A
T	H
H	M
H	A
H	T

Since M, A, T, or H may be selected on the first draw, the tree diagram requires four branches extending from START, one to each of these four letters. After the first draw, three branches are required for each of the second draws since any one of the remaining three letters may be selected.

The sample space requires one line for each possible combination of a first draw followed by a second draw. Since there are four possible letters to choose on the first draw, each followed by three possible second draws, the sample space needs 4×3, or 12, lines.

(2)

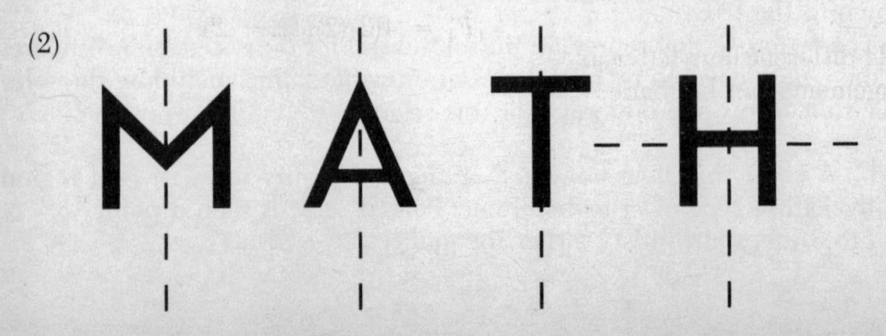

The diagram shows the lines of symmetry as broken lines for each of the four letters. M, A, and T each have vertical lines of symmetry. H has both horizontal and vertical lines of symmetry.

$$\text{Probability of an event occurring} = \frac{\text{number of successful outcomes}}{\text{total possible number of outcomes}}.$$

The probability of selecting a letter having both horizontal and vertical line symmetry is the same as the probability of selecting an H.

There are 6 branch paths on the tree diagram that include an H as one of the letters: M-H, A-H, T-H, H-M, H-A, and H-T. Similarly, the sample space contains 6 lines with these same combinations. Thus the number of successful outcomes is 6. The total number of possible outcomes is the total number, 12, of complete branch paths on the tree diagram or the total number, 12, of lines in the sample space.

The probability of selecting one letter with both horizontal and vertical line symmetry is $\frac{6}{12}$.

The probability is $\frac{6}{12}$.

(3) All of the letters, M, A, T, and H, have at least one line of symmetry. Therefore the probability of selecting both letters with at least one line of symmetry is certainty, which is expressed as a probability of 1.

The probability definition itself may also be used to obtain the result. Since there are 12 branch paths or 12 sample space lines, all of which have two letters with at least one line of symmetry, there are 12 successful outcomes. The total possible number of outcomes is also 12, so the probability is $\frac{12}{12}$, or 1.

The probability is **1**.

b. The number of different arrangements, or permutations, of n letters taken n at a time is given by this formula:

$${}_nP_n = n(n - 1)(n - 2) \ldots (3)(2)(1)$$

For letters M, A, T, and H, $n = 4$:

$${}_4P_4 = 4(3)(2)(1) = 24$$

24 different four-letter arrangements can be made.

Topic	Question Numbers	Number of Points	Your Points	Your Percentage
1. Numbers (rat'l, irrat'l); Percent	36c	2		
2. Properties of No. Systems	11	2		
3. Operations on Rat'l Nos. and Monomials	23	2		
4. Operations on Multinomials	—	0		
5. Square Root; Operations involving Radicals	28	2		
6. Evaluating Formulas and Expressions	6	2		
7. Linear Equations (simple cases incl. parentheses)	3	2		
8. Linear Equations Containing Decimals or Fractions	5, 9	2 + 2 = 4		
9. Graphs of Linear Functions (slope)	26, 30	2 + 2 = 4		
10. Inequalities	24, 27	2 + 2 = 4		
11. Systems of Eqs. & Inequalities (alg. & graphic solutions)	41a, b	8 + 2 = 10		
12. Factoring	—	0		
13. Quadratic Equations	32	2		
14. Verbal Problems	37, 40	10 + 10 = 20		
15. Variation	19	2		
16. Literal Eqs.; Expressing Relations Algebraically	20, 22	2 + 2 = 4		
17. Factorial *n*	—	0		
18. Areas, Perims., Circums., Vols. of Common Figures	8, 15, 16, 18, 31, 39b, c, d	2 + 2 + 2 + 2 + 2 + 2 + 2 + 3 = 17		
19. Geometry (congruence, parallel lines, compls., suppls.)	14, 39a	2 + 3 = 5		
20. Ratio & Proportion (incl. similar triangles & polygons)	13	2		
21. Pythagorean Theorem	29	2		
22. Logic (symbolic rep., logical forms, truth tables)	2, 12, 25, 38a, b	2 + 2 + 2 + 9 + 1 = 16		
23. Probability (incl. tree diagrams & sample spaces)	1, 35, 36d, 42a	2 + 2 + 2 + 8 = 14		
24. Combinations, Arrangements, Permutations	42b	2		
25. Statistics (central tend., freq. dist., histogr., quartiles, percentiles)	4, 7, 36a, b	2 + 2 + 4 + 2 = 10		

Topic	Question Numbers	Number of Points	Your Points	Your Percentage
26. Properties of Triangles and Quadrilaterals	10, 33	2 + 2 = 4		
27. Transformations (reflect., translations, rotations, dilations)	34	2		
28. Symmetry	21	2		
29. Area from Coordinate Geometry	—	0		
30. Dimensional Analysis	—	0		
31. Scientific Notation; Negative & Zero Exponents	17	2		

Examination January 1993

Three-Year Sequence for High School Mathematics—Course I

PART ONE

DIRECTIONS: *Answer 30 questions from this part. Each correct answer will receive 2 credits. No partial credit will be allowed. Write your answers in the spaces provided. Where applicable, answers may be left in terms of π or in radical form.* [60]

1 A bag contains three red marbles, two green marbles, and four orange marbles. If one marble is selected at random, what is the probability that it is *not* orange? 1_____

2 In the accompanying diagram, $\triangle ABC \sim \triangle DEF$, $AB = 12$, $AC = 8$, $DE = x$, and $DF = 2$. Find the value of x.

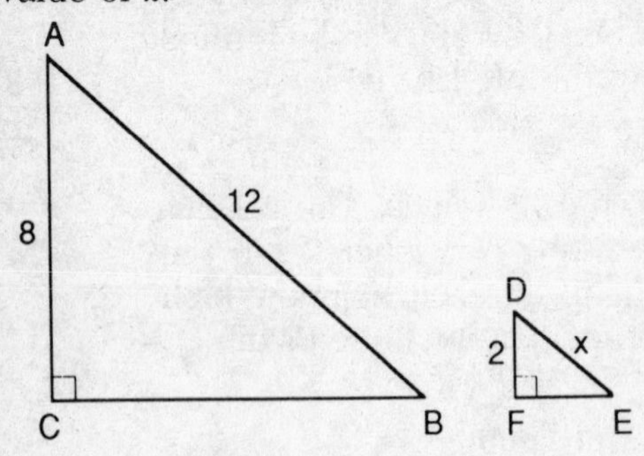

2_____

3 Let p represent "I go to the movies" and let q represent "I study my math." Using p and q, write in symbolic form: "If I do not study my math, then I do not go to the movies." 3_____

 ISBN 0-8120-1665-3

4 Express the sum of $3x^2 - 2x + 5$ and $x^2 + 2x - 8$ in simplest form. 4_____

5 Solve for x: $7x - 2 = 5x + 3$ 5_____

6 Express $4x^2 - 25$ as the product of two binomials. 6_____

7 In the accompanying diagram, $\overrightarrow{DH} \parallel \overline{EF}$. If $m\angle HDF = 30$ and $m\angle EDF = 55$, find $m\angle E$. 7_____

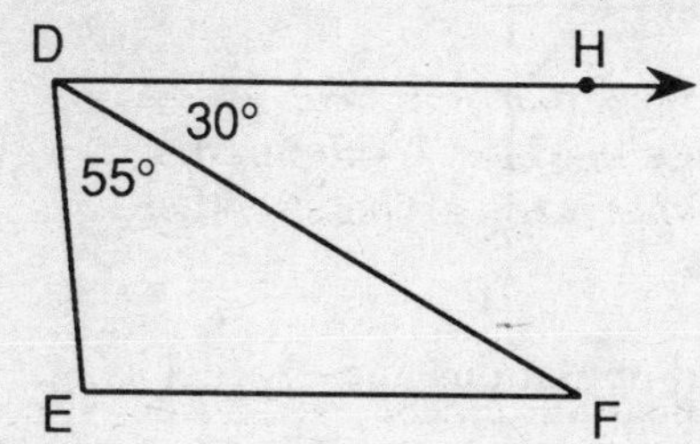

8 Without repetition of digits, how many different three-digit numbers less than 800 can be formed using the numbers in the set $\{2,3,9\}$? 8_____

9 A census taker visited 100 homes. The accompanying table shows the frequencies for the number of people living in each home. Which interval contains the median for these data?

Number of People
Living in One Home

Interval	Frequency
1–2	40
3–4	39
5–6	16
7–8	5

9_____

10 Find, in terms of x, the mean of $3x - 5$, $5x - 6$, and $4x + 11$. 10_____

11 What is the product of $\frac{1}{2}x^3y$ and $\frac{1}{4}xy^2$? 11_____

12 Solve for m: $\frac{m + 3}{2m} = \frac{4}{5}$, $m \neq 0$ 12_____

13 In the accompanying diagram, the length of the diameter of circle O equals the length of a side of the square. If the circumference of the circle is 6π, what is the perimeter of the square?

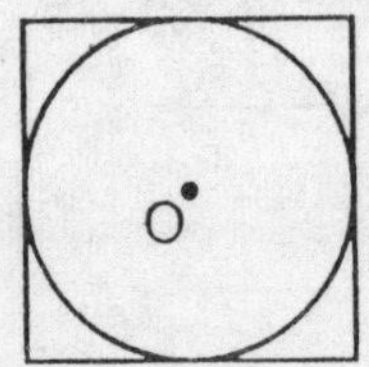

13_____

14 Perform the indicated operations and express as a binomial:

$$-3(x + 2) - x$$

14_____

15 Express the product of $2a + 1$ and $a + 5$ as a trinomial. 15_____

16 In the accompanying figure, $\triangle ABC$ has coordinates $A(0,3)$, $B(7,3)$, and $C(7,7)$. Find the area of $\triangle ABC$.

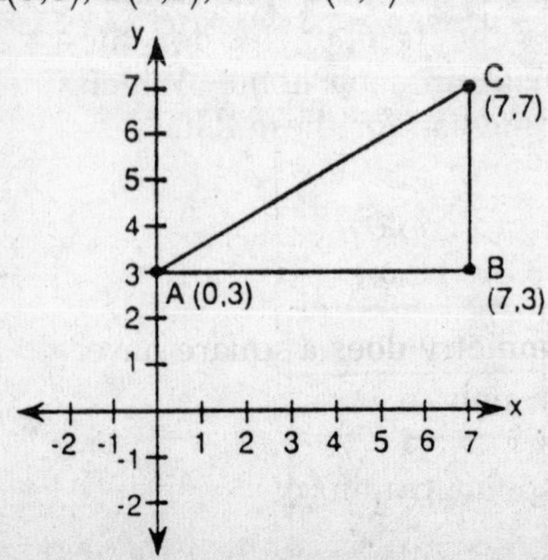

16_____

17 In the accompanying diagram of rectangle $ABCD$, $AD = 12$ and $AB = 5$. What is the length of diagonal $\overline{AC}$?

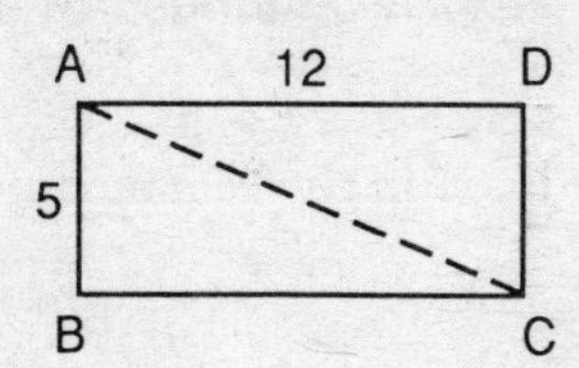

17_____

18 Add: $\frac{3}{2x} + \frac{2}{5x}$, $x \neq 0$ 18_____

19 Solve for x: $0.3x - 2 = 0.7$ 19_____

DIRECTIONS (**20–35**): *For* each *question chosen, write the* numeral *preceding the word or expression that best completes the statement or answers the question.*

20 When $12x^4 - 3x^3 + 6x^2$ is divided by $3x^2$, the quotient is

(1) $4x^2 - 3x^3 + 6x^2$
(2) $12x^4 - 3x^3 + 2$
(3) $9x^2 - x + 2$
(4) $4x^2 - x + 2$

20_____

21 If a regular pentagon has a side of length $2x + 1$, what is the perimeter of the pentagon?

(1) $4x^2 + 1$
(2) $10x + 5$
(3) $10x + 1$
(4) $10x^2 + 5$

21_____

22 Which type of symmetry does a square have?

(1) line symmetry, only
(2) point symmetry, only
(3) both line and point symmetry
(4) no symmetry

22_____

23 The interest on a loan varies directly as the rate. If the rate is halved, then the interest

(1) is halved
(2) is doubled
(3) remains the same
(4) is multiplied by 4

23_____

24 Which is an equation for line ℓ in the accompanying diagram?

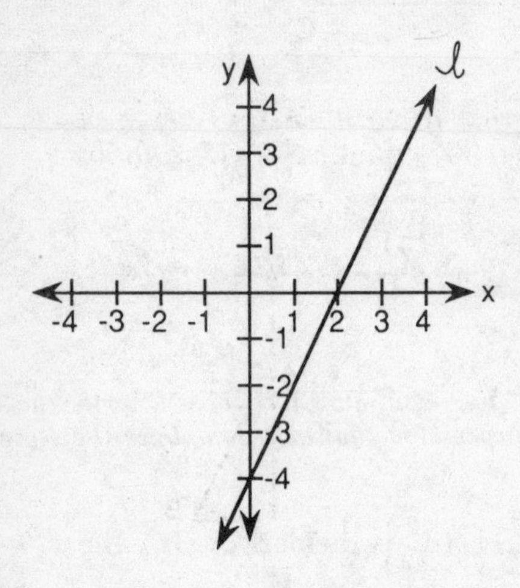

(1) $y = 2x + 2$
(2) $y = 2x - 4$
(3) $y = -2x - 4$
(4) $y = -2x + 2$

24_____

25 One member of the solution set of $3x - 1 \geq 4$ is

(1) 1
(2) $\frac{2}{3}$
(3) $\frac{5}{3}$
(4) $-\frac{4}{3}$

25_____

26 If two angles of a triangle are complementary, then the triangle *must* be

(1) obtuse
(2) equilateral
(3) isosceles
(4) right

26_____

27 In the accompanying diagram, $\triangle ABC$ is a right triangle.

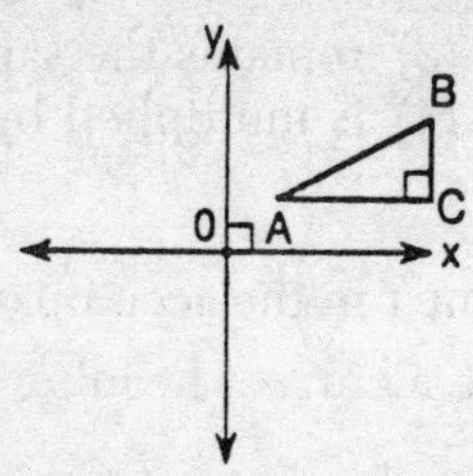

Which diagram below represents the image of $\triangle ABC$ when rotated 90° counterclockwise about the origin?

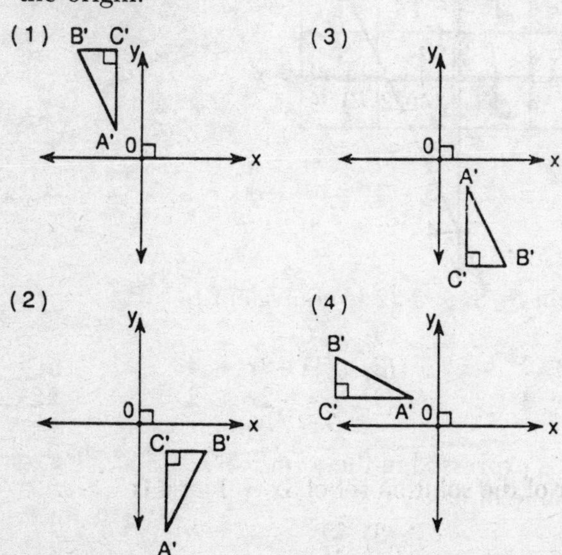

27_____

28 Which figure has the largest area?

(1) a square whose side measures 6
(2) a circle whose diameter measures 6
(3) a triangle whose base and height each measure 6
(4) an equilateral triangle whose side measures 6

28_____

29 Written in factored form, the trinomial $3x^2 + 5x - 2$ is equivalent to

(1) $(3x + 1)(x - 2)$
(2) $(3x - 1)(x + 2)$
(3) $(3x + 2)(x - 1)$
(4) $(3x - 2)(x + 1)$

29_____

30 Which statement would be a correct heading for the last column of the table?

p	q	$\sim q$	?
T	T	F	F
T	F	T	T
F	T	F	T
F	F	T	F

(1) $q \rightarrow \sim q$
(2) $p \wedge \sim q$
(3) $p \leftrightarrow \sim q$
(4) $p \vee \sim q$

30_____

31 The expression $5\sqrt{8} - 3\sqrt{2}$ is equivalent to

(1) 7
(2) $7\sqrt{2}$
(3) $2\sqrt{6}$
(4) $\sqrt{34}$

31_____

32 If 0.0000055 is expressed in the form 5.5×10^n, what is the value of n?

(1) 5
(2) 6
(3) −5
(4) −6

32_____

33 The probability of throwing two fours on a single toss of a pair of dice is

(1) $\frac{1}{36}$
(2) $\frac{1}{12}$
(3) $\frac{1}{6}$
(4) $\frac{1}{3}$

33_____

34 If $x = -2$ and $y > 0$, in which quadrant of the coordinate plane is point $P(x,y)$ located?

(1) I
(2) II
(3) III
(4) IV

34______

35 The converse of $\sim p \rightarrow q$ is

(1) $p \rightarrow \sim q$
(2) $\sim q \rightarrow p$
(3) $q \rightarrow \sim p$
(4) $\sim p \rightarrow \sim q$

35______

PART TWO

DIRECTIONS: *Answer* four *questions from this part. Clearly indicate the necessary steps, including appropriate formula substitutions, diagrams, graphs, charts, etc. Calculations that may be obtained by mental arithmetic or the calculator do not need to be shown.* [40]

36 *a* On the same set of coordinate axes, graph the following system of inequalities and label the solution set *S*.

$$y \geq 3x + 1$$
$$x + 2y < 5$$
[8]

b Write the coordinates of a point in the solution set of the system of inequalities graphed in part *a*. [2]

37 A fair die and a fair coin are tossed.

a Draw a tree diagram or list the sample space for all possible outcomes. [2]

b Find the probability of getting

(1) an odd number and a head [2]
(2) a number greater than 5 and a tail [2]
(3) a number greater than 4 or a head [2]
(4) a prime number and either a head or a tail [2]

38 The denominator of a fraction is four less than twice the numerator. If three is added to both the numerator and the denominator, the new fraction is equal to $\frac{2}{3}$. Find the original fraction. [*Only an algebraic solution will be accepted.*] [5,5]

39 A taxi ride costs \$1.25 for the first quarter of a mile and \$0.30 for each subsequent quarter of a mile.

a A ride that costs \$5.45 is how many miles longer than a ride that costs \$3.95? [Show or explain the procedure used to obtain your answer.] [6]

b If n represents the number of quarter miles traveled and C represents the total cost of the trip, which formula could be used to find C? [2]

(1) $C = 1.25n$
(2) $C = 1.25 + 0.30n$
(3) $C = 1.25 + 0.30(n - 1)$
(4) $C = 1.25n + 0.30n$

c Using your answer from part *b*, find the value of C if $n = 20$. [2]

40 Solve the following system of equations algebraically and check:

$$3x + 5y = 4$$
$$4x + 3y = -2$$

[8,2]

41 A small company recorded the number of hours each of their 21 part-time employees worked during one week. The hours employees worked were: 14, 22, 13, 2, 7, 13, 18, 29, 19, 15, 9, 16, 23, 12, 17, 28, 20, 4, 18, 8, 24.

a *On your answer paper*, copy and complete the tables below using the given data. [4]

Interval (in hours)	Frequency
26–30	
21–25	
16–20	
11–15	
6–10	
0–5	2

Interval (in hours)	Cumulative Frequency
0–30	
0–25	
0–20	
0–15	
0–10	
0–5	2

b Which interval in the frequency table contains the median number of hours worked? [2]

c Construct a cumulative frequency histogram using the data from the appropriate table. [4]

42 *a* *On your answer paper*, copy and complete the truth table for the statement $\sim(p \rightarrow q) \leftrightarrow (p \wedge \sim q)$. [8]

p	q	$p \rightarrow q$	$\sim(p \rightarrow q)$	$\sim q$	$p \wedge \sim q$	$\sim(p \rightarrow q) \leftrightarrow (p \wedge \sim q)$
T	T					
T	F					
F	T					
F	F					

b Is the statement $\sim(p \rightarrow q) \leftrightarrow (p \wedge \sim q)$ a tautology? [1]

c Justify the answer given in part *b*. [1]

Answers January 1993

Three-Year Sequence for High School Mathematics—Course I

ANSWER KEY

PART ONE

1. $\frac{5}{9}$

2. 3

3. $\sim q \rightarrow \sim p$

4. $4x^2 - 3$

5. $\frac{5}{2}$

6. $(2x + 5)(2x - 5)$

7. 95

8. 4

9. 3–4

10. $4x$

11. $\frac{1}{8}x^4y^3$

12. 5

13. 24

14. $-4x - 6$

15. $2a^2 + 11a + 5$

16. 14

17. 13

18. $\frac{19}{10x}$

19. 9

20. (4)

21. (2)

22. (3)

23. (1)

24. (2)

25. (3)

26. (4)

27. (1)

28. (1)

29. (2)

30. (3)

31. (2)

32. (4)

33. (1)

34. (2)

35. (3)

PART TWO *See answers explained section.*

ANSWERS EXPLAINED

PART ONE

1. To find the probability that a marble selected at random is *not* orange, divide the number of successful outcomes by the total possible number of outcomes:

$$\text{Probability of an event occuring} = \frac{\text{Number of successful outcomes}}{\text{Total possible number of outcomes}}$$

Since the bag contains three red marbles, two green marbles and four orange marbles, the total possible number of outcomes is $3 + 2 + 4 = 9$. The number of successful outcomes is the sum of the number of red marbles and the number of green marbles, which is $3 + 2 = 5$. Therefore, the probability of *not* picking an orange marble is $\frac{5}{9}$.

The probability is $\mathbf{\frac{5}{9}}$.

2. Given: $\Delta ABC \sim \Delta DEF$. In similar triangles, the lengths of corresponding sides are in proportion.

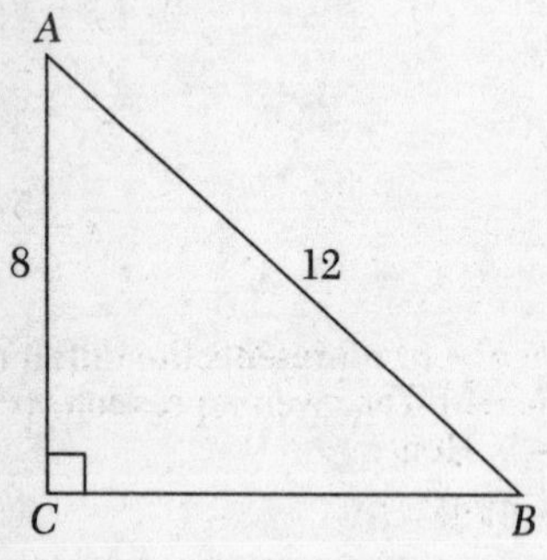

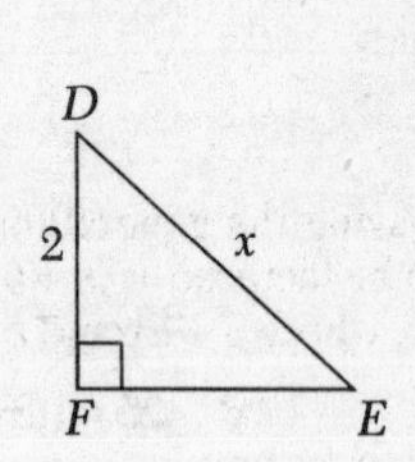

Hence: $$\frac{DE}{DF} = \frac{AB}{AC}$$

Also given: $AB = 12$, $AC = 8$, $DE = x$, and $DF = 2$. Then: $$\frac{x}{2} = \frac{12}{8}$$

In a proportion the product of the means equals the product of the extremes (cross-multiply): $$8x = 24$$

The value of x is **3**. $$x = \frac{24}{8} = 3$$

3. If p represents "I go to the movies," then the statement "I do not go to the movies" is the negation of p, written as $\sim p$.

If q represents "I study my math," then the statement "I do not study my math" is the negation of q, written as $\sim q$.

"If I do not study my math, then I do not go to the movies." is a conditional statement that may be represented as $\sim q \rightarrow \sim p$.

The statement written in symbolic form is $\boldsymbol{\sim q \rightarrow \sim p}$.

4. Write the polynomials to be added so that like terms appear in the same vertical column:

$$\begin{array}{r} 3x^2 - 2x + 5 \\ x^2 + 2x - 8 \\ \hline 4x^2 + 0x - 3 \end{array}$$

Combine like terms in each column:

The sum in simplest form is $\mathbf{4x^2 - 3}$.

5. To solve the given equation, $7x - 2 = 5x + 3$, isolate the variable on one side of the equation by adding 2 to each side and subtracting $5x$ from each side.

On each side, add 2 and subtract $5x$:

$$\begin{array}{r} 7x - 2 = 5x + 3 \\ -5x + 2 = -5x + 2 \\ \hline 2x \quad = \quad 5 \end{array}$$

Divide both sides by 2:

$$\frac{2x}{2} = \frac{5}{2}$$

The value of x is $\mathbf{\frac{5}{2}}$.

$$x = \frac{5}{2}$$

6. A binomial having the general form $a^2 - b^2$ represents the difference of two squares and can be factored as $(a+b)(a-b)$. The given expression, $4x^2 - 25$, has the form $a^2 - b^2$, where $a = 2x$ and $b = 5$. Hence:

$$4x^2 - 25 = (2x + 5)(2x - 5)$$

The given expression may be written as $\mathbf{(2x + 5)(2x - 5)}$.

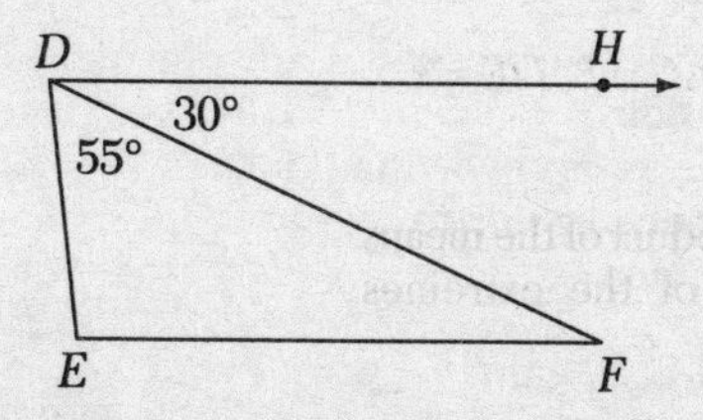

7. Given: $\overline{DH} \parallel \overline{EF}$. Since interior angles on the same side of a transversal intersecting parallel lines are supplementary,

$$m\angle E + m\angle EDH = 180$$

The measure of $\angle EDC$ is the sum of the measures of angles EDF and FDH. Hence:

$$m\angle E + (55 + 30) = 180$$
$$m\angle E + 85 = 180$$
$$m\angle E + 85 = 180 - 85$$
$$m\angle E + 85 = 95$$

The measure of $\angle E$ is **95**.

8. The problem of forming a three-digit number less than 800 using the digits 2, 3, and 9 is equivalent to making choices to fill each of three ordered positions: __ __ __. Since the three-digit number must be less than 800, the first position may be filled only with the digit 2 or 3, so there are two possible ways the first position can be filled. The second position can be filled with the digit 9 or, since repetition of digits is not allowed, the remaining digit, 2 or 3, that was not used to fill the first position. Hence, there are two ways in which the second position can be filled. Without repeating digits, there is only one remaining digit that can be used to fill the last slot.

The pairings of the three digits may be performed in $\underline{2} \times \underline{2} \times \underline{1} = 4$ ways.

Therefore, **4** different three-digit numbers less than 800 can be formed.

9. The median is the middle observation when the data are arranged in size order. Half the scores are less than the median, and half the scores are above the median.

Number of People Living in One Home

Interval	Frequency
1–2	40
3–4	39
5–6	16
7–8	5

The sum of the frequencies in the given table is 100. Since one-half of 100 is 50, 50 scores will lie below the median and 50 scores will lie above it. The median will lie halfway between the 50th and 51st observations.

Count down from the top of the frequency column until the interval that contains the 50th and 51st observations is found. There are 40 observations in the first interval. The next interval, 3–4, contains the next 39 observations. Thus, it must include the 50th and 51st observations.

The interval **3–4** contains the median.

10. The mean of a set of expressions is the sum of the expressions divided by the number of expressions.

Add the given expressions by writing them underneath one another, aligning like terms in the same vertical column:

$$\begin{array}{r} 3x - 5 \\ 5x - 6 \\ 4x + 11 \\ \hline 12x + 0 \end{array} = 12x$$

To find the mean, divide the sum by the number of expressions added:

$$\text{Mean} = \frac{\text{Sum}}{\text{Number of expressions}} = \frac{12x}{3} = 4x$$

The mean is $\mathbf{4x}$.

11. To find the product of $\frac{1}{2}x^3y$ and $\frac{1}{4}xy^3$, multiply their numerical coefficients and then multiply powers of the same base by keeping the base and adding the exponents.

Given: $\left(\frac{1}{2}x^3y\right)\left(\frac{1}{4}xy^2\right)$

Multiply numerical coefficients and group like terms, using the commutative property. Replace x by x^1 and y by y^1: $\frac{1}{8}\left(x^3 \cdot x^1\right)\left(y^1 \cdot y^2\right)$

To multiply powers having the same base, add the exponents: $\frac{1}{8}x^4y^3$

The product is $\mathbf{\frac{1}{8}x^4y^3}$.

12. Given: $\frac{m+3}{2m} = \frac{4}{5} \quad (m \neq 0)$

In a proportion the product of the means equals the product of the extremes (cross-multiply): $5(m+3) = 4(2m)$

Simplify: $5m + 15 = 8m$

Add $-5m$ to each side of the equation:

$$\begin{array}{r} 5m + 15 = 8m \\ -5m \quad\;\; = -5m \\ \hline 15 = 3m \end{array}$$

Divide each side of the equation by 3: $\frac{15}{3} = \frac{3m}{3}$

The value of m is **5**. $5 = m$

13. The circumference C of a circle is given by the formula $C = \pi D$, where D represents the diameter of the circle. If $C = 6\pi$, then $\pi D = 6\pi$, or $D = 6$.

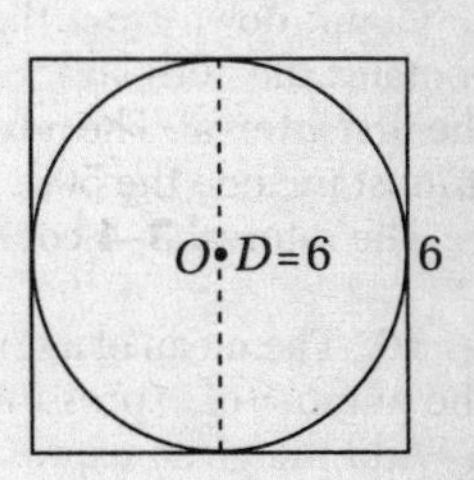

Since the length of a side of the square is equal to the diameter of the circle, each side of the square measures 6.

Therefore, the perimeter of the square is 4×6, or 24.

The perimeter of the square is **24**.

14. Given: $-3(x+2) - x$

Use the distributive property: $-3x + (-3)(2) - x$

Simplify: $-4x - 6$

The binomial is $\mathbf{-4x - 6}$.

15. To find the product of $2a + 1$ and $a + 5$, use FOIL:

Given:	$(2a+1)(a+5)$
Write the product of the ***F***irst terms:	$(\underline{2a}+1)(\underline{a}+5) = 2a^2 +$
Write the product of the ***O***uter terms:	$(\underline{2a}+1)(a+\underline{5}) = 2a^2 + (2a)(5) +$
Write the product of the ***I***nner terms:	$(2a+\underline{1})(\underline{a}+5) = 2a^2 + 10a + (1)(a) +$
Write the product of the ***L***ast terms:	$(2a+\underline{1})(a+\underline{5}) = 2a^2 + 10a + a + (1)(5)$
Simplify:	$(2a+1)(a+5) = 2a^2 + 11a + 5$

The product is $\mathbf{2a^2 + 11a + 5}$.

16.

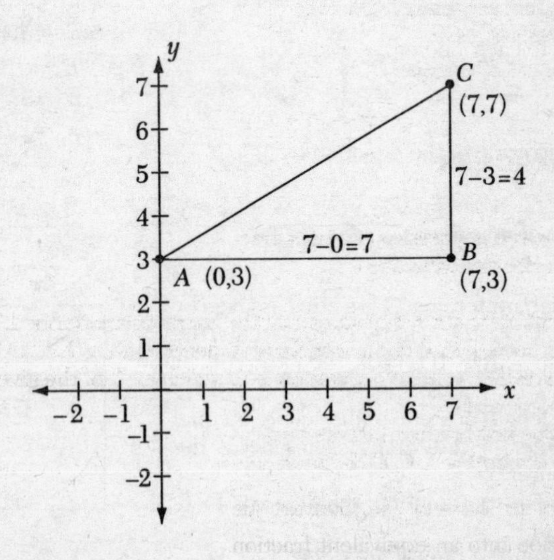

The area of right triangle ABC is equal to one-half of the product of the lengths of its two legs, $\overline{AB}$ and $\overline{BC}$.

Since $\overline{AB}$ is a horizontal line segment, its length is equal to the difference between the x-coordinates of its endpoints, which is $7 - 0 = 7$.

The length of vertical segment BC is equal to the difference between the y-coordinates of its endpoints, which is $7 - 3 = 4$.

Thus, $AB = 7$ and $BC = 4$. Then:

$$\text{Area of right } \Delta ABC = \frac{1}{2}(AB)(BC)$$
$$= \frac{1}{2}(7)(4)$$
$$= \frac{1}{2}(28)$$
$$= 14$$

The area of ΔABC is **14**.

17. A rectangle contains four right angles and has all of the properties of a parallelogram.

In the given rectangle, $BC = AD = 12$. Diagonal $\overline{AC}$ is the hypotenuse of right triangle ABC.

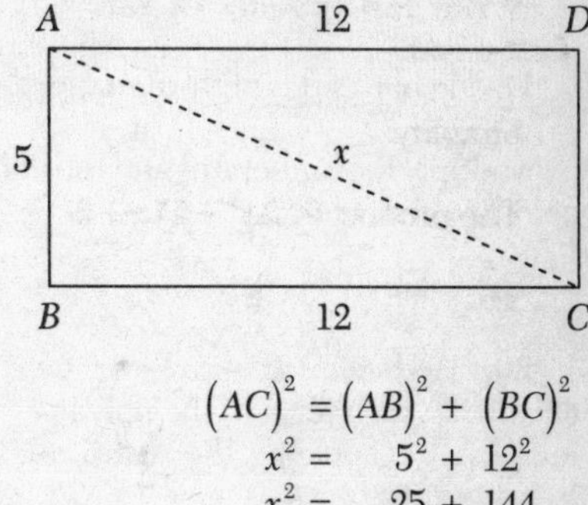

To find the length of $\overline{AC}$, use the Pythagorean theorem, where x represents the length of $\overline{AC}$:

Let $AC = x$, $AB = 5$, and $BC = 12$:

$$(AC)^2 = (AB)^2 + (BC)^2$$
$$x^2 = 5^2 + 12^2$$
$$x^2 = 25 + 144$$
$$x^2 = 169$$
$$x = \sqrt{169} = 13$$

The length of $\overline{AC}$ is **13**.

18. The given expression contains fractions with unlike denominators:

$$\frac{3}{2x} + \frac{2}{5x} \quad (x \neq 0)$$

The restriction that $x \neq 0$ ensures that the fractions are defined. To add these fractions, we need to find the least common denominator (L.C.D.) The L.C.D. is $10x$, since it is the smallest expression into which each of the given denominators will divide evenly.

Convert the first fraction into an equivalent fraction having the L.C.D. by multiplying it by 1 in the form of $\frac{5}{5}$. Convert the second fraction into an equivalent fraction having the L.C.D. by multiplying it by 1 in the form of $\frac{2}{2}$:

$$\frac{(5)3}{(5)2x} + \frac{(2)2}{(2)5x}$$

Simplify:

$$\frac{15}{10x} + \frac{4}{10x}$$

Add the fractions by writing the sum of their numerators over their common denominator:

$$\frac{15+4}{10x}$$

$$\frac{19}{10x}$$

The sum is $\mathbf{\frac{19}{10x}}$.

19. Given:

$$0.3x - 2 = 0.7$$

Add 2 to both sides of the equation:

$$\begin{array}{r} 0.3x - 2 = 0.7 \\ 2 = 2 \\ \hline 0.3x \quad = 2.7 \end{array}$$

Divide both sides of the equation by 0.3:

$$\frac{0.3x}{0.3} = \frac{2.7}{0.3}$$

Multiply the numerator and the denominator by 10:

$$x = \frac{27}{3}$$

The value of x is **9**.

$$x = 9$$

20. To divide $12x^4 - 3x^3 + 6x^2$ by $3x^2$, divide each term of the polynomial by $3x^2$, applying the distributive property:

$$\frac{12x^4 - 3x^3 + 6x^2}{3x^2} = \frac{12x^4}{3x^2} - \frac{3x^3}{3x^2} + \frac{6x^2}{3x^2}$$

Divide the numerical coefficients of each term to obtain the numerical coefficient of each quotient. To obtain the quotient of powers of x, subtract the exponents:

$$\frac{12x^4 - 3x^3 + 6x^2}{3x^2} = 4x^2 - x + 2$$

The correct choice is **(4)**.

21. A pentagon is a five-sided polygon. In a *regular* pentagon each of the five sides has the same length. Therefore, the perimeter of a regular pentagon is found by multiplying the length of a side by 5.

If the length of a side of a regular pentagon is $2x + 1$, then the perimeter of the pentagon is $5(2x + 1) = 10x + 5$.

The correct choice is **(2)**.

22. A figure has line symmetry if there exists a line that can be drawn through the figure so that it divides the figure into two parts that are "mirror images" of each other.

A vertical line that bisects the two sides of a square it intersects, as shown in the accompanying diagram, is a line of symmetry for a square.

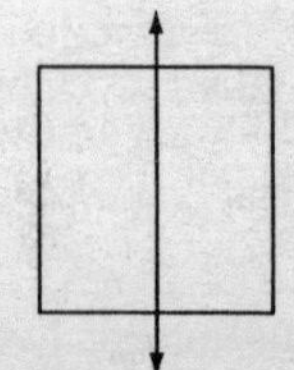

Vertical line of symmetry

A figure has point symmetry if the figure coincides with itself when rotated in either direction 180° about some fixed point.

A square has point symmetry since, if square $ABCD$ is rotated 180° about its center (the point at which diagonals $\overline{AC}$ and $\overline{BD}$ intersect) in the clockwise direction, the square will coincide with itself, with vertex A mapped onto vertex C, vertex B mapped onto vertex D, vertex C mapped onto vertex A, and vertex D mapped onto vertex B.

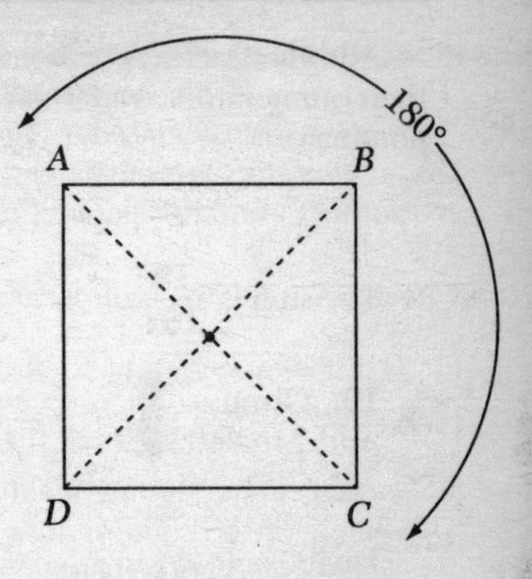

Thus, a square has both line and point symmetry.

The correct choice is **(3)**.

23. If a quantity varies directly with another quantity, then the ratio of the two quantities remains constant when the values of the variables change.

Let I = the interest on a loan, and let R = the rate of the loan. Since I varies directly as R, then $\frac{I}{R} = k$, where k is some nonzero constant.

Halving R requires halving I in order for their ratio to remain equal to the same nonzero constant, k:

$$\frac{\frac{1}{2}I}{\frac{1}{2}R} = \frac{I}{R} = k$$

The correct choice is **(1)**.

24.

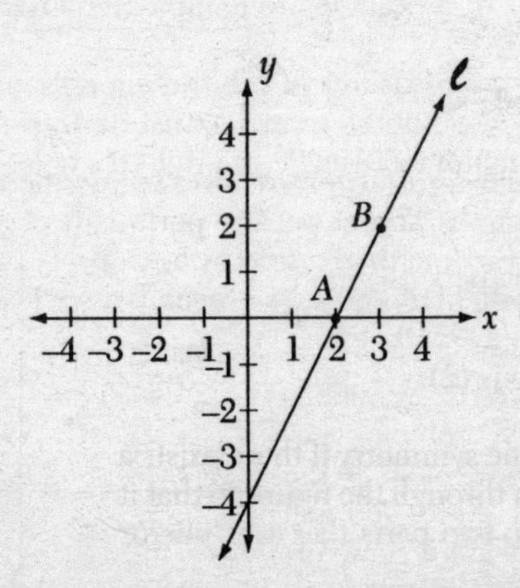

In an equation having the form $y = mx + b$, m represents the slope of the line and b is the y-intercept of the line.

In the given graph, line l cuts the y-axis at the point whose y-coordinate is -4, so $b = -4$. The slope of the line is defined as the change in y over the change in x, and is calculated using any two points on the line.

Pick any two convenient points on the line, say A (2,0) and B (3,2). Use these points to find the slope of the line:

$$m = \text{slope} = \frac{\text{change in } y}{\text{change in } x} = \frac{2-0}{3-2} = \frac{2}{1} = 2$$

Since $m = 2$ and $b = -4$, an equation of line l is $y = 2x - 4$.
The correct choice is **(2)**.

25. Solve the inequality $3x - 1 \geq 4$ for x. An equivalent inequality results whenever the same number is added to both sides of the inequality, or both sides of an inequality are divided by the same positive number.

Given: $3x - 1 \geq 4$

Add 1 to each side: $1 = 1$

$$3x \geq 5$$

Divide each side by 3: $\frac{3x}{3} \geq \frac{5}{3}$

$$x \geq \frac{5}{3}$$

Consider each choice in turn. The value $\frac{5}{3}$ is the only replacement value for x that makes the inequality $x \geq \frac{5}{3}$ a true statement.
The correct choice is **(3)**.

26. If two angles of a triangle are complementary, then the sum of their measures is 90.

Since the sum of the measures of the three angles of a triangle is 180, the measure of the third angle of the triangle must be $180 - 90 = 90$.

A triangle that contains a 90° angle is a right triangle.
The correct choice is **(4)**.

27. The accompanying diagram, in which right triangle ABC is located in Quadrant I, is given.

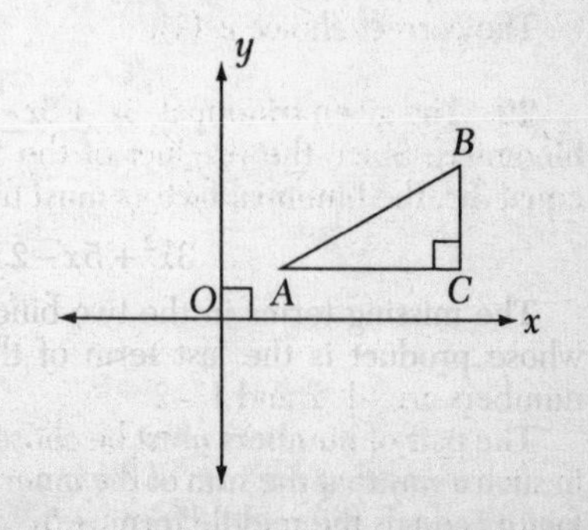

If you turn the page that contains the diagram for this problem one-quarter of a full rotation, or 90° in the counterclockwise direction, you will see that the positive x-axis is now vertical, the positive y-axis is now horizontal, and the image of ΔABC is now located in Quadrant II, so that the image of side $\overline{AC}$ is vertical and the image of $\overline{BC}$ is horizontal.

This situation corresponds to the diagram shown in choice (1).

The correct choice is **(1)**.

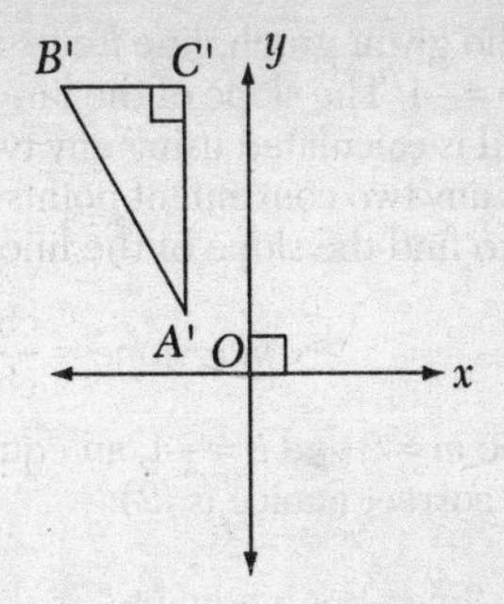

28. Consider each choice in turn:

(1) The area of a square is equal to the square of the measure of a side. A square whose side measures 6 has an area of 6^2, or 36.

(2) The area of a circle is equal to πr^2. In a circle whose diameter measures 6, the radius is 3 and the area is $\pi(3^2)$, or 9π. Since π is less than 4 (approximately equal to 3.14), the area of the circle is less than 36. Thus, the area of this circle is less than the area calculated in choice (1).

(3) The area of a triangle is equal to one-half the product of the lengths of the base and the height. In a triangle whose base and height each measure 6, the area is $\frac{1}{2}(6)(6) = 18$, which is less than the area calculated in choice (1).

(4) In an equilateral triangle each of whose sides measures 6, the base is 6 and the altitude must be less than 6, since it is a leg of a right triangle in which the hypotenuse is 6 (see diagram). The area of this equilateral triangle must be less than $\frac{1}{2}(36)$, so its area is less than the area calculated in choice (1).

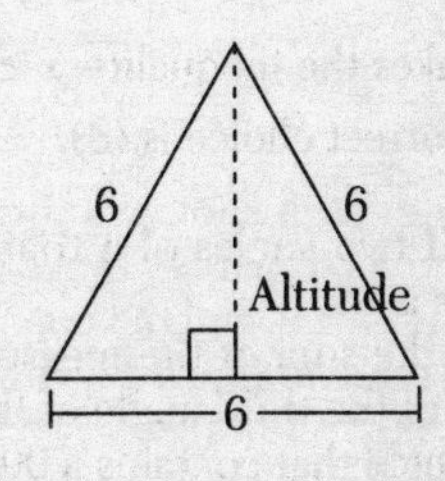

The correct choice is **(1)**.

29. The given trinomial, $3x^2 + 5x - 2$, can be factored as the product of two binomials. Since the product of the first terms of the binomial factors must equal $3x^2$, the binomial factors must take the following form:

$$3x^2 + 5x - 2 = (3x + ?)(x + ?)$$

The missing terms in the two binomial factors must be a pair of numbers whose product is the last term of the trinomial, –2. Possible pairs of such numbers are –1, 2 and 1, –2.

The pair of numbers must be chosen and placed within the binomial factors in such a way that the sum of the inner and outer cross-products of the binomial factors equals the middle term, + 5x, of the trinomial.

Try –1 and 2, and place them within the binomials as follows

inner product = $-x$

$(3x - 1)(x + 2)$

outer product = $6x$

Since $-x + 6x = +5x$, $(3x - 1)$ and $(x + 2)$ are the correct binomial factors of $3x^2 + 5x - 2$.

Hence, $3x^2 + 5x - 2 = (3x - 1)(x + 2)$.

Alternative Solution: The problem can also be solved by using FOIL to express the product in each of the four choices as a trinomial. The correct choice is the one that gives a product that matches the given trinomial.

The correct choice is **(2)**.

30.

p	q	$\sim q$	?
T	T	F	F
T	F	T	T
F	T	F	T
F	F	T	F

Consider each choice in turn:

(1) A conditional is always true *except* in the single instance in which the hypothesis is true, and the conclusion is false. In row 4, q is false, $\sim q$ is true, but the truth value in the last column is False, so the last column cannot represent the truth values for $q \rightarrow \sim q$.

(2) A conjunction is true only when each of the conjuncts is true. In row 3, p is false, $\sim q$ is true, but the truth value in the last column is True, so the last column cannot represent the truth values for $p \wedge \sim q$.

(3) A biconditional is true when the statements on both sides of the symbol $\leftrightarrow$ have the same truth value. In rows 2 and 3, p and $\sim q$ have the same truth value, and the last column has the truth value True for each of these rows. In rows 1 and 4, p and $\sim q$ have different truth values, and the last column has the truth value False for each of these rows. Hence, the last column represents the truth values for the biconditional of p and $\sim q$. The correct choice is (3).

Note: Since a disjunction is true whenever at least one of the disjuncts is true, the truth values in the first row rule out choice (4), $p \vee \sim q$, since the truth value of the last column is False, although p is true.

Hence, the correct heading for the last column of the table is $p \leftrightarrow \sim q$.

The correct choice is **(3)**.

31. The four choices suggest that it is necessary to combine the radicals in the expression given, $5\sqrt{8}-3\sqrt{2}$. Square root radicals can be combined only if they have the same radicand.

Given:	$5\sqrt{8} - 3\sqrt{2}$
Factor the radicand 8:	$5\sqrt{4\cdot 2} - 3\sqrt{2}$
Use the multiplication property of radicals:	$5\sqrt{4}\sqrt{2} - 3\sqrt{2}$
Simplify:	$5\cdot 2\sqrt{2} - 3\sqrt{2}$
	$10\sqrt{2} - 3\sqrt{2}$

The radicals are now like radicals since they have the same index (both are square roots) and the same radicand, 2. Like radicals are combined by combining their numerical coefficients: $7\sqrt{2}$

The correct choice is **(2)**.

32. The effect of multiplying a number by a power of 10 is to move the position of the decimal point of the original number. To have the decimal point in the number 0.0000055 move to the right so that the number becomes 5.5, the decimal point must move six decimal positions.

Moving one decimal position to the right is equivalent to dividing the number by 10 or, equivalently, multiplying the number by 10^{-1}.

Therefore, 5.5 must be multiplied by 10^{-6} on order to equal 0.0000055. The value of n is –6.

The correct choice is **(4)**.

33. Since a die has six faces, numbered 1, 2, 3, 4, 5, and 6, respectively, the probability of obtaining a 4 on a single toss of a die is $\frac{1}{6}$. In a single toss of a *pair* of dice, the probability of throwing two 4's is obtained by multiplying the probabilities of obtaining a 4 on each die:

$$P\text{ (Tossing two 4's)} = \frac{1}{6}\times\frac{1}{6} = \frac{1}{36}$$

The correct choice is **(1)**.

34. Since it is given that $x = -2$, the point $P(x,y)$ must lie in either Quadrant II or in Quadrant III, where x is negative.

In Quadrant II y is positive, and in Quadrant III y is negative.

Since it is given that $y > 0$, y is positive, so $P(x,y)$ must lie in Quadrant II.

The correct choice is **(2)**.

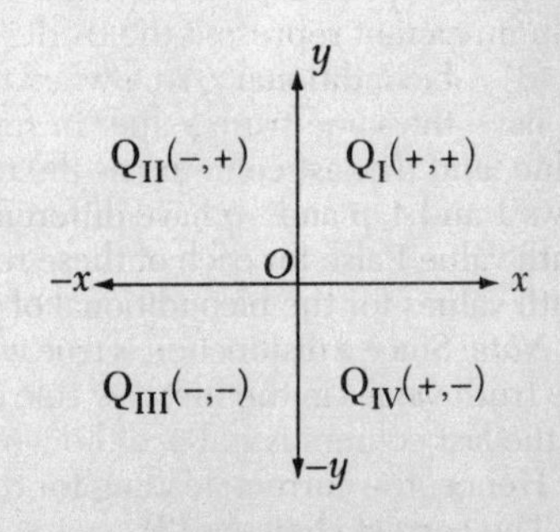

35. The converse of a conditional statement is formed by interchanging its parts.

Therefore, the converse of $\sim p \rightarrow q$ is the conditional $q \rightarrow \sim p$.

The correct choice is **(3)**.

PART TWO

36. a. STEP 1. Draw the graph of $y \geq 3x + 1$ by first graphing $y = 3x + 1$.

Prepare a table of pairs of values of x and y by selecting three convenient values for x (say, –2, 0, and 2) and substituting them into the equation $y = 3x + 1$ to find the corresponding values of y.

x	$3x + 1$	$= y$
–2	$3(-2) + 1 = -6 + 1$	$= -5$
0	$3(0) + 1 = 0 + 1$	$= 1$
2	$3(2) + 1 = 6 + 1$	$= 7$

Plot points (–2,–5), (0,1), and (2,7), and draw a *solid* line through them. This line is the graph of $y = 3x + 1$. It is drawn as a solid line to show that points on this line are contained in the solution set of $y \geq 3x + 1$.

The solution set of $y > 3x + 1$ lies on one side of the line $y = 3x + 1$. To find out on which side of the line the solution set lies, choose a convenient test point, say (0,0), and substitute its coordinates in the inequality:

$$y > 3x + 1$$
$$0 > 3(0) + 1$$
$$0 > 1$$

Since (0,0) does not satisfy the inequality, it does *not* lie in the solution set of $y > 3x + 1$. Therefore, shade the *opposite* side of the line by drawing lines extending up and to the left as indicated in the accompanying diagram.

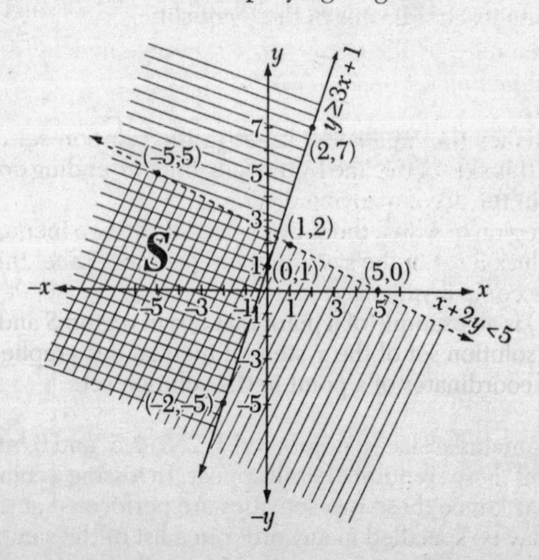

STEP 2. To draw the graph of $x + 2y < 5$, first solve for y by adding $-x$ to each side and then dividing each side by 2, the coefficient of y:

$$\begin{aligned} x + 2y &< 5 \\ \underline{-x\quad} &\underline{= -x} \\ 2y &< -x + 5 \\ \frac{2y}{2} &< \frac{-x+5}{2} \\ y &< \frac{-x+5}{2} \end{aligned}$$

Graph $y = \frac{-x+5}{2}$ by following the procedure used in Step 1:

x	$\frac{-x+5}{2}$	$= y$
−5	$\frac{-(-5)+5}{2} = \frac{5+5}{2} = \frac{10}{2}$	$= 5$
1	$\frac{-1+5}{2} = \frac{-1+5}{2} = \frac{4}{2}$	$= 5$
5	$\frac{\_5+5}{2} = \frac{-5+5}{2}$	$= 0$

Plot points (−5,5), (1,2), and (5,0), and draw a *broken* line through them. This broken line is the graph of $x + 2y < 5$. It is drawn as a broken line to show that points on this line are *not* contained in the solution set of $x + 2y < 5$.

The solution set of $x + 2y < 5$ lies on one side of the line $x + 2y < 5$. To find out on which side of the line the solution set lies, choose a convenient test point, say (0,0), and substitute its coordinates in the inequality:

$$\begin{aligned} x + 2y &< 5 \\ 0 + 2(0) &< 5 \\ 0 &< 5 \end{aligned}$$

Since (0,0) satisfies the inequality, it lies in the solution set of $x + 2y < 5$. Therefore, shade this side of the line by drawing lines extending down and to the left, as indicated in the accompanying diagram.

STEP 3. The region in which the solution sets of the two inequalities overlap represents the solution set of the system of inequalities. Label this region **S**, as indicated in the accompanying diagram.

b. Point (−3,1) is an example of a point that lies in region S and is, therefore, a member of the solution set of the system of inequalities graphed in part **a.**

(−3,1) are the coordinates of a point in the solution set.

37. a. A die contains six faces, numbered 1, 2, 3, 4, 5, and 6, respectively. In tossing a die, one of these six numbers will appear. In tossing a coin either a head or a tail will appear. Since these two activities are performed at the same time, their outcomes may be specified in any order in a list of the sample space.

Here is a listing of the sample space in which the possible result of tossing a die is given first, and the possible result of tossing a coin at the same time is indicated second:

(1, H) (2, H) (3, H) (4, H), (5, H), (6, H)
(1, T), (2, T), (3, T), (4, T), (5, T), (6, T)

b. In each part, use the relationship

$$\text{Probability of an event occurring} = \frac{\text{number of successful outcomes}}{\text{total possible number of outcomes}}$$

In each part, the total possible number of outcomes is 12 since 12 different outcomes are listed in the sample space obtained in part **a**.

(1) From part **a**, there are three outcomes in which an odd number and a head may occur: (1, H), (3, H), and (5, H). Thus:

$$P\,(\text{Odd number and head}) = \frac{3}{12} \text{ or } \frac{1}{4}$$

The probability of getting an odd number and a head is $\mathbf{\frac{3}{12}}$.

(2) From part **a**, there is one outcome in which a number greater than 5 and a tail may occur: (6, T). Thus:

$$P\,(\text{Number} > 5 \text{ and tail}) = \frac{1}{12}$$

The probability of getting a number greater than 5 and a tail is $\mathbf{\frac{1}{12}}$.

(3) From part **a**, there are four outcomes that include a number greater than 4: (5, H), (6, H), (5, T), and (6, T).

There are six outcomes that include a head: (1, H), (2, H), (3, H), (4, H), (5, H), (6, H). However, two of these outcomes, (5, H) and (6, H), have already been counted in the number of outcomes that include a number greater than 4.

Hence, without counting the same outcome twice, there is a total of eight different outcomes that satisfy both conditions: (1, H), (2, H), (3, H), (4, H), (5, H), (6, H), (5, T), and (6, T). Thus:

$$P\,(\text{Number} > 4 \textit{ or } \text{head}) = \frac{8}{12} \text{ or } \frac{2}{3}$$

The probability of getting a number greater than 4 or a head is $\mathbf{\frac{8}{12}}$.

(4) A prime number is any integer greater than 1 that is evenly divisible only by itself and 1. In the set of numbers 1, 2, 3, 4, 5, and 6, the prime numbers are 2, 3, and 5.

There are six outcomes that include a prime number and either a head *or* a tail: (2, H), (2, T), (3, H), (3, T), (5, H), and (5, T). Thus:

$$P\,(\text{Prime number and } \textit{either} \text{ head } \textit{or} \text{ tail}) = \frac{6}{12} \text{ or } \frac{1}{2}$$

The probability of getting a prime number and either a head or a tail is $\mathbf{\frac{6}{12}}$.

38. Let x = the numerator of the original fraction.
Then $2x - 4$ = the denominator of the original fraction.

Hence, the original fraction is $\frac{x}{2x-4}$.

Form an equation by using the given condition that, if 3 is added to both the numerator and the denominator of the original fraction, the new fraction that results is equal to $\frac{2}{3}$:

$$\frac{x+3}{(2x-4)+3} = \frac{2}{3}$$

Simplify the denominator:

$$\frac{x+3}{2x-1} = \frac{2}{3}$$

In a proportion the product of the means equals the product of the extremes (cross-multiply):

$$3(x+3) = 2(2x-1)$$
$$3x+9 = 4x-2$$

Add $-3x$ to both sides, and add 2 to both sides:

$$\begin{aligned} 3x+9 &= 4x-2 \\ -3x+2 &= -3x+2 \\ \hline 11 &= x \end{aligned}$$

Since $x = 11$, replace x by 11 in the expression, $\frac{x}{2x-4}$ for the original fraction:

$$\frac{11}{2(11)-4}$$
$$\frac{11}{22-4} = \frac{11}{18}$$

The original fraction is $\mathbf{\frac{11}{18}}$.

39. a. The difference in the given fares, \$5.45 – \$3.95 = \$1.50, represents the cost of traveling the additional distance covered by the ride that costs the greater amount.

Each quarter mile of the additional distance that is traveled costs \$0.30. Since 5 × \$0.30 = \$1.50, the difference of \$1.50 represents an additional distance traveled of 5 quarter miles, or $1\frac{1}{4}$ miles.

A taxi ride that costs \$5.45 is $\mathbf{1\frac{1}{4}}$ miles longer than a ride that costs \$3.95

b. Let C = the cost of the taxi ride.
Let n = the number of quarter miles of the taxi ride.
Construct a table that shows the cost of a taxi ride for trip lengths given in quarter miles.

Taxi Ride (No. of Quarter Miles)	Cost of Taxi Ride
1	\$1.25
2	\$1.25 + 0.30
3	\$1.25 + 0.30 + 0.30 = \$1.25 + 2(0.30)
4	\$1.25 + 0.30 + 0.30 + 0.30 = \$1.25 + 3(0.30)
•	•
•	•
•	•
n	\$1.25 + (n – 1) 0.30

Observing the pattern in the first four lines of the table, we may generalize that $C = 1.25 + (n - 1)\ 0.30$. This formula can also be written as $C = 1.25 + 0.30(n - 1)$.

The correct choice is **(3)**.

c. Using the answer from part **b,** let $n = 20$, and find the value of C.

From part **b:** $C = 1.25 + 0.30(n - 1)$

Let $n = 20$:

$$C = 1.25 + 0.30(20 - 1)$$
$$C = 1.25 + 0.30(19)$$
$$C = 1.25 + 5.70$$
$$C = 6.95$$

The value of C is **\$6.95.**

40. The given system of equations is:

Work toward eliminating the variable y.

$$\begin{cases} 3x + 5y = 4 \\ 4x + 3y = -2 \end{cases}$$

Begin by multiplying each term of the first equation by –3, and each term of the second equation by 5:

$$\begin{cases} (-3)3x + (-3)5y = (-3)4 \\ (5)4x + (5)3y = (5)(-2) \end{cases}$$

$$-9x - 15y = -12$$
$$20x + 15y = -10$$

Add the two equations:

$$11x \quad = -22$$

Divide both sides of the equation by 11:

$$\frac{11x}{11} = \frac{-22}{11}$$
$$x = -2$$

Substitute –2 for x in the first equation of the original system of equations and solve for y:

$$3(-2)+5y=4$$
$$-6+5y=4$$

Add 6 to each side of the equation:

$$\begin{array}{r} -6+5y=4 \\ \underline{6 \quad\;\; =6} \\ 5y=10 \end{array}$$

Divide each side of the equation by 5:

$$\frac{5y}{5}=\frac{10}{5}$$

The solution is **$x = -2$ and $y = 2$.**

$$y=2$$

CHECK: The solution $x = -2$ and $y = 2$ must satisfy each equation in the original system of equations. Substitute –2 for x and 2 for y in each of these equations to determine whether it is satisfied:

(1)
$$\begin{aligned} 3x+5y&=4 \\ 3(-2)+5(2)&\overset{?}{=}4 \\ -6+10&\overset{?}{=}4 \\ 4&\overset{\checkmark}{=}4 \end{aligned}$$

(2)
$$\begin{aligned} 4x+3y&=-2 \\ 4(-2)+3(2)&\overset{?}{=}-2 \\ -8+6&\overset{?}{=}-2 \\ -2&\overset{\checkmark}{=}-2 \end{aligned}$$

41. a. The given set of hours is 14, 22, 13, 2, 7, 13, 18, 29, 19, 15, 9, 16, 23, 12, 17, 28, 20, 4, 18, 8 and 24.

Enter a tally mark in the frequency column that corresponds to the interval in which each of the given data values falls.

Check that you have accounted for each of the data values by verifying that the sum of the frequencies equals 21, the number of data values in the given list.

Interval (in hours)	Frequency
26–30	// = 2
21–25	/// = 3
16–20	~~////~~ / = 6
11–15	~~////~~ = 5
6–10	/// = 3
0–5	2

Sum = 21

The first entry (bottom row) in the cumulative frequency table is 2. The cumulative frequency for each interval after this is obtained by adding the frequency for the next interval in the frequency table to the entry in the cumulative frequency table for the interval immediately below it.

Interval (in hours)	Cumulative Frequency
0–30	19 + 2 = 21
0–25	16 + 3 = 19
0–20	10 + 6 = 16
0–15	5 + 5 = 10
0–10	2 + 3 = 5
0–5	2

b. The median of a list of data values is the middle value when the data values are arranged in order of size. Half the data values are less than the median, and half the data values are greater than the median.

In an ordered list having 21 data values, there are 10 data values below the 11th data value and 10 data values above the 11th data value. Hence, the median is the 11th data value.

Counting up from the bottom of the frequency table, we find that there are $2 + 3 + 5 = 10$ data values in the first three intervals. Since the next higher interval in the frequency, 16–20, contains six data values, it must include the 11th data value.

Thus the median lies in the interval **16–20**.

c.

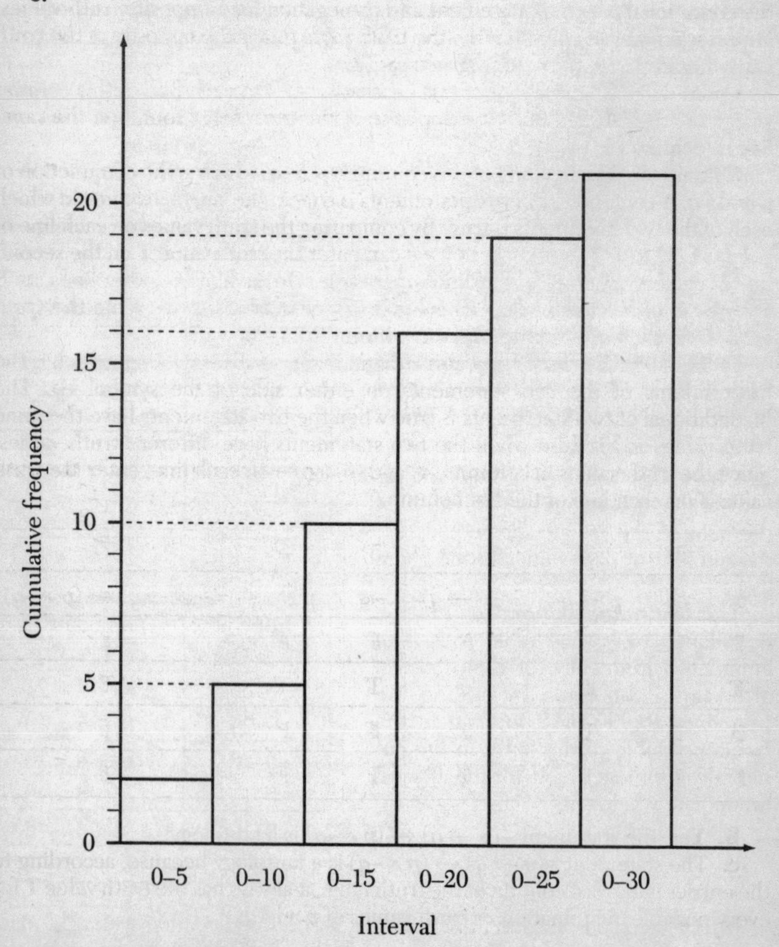

42. a. For each statement that appears in the headings of columns (3) to (7) of the given truth table, the truth values, T (true) and F (false), on each line are determined in accordance with the truth values of the component statements that appear on the same line in the preceding column(s).

Column (3): the heading of this column is $p \rightarrow q$, which is the implication, or conditional, if p, then q. The conditional is always true *except* in the single instance (line 2) in which p is true and q is false. Hence, enter the truth value F on line 2 and the truth value T on each of the other lines of this column.

Column (4): The heading of this column is $\sim(p \rightarrow q)$, which is the negation of the conditional $p \rightarrow q$. A statement and its negation have opposite truth values. On each line of this column write the truth value that is the opposite of the truth value found on the same line of column (3).

Column (5): The heading of this column is $\sim q$. On each line of this column write the truth value that is the opposite of the truth value found on the same line of column (2).

Column (6): The heading of this column is $p \wedge \sim q$, which is the conjunction of p and $\sim q$. A conjunction of two statements is true in the single instance in which each of the two statements is true. By comparing the truth values on each line of columns (1) and (5), we know that we can enter the truth value T on the second line of column (6) since both p and $\sim q$ are true. On each of the other lines, an F appears in either column (1) or column (5), or in both, so we write the truth value F on each of the other lines in column (6).

Column (7): The heading of this column is $\sim(p \rightarrow q) \leftrightarrow (p \wedge \sim q)$ which is the biconditional of the two statements on either side of the symbol $\leftrightarrow$. The biconditional of two statements is true when the two statements have the same truth value, and is false when the two statements have different truth values. Since the truth values in columns (4) and (6) agree on each line, enter the truth value T on each line of the last column.

(1)	(2)	(3)	(4)	(5)	(6)	(7)
p	q	$p \rightarrow q$	$\sim(p \rightarrow q)$	$\sim q$	$p \wedge \sim q$	$\sim(p \rightarrow q) \leftrightarrow (p \wedge \sim q)$
T	T	T	F	F	F	T
T	F	F	T	T	T	T
F	T	T	F	F	F	T
F	F	T	F	T	F	T

b. Yes, the statement $\sim(p \rightarrow q) \leftrightarrow (p \wedge \sim q)$ is a tautology.

c. The statement $\sim(p \rightarrow q) \leftrightarrow (p \wedge \sim q)$ is a tautology because, according to the entries in the last column of the truth table, it always has the truth value T for every possible combination of truth values of p and q.

Topic	Question Numbers	Number of Points	Your Points	Your Percentage
1. Numbers (rat'l, irrat'l); Percent	—	0		
2. Properties of No. Systems	—	0		
3. Operations on Rat'l Nos. and Monomials	11,18	2 + 2 = 4		
4. Operations on Multinomials	4,10,14,15, 20	2 + 2 + 2 +2 + 2 = 10		
5. Square Root; Operations Involving Radicals	31	2		
6. Evaluating Formulas and Expressions	—	0		
7. Linear Equations (simple cases incl. parentheses)	5	2		
8. Linear Equations Containing Decimals or Fractions	19	2		
9. Graph of Linear Functions (slope)	24,34	2 + 2 = 4		
10. Inequalities	25	2		
11. Systems of Eqs. & Inequalities (alg. & graphic solutions)	36,40	10 + 10 = 20		
12. Factoring	6,29	2 + 2 = 4		
13. Quadratic Equations				
14. Verbal Problems	38,39	10 + 10 = 20		
15. Variation	23	2		
16. Literal Eqs.; Expressing Relations Algebraically	—	0		
17. Factorial *n*	—	0		
18. Areas, Perims., Circums., Volumes of Common Figures	13,21,28	2 + 2 + 2 = 6		
19. Geometry (congruence, parallel lines, compls., suppls.)	7,26	2 + 2 = 4		
20. Ratio & Proportion (incl. similar triangles and polygons	2,12	2 + 2 = 4		
21. Pythagorean Theorem	17	2		
22. Logic (symbolic rep., logical forms, truth tables)	3,30,35,42	2 + 2 + 2 + 10 = 16		

Topic	Question Numbers	Number of Points	Your Points	Your Percentage
23. Probability (incl. tree diagrams & sample spaces)	1 ,33,37	2 + 2 + 10 = 14		
24. Combinations, Arrangements, Permutations	8	2		
25. Statistics (central tend., freq. dist., histogr., quartiles, percentiles)	9,41	2 + 10 = 12		
26. Properties of Triangles and Quadrilaterals	—	0		
27. Transformations (reflect., translations, rotations, dilations)	27	2		
28. Symmetry	22	2		
29. Area from Coordinate Geometry	16	2		
30. Dimensional Analysis	—	0		
31. Scientific Notation; Negative & Zero Exponents	32	2		

Notes